PropTech and Real Estate Innovations

This textbook serves as a guide to real estate students and educators on the various property innovations and digital technologies that continue to shape the property industry. The advancement of PropTech in the last few decades has led to significant changes in real estate systems, operations, and practice, and this new textbook provides insight on the past, present, and future of PropTech innovations that have spread across the value chain of real estate through planning, development, management, finance, investment, operations, and transactions. The textbook approaches this subject from the real estate components, asset classes, and submarkets and links them to the associated innovations and digital technologies. It concludes by reviewing the role of education, innovation, skill development, and professionalism as major elements of the future of real estate operations and practice.

This book's unique contributions are in putting the "property" element at the forefront and then illustrating how technology can enhance the various areas of real estate; the focus on how the different innovations and technologies can enhance the economic, environmental, social, and physical efficiency of real estate; and its coverage of some non-technological innovations like flexible working and more practical areas of real estate innovation such as skills, employability, creativity, and education. It contains 21 case studies and 29 case summaries, which can serve as practice exercises for students.

This book will be useful to students in helping them build a knowledge base and understanding of innovation and digital technologies in the industry. Real estate educators can use the textbook as a guide to incorporate real estate innovation and digital technologies into their current teaching and also to develop their real estate curricula through PropTech-related modules and courses where necessary. It will also be valuable to real estate researchers in search of the theoretical and conceptual linkages, as well as industry practitioners who seek insight into the current and future potential of digital technologies and their applications to real estate operations and practice.

Olayiwola Oladiran is currently a Lecturer (Assistant Professor) in Real Estate at the University of Sheffield, Lead of Shef.AI Urban Interest Group and Co-Lead of the Shef.AI Innovation and Pedagogy Interest Groups. He is also the Co-Founder of Bloomshill. He holds a PhD in Real Estate Economics, an MSc in International Real Estate Investment and Finance, and a BTech in Estate Management. Olayiwola is an RICS-Chartered Surveyor (MRICS) and a Fellow of the UK Higher Education Academy (FHEA). He is also a Board Member of the RICS Yorkshire and the Humber Regional Board. He won the UK PropTech Association Special Achievement Award in 2022 and the University College of Estate Management Harold Samuel Research Prize in 2020.

Louisa Dickins is the Co-Founder of LMRE, the market-leading global built environment recruitment platform and search consultancy with operations across North America, the United Kingdom, Europe, Asia-Pacific, and the Middle East and North Africa (MENA). She hosts "The Propcast", a built environment technology podcast; mentors for REACH, a global early-stage technology fund; and is a Committee Member of Real Estate Women, a non-profit association whose mission is to create an equitable and inclusive industry, advocating for fair representation of women and minorities in leadership positions. She was also previously a Board Director for the UK PropTech Association, and was awarded the Young Property Personality of the Year Award at the Property Awards.

"This Textbook is a great way for real estate professionals, students, innovators, and other stakeholders to deep dive into the innovation in the built environment. It offers great global perspectives and insight into a wide range of digital tools and technologies across the real estate value chain from planning and urban management to development, investment, and brokerage across all asset classes. The authors also introduce new concepts and theories that have the potential to shape future innovation and digitisation in real estate and they provide empirical-based insights into future trends and growth areas such as climate tech, AI, and blockchain technologies. This education is a valuable and important contribution to the further development of the PropTech ecosystem."

– Aaron Block, Co-founder & Managing Partner MetaProp

"Too little has been written about innovation in the real estate industry. Louisa and Olayiwola have done something about this by producing a scholarly yet at the same time pragmatic and comprehensive book focussed in digital transformation and PropTech."

– Professor Andrew Baum, Chairman, Newcore Capital. Senior Research Fellow, Green Templeton College and Emeritus Professor, University of Oxford

"This timely publication will help people navigate the huge variety and number of digital solutions that are now available to support the built and natural environment right across the property lifecycle. It can be argued that the sector still needs to reap the full benefits of digitisation by replacing inefficient 'analogue' processes with new and more productive digital equivalents to address long-standing issues around productivity, to support the newer net zero and sustainability agendas, and to deal with the effects of the pandemic and the changing ways in which real estate is being used – all the while showing a positive return on investment.

The continued and dramatic rise in the use of artificial intelligence in all its forms is now presenting the sector with both opportunities and risks, and in tandem with the drive to digitise, the sector will have a profound impact on skills, job roles, and how we train, develop, and nurture the next generation of property professionals."

– Andrew Knight, Global Data and Tech Lead, Thought Leadership and Analytics, RICS

"This book is a gold mine of information detailing the a-z of the real estate tech sector. A must-read for students and those looking to break into PropTech. It reads like a novel… you will not be able to put it down!"

– Anouk Khan, Chief Operating Officer Real Estate Women

"I wish this book was available when I started in PropTech! Louisa and Olayiwola demystify PropTech, offer valuable insights into current practices in real estate and construction, and present case studies to demonstrate how technology and innovation are reshaping the way we design, build, and live. Sharing these lessons is essential for innovators, professionals, and students who want to learn from the past and drive the industry forward."

– Courtney Cooper, Principal at Alate Partners and Co-founder of PropTech Collective

"The transformation of real estate through digitization, well presented in this book, is a testament to the evolving dynamics of the industry. As global PropTech actors (owners/

operators, startups, venture capitalists, etc) proliferate, the discourse around them has notably shifted, no longer merely focusing on the technological tools but emphasizing the genuine integration of these tools into the value chain of real estate. This evolution is cogently showcased through Dickins & Oladiran's writing weaving a seamless narrative around the profound marriage of technology and real estate. The expansion of the ESEP (Environmental, Social, Economic, and Physical) efficiency framework provides readers with a structured lens to analyze both the antecedents and outcomes of real estate digital tools and platforms.

Not only does it demystify complex technologies; but it also showcases their practical applications, paving the way for future innovations in real estate. Each chapter is tailored to ensure that students, educators, professionals, and even PropTech enthusiasts are well-equipped with a robust understanding of the sector's evolution. This work resonates with both the novice seeking foundational knowledge and the expert yearning for a deeper understanding. Through a meticulous blend of historical analysis, technological elucidation, and real-world application, this book is an indispensable guide for anyone navigating the ever-evolving corridors of real estate in the digital age."

– Jeffrey Berman, Partner Camber Creek

"Proptech and Real Estate Innovations' is an absolute must-read not just for professionals in the real estate technology sector, but for anyone who is interested in where the future of the built world is heading. The authors absolutely capture the pulse of this growing industry and the book is a fantastic way to understand the PropTech sector, gain insights on the major tech adoption trends happening today and provide a terrific lens into where innovation and technology are heading in the future in the built world. I have no doubt that technology will profoundly change the way the real estate industry invests, builds and operates in the future and this book is a terrific primer to gain great insight into the sector in a very digestible manner. I therefore highly recommend this book and appreciate the authors' outstanding leadership in being catalysts for reimagining real estate".

– Michael Beckerman, CEO of CREtech

"PropTech has taken centre stage in disrupting and redefining the value chain of real estate markets. However, in-depth knowledge of PropTech is lacking. This book by Olayiwola and Louisa addresses the many unknowns and myths surrounding PropTech. There is a careful and thorough unpacking of the PropTech ecosystem from the concepts to existing deployments to solve real estate problems. This book is a game changer in documenting advancements in the real estate markets through innovation and digital technology. It pioneers a new strand of innovativeness in the property industry. This book is a must-have on the shelves of real estate academics, professionals, students, and any keen enthusiasts interested in the changing dynamics of real estate".

– Professor Omokolade Akinsomi (PhD, MRICS), President-Elect-International Real Estate Society (2024), Associate Professor of Real Estate Finance, University of Witwatersrand, South Africa

"We have experienced exceptional advancement in real estate innovation and digital technology in the last decade and it is important that built environment students are exposed to the drivers, concepts, principles, theories and applications of these innovations and digital technologies. This textbook is a valuable resource that provides students and educators with a solid knowledge base for the application of digital technology in the built environment. Olayiwola and Louisa cover a wide range of digital tools, their applications to all segments of the real estate value chain and the emergent digital systems within these segments. They also present case studies that can be used as springboards for further innovation and digitisation in the industry to enhance environmental, social, economic, and physical/spatial efficiency. I, therefore, highly recommend this book to built environment academics, professionals, students, and other enthusiasts who are interested in transforming the use, operations, and services in the built environment".

– **Sammy Pahal, UK PropTech Association Managing Director**

PropTech and Real Estate Innovations

A Guide to Digital Technologies and Solutions in the Built Environment

Olayiwola Oladiran and Louisa Dickins

Routledge
Taylor & Francis Group

LONDON AND NEW YORK

Designed cover image: © Getty Images

First published 2025
by Routledge
4 Park Square, Milton Park, Abingdon, Oxon OX14 4RN

and by Routledge
605 Third Avenue, New York, NY 10158

Routledge is an imprint of the Taylor & Francis Group, an informa business

British Library Cataloguing-in-Publication Data
A catalogue record for this book is available from the British Library

ISBN: 978-1-032-18857-7 (hbk)
ISBN: 978-1-032-18713-6 (pbk)
ISBN: 978-1-003-26272-5 (ebk)

DOI: 10.1201/9781003262725

Typeset in Galliard
by codeMantra

*In the creation of this textbook, generative AI (Chat GPT-3) was utilised to assist
in certain aspects. These include suggesting sub-topics for further research and
identifying points of interest for chapters 4–8, compiling a glossary of terms, and
providing initial drafts for abstracts, and for section summaries in chapter 10. All
AI-generated content was thoroughly reviewed and edited to ensure it accurately
reflects the authors' thoughts and perspectives.*

Contents

About the Authors

Olayiwola Oladiran

Olayiwola is a real estate market analyst, researcher, and educator. He is currently a Lecturer in Real Estate at the University of Sheffield. Lead of Shef.AI Urban Interest Group and Co-Lead of Shef.AI Innovation and Pedagogy Interest Groups. He is also the Co-Founder of Bloomshill. Olayiwola holds a PhD in Real Estate Economics from the Henley Business School, University of Reading; an MSc in International Real Estate Investment and Finance from Nottingham Trent University; and a BTech in Estate Management from the Federal University of Technology, Minna, Nigeria. He is an RICS-Chartered Surveyor (MRICS) and a Fellow of the UK Higher Education Academy (FHEA). He is also a Board Member of the RICS Yorkshire and the Humber Regional Board. Previously, he worked as a Lecturer in Real Estate and Property Management at Leeds Beckett University and a Teaching Fellow at the Henley Business School, University of Reading. Prior to joining academia, he built his career in international real estate professional practice in the private and public sectors.

Olayiwola specialises in Real Estate Economics with core interests in real estate digital applications (PropTech), operational real estate, residential real estate, and urban markets. He has published his research output in several highly-ranked journals, and his research has been presented at several international conferences in France, Italy, China, the United States of America, the Netherlands, Zambia, Singapore, Germany, and the United Kingdom. He is currently a co-editor Journal of European Real Estate Research Special Issue on Digital Technology and Innovation in Real Estate Special and a reviewer for international peer reviewed journals within his area of interest. Over the years, he has led and collaborated on several research projects. He was the lead investigator on the PropTech Education Integration Framework development following the award of the University College of Estate Management Harold Samuel Research Prize in 2020. He also won the 2022 UK PropTech Association Special Achievement Award. He is enthusiastic about developing and promoting innovation in the real estate sector.

Louisa Dickins (LMRE)

Louisa is the Co-Founder of LMRE, which has rapidly become the market-leading global built environment recruitment platform and search consultancy with operations across North America, the United Kingdom, Europe, Asia-Pacific, and MENA.

To promote the industry she is so passionate about, Louisa set up a global podcast "The Propcast", where she hosts and invites guests from the built environment space to join her in conversation about innovation. In addition, Louisa mentors for REACH, a global scale-up program created by Second Century Ventures, an early-stage technology fund. Louisa is also a Committee Member of Real Estate Women (RE:WOMEN), a non-profit association whose mission is to create an equitable and inclusive industry, advocating for fair representation of women and minorities in leadership positions. Louisa was also a Board Director for the UK PropTech Association.

In recognition of Louisa's passion for PropTech, she was awarded the Young Property Personality of the Year award as part of the 2023 Property Awards, commending her entrepreneurialism, tenacity, and vision in everything she does at LMRE. Louisa is constantly inspiring the LMRE team to grow, expand, and create long-lasting relationships with both clients and candidates alike. This passion inspires not only her team but the whole PropTech world, where she continues to bridge the gap between real estate and technology, promoting PropTech externally and internally and bringing the top talent into the industry.

Louisa has spoken on several panels, including at the BuiltWorlds' Women in ConTech Luncheon 2022 and at MIPIM 2023, where she spoke about using PropTech to explore and improve the most interesting, impactful, and profitable issues in our sector. In September 2023, Louisa also moderated a panel at London PropTech Connect, discussing Mastering Leadership Challenges: Errors that are Losing the Modern Workforce, as well as a number of speaking engagements at Cityscape Global in Riyadh.

Outside of mentoring, speaking, and running LMRE, in September 2023 Louisa launched the second global built environment and innovation salary report. The purpose of this guide was to provide up-to-date data on salaries within PropTech and the Built Environment, whilst looking at the push and pull factors around hiring and retaining the top talent globally.

Preface

The acceleration of innovation and digitisation in the built environment has led to the creation of a multi-billion-dollar industry within the real estate ecosystem and the emergence of the buzzword "PropTech". Despite the increasing global interest, attention, and investment in this emerging sector, the knowledge base is fragmented. This has therefore made it difficult for built environment students, academics, and professionals to conceptualise this emerging phenomenon. This book provides a comprehensive and integrated knowledge base that students, scholars, and enthusiasts, particularly those whose interests are in real estate, and the built environment more broadly can draw on to conceptualise the "PropTech" phenomenon.

This book has three sections. The first section, comprising three chapters, attempts to demystify the ideas and concepts around PropTech, thereby laying the foundation for knowledge development. This positions the readers to understand PropTech from both conceptual and practical angles. This section also provides an overview of various digital tools and technologies that have transformed the real estate sector and concludes with a review of the PropTech evolution and investment capital over time. These insights enable readers to understand the process of real estate innovation and the key enablers of the transformation that has accompanied the PropTech evolution.

The second section provides an extensive analysis of the key drivers and the impact of digitisation and innovation across the real estate value chain. Each chapter focuses on a specialised area (or conglomerate of similar specialisations) of the real estate value chain. The first chapter of this section focuses on planning, land management, and urban management, while the second chapter focuses on real estate development, construction, and maintenance. The other three chapters focus on real estate investment, finance, and funding; valuation and appraisal; and agency, brokerage, marketing, and other allied services. Each chapter in this section begins by providing an overview of the real estate services and operations within the specialist area and highlighting the efficiencies and challenges within that segment of the real estate value chain. This is then followed by a critical analysis of the emergent tools and systems that have been driven by digital technology and concludes with a case study analysis of specific tools and platforms within that area using the ESEP efficiency framework.

Finally, the last section provides a range of insights, discussions, and perspectives on the current state of real estate innovation, its prospects, and its future trajectory. It highlights core areas of growth, enablers, and support systems that can further enhance innovation and digitisation in the built environment. Furthermore, it identifies relevant stakeholders, processes, and support systems and provides a review of their roles in deepening and strengthening innovation and digitisation in real estate.

Although the built environment is gradually embracing innovation and digitisation, there are still several "unknowns" and potential areas of growth. This book therefore concludes by highlighting the evolutionary prospects of PropTech and the potential for a heightened adoption of digital technology and innovation in the built environment sector. Readers are encouraged to transcend limitations, think out of the box, and actively contribute to the ongoing transformation of the built environment.

Foreword

Professor Saheed Ajayi

In a world where change is the only constant, the real estate industry, just like the wider construction and the built environment, has often been regarded as being traditional, inflexible, and slow to adapt. The last decade has, however, witnessed a great transformation that is driven by the rapid advancement of digital technologies. This transformation has given rise to a new buzzword, "PropTech," which captivates the imagination of real estate professionals, investors, entrepreneurs, academics, and other stakeholders. PropTech has become a major catalyst for change in real estate, heralding an era of innovation and digitisation.

PropTech and Real Estate Innovations: A Guide to Real Estate Innovations and Digital Technologies" is a pioneering work that unravels the complexities of PropTech and offers readers a comprehensive and integrated knowledge base to navigate this dynamic landscape. As a Professor in Digital Construction, I am excited to endorse this book, which is an invaluable resource for students, educators, professionals, and anyone interested in the future of real estate.

The authors, Dr Olayiwola Oladiran and Louisa Dickins, are leading figures in the field, with Olayiwola's extensive expertise in real estate economics and PropTech and Louisa's expertise in the global built environment recruitment. Their combined knowledge and passion for innovation shine through in this handbook.

Section One of the book serves as a strong foundation, demystifying the concepts and ideas surrounding real estate innovation and digital technology. It also defines and synthesises core concepts, provides a comprehensive overview of its enablers, and takes us on a journey through the evolution of PropTech and its significant players, providing essential historical context. In Section Two, the book delves into various segments of the real estate value chain, from planning to brokerage, showcasing how digital technology has revolutionised each area. The chapters present a blend of theoretical insights and real-world applications through case studies, providing readers with an understanding of how digital innovations are addressing inefficiencies and propelling the industry forward. Section Three explores the prospects and the future of PropTech, emphasising the key enablers, stakeholders, and systems that will further drive innovation and digitisation in real estate. The book concludes by summarising the key themes and inviting readers to embrace the transformative mindset that innovation brings to the real estate industry.

PropTech and Real Estate Innovations is more than just a handbook; it is a guide to navigating the ever-changing landscape of the built environment. It equips readers with the knowledge and insights they need to understand the past, present, and future of PropTech and its profound impact on real estate. I believe that this book is an essential addition to the library of anyone seeking to understand the intersection of technology and real estate. Olayiwola Oladiran and Louisa Dickins have provided a comprehensive and insightful resource that will be invaluable to students, educators, professionals, and industry leaders. It is a must-read for anyone looking to stay ahead in the world of PropTech and real estate innovation.

Professor Saheed Ajayi, BSc (Hons), MSc, PhD, PGCert AP, FHEA, MCIOB
Professor of Digital Construction and Project Management
and the Director of Construction Informatics and Digital
Enterprise Laboratory (CIDEL) within the School
of Built Environment, Engineering and Computer,
Leeds Beckett University, United Kingdom

Foreword

Professor Paloma Taltavull

This book is designed to cultivate integrated knowledge in PropTech, intended to assist both students and educators in comprehensively understanding this phenomenon and structuring its fundamental principles for learning. It serves as a handbook for PropTech students, organised from an educational perspective into three main sections; the first section introduces fundamental concepts; the second section delves deep into the key issues in real estate and how digital technologies are reshaping traditional market dynamics; and the third section provides a summary where the author applies this knowledge to new technological initiatives (such as the metaverse) and engages in discussions about professional roles, ethical concerns, curriculum development, and the future of the profession.

This book commences with the premise that the knowledge base in PropTech is fragmented, and it dedicates three sections to establishing connections between different aspects of real estate and their associated digital tools. This approach demonstrates their interrelationships along a logical path, covering some of the most relevant market issues, including supply, demand, and marketing transactions.

This book comprehensively addresses the multifaceted domain of PropTech, offering insights into its present state and future potential. It explores the key stakeholders, processes, and systems that drive innovation and digitalisation within the real estate industry.

It provides a comprehensive result of an in-depth literature review covering various aspects related to real estate. Given the diverse factors influencing the real estate market, the content, summaries, and organisation offer an encompassing understanding of how digital technology has already influenced and continues to impact the real estate market. This, in itself, is a valuable contribution.

A second significant contribution comes from the case studies presented in each section, which incorporate a majority of real-world cases, companies, innovations, and more that are currently active in the markets. These cases are meticulously analysed and compared, enhancing and complementing the underlying conceptual structure. This component effectively illustrates the point that certain theoretical concepts have yet to be unified.

The book's structure, organised into three parts, with the second part delving into the most relevant technical aspects, the first defining, and the third providing a future perspective, successfully achieves its objective as a study handbook. It introduces fundamental concepts, addresses all technical issues of interest, and concludes by rationalising the future evolution of the real estate market in the context of digital technology.

The book's commitment to education is apparent through precise definitions, clear writing, and the presentation of ideas in various tables. Each chapter includes a list of learning objectives, which is especially valuable for teachers and educators. The inclusion of multiple cases in each chapter further enriches the book's content, offering real-life examples of situations within the real estate markets.

Professor Paloma Taltavull

Professor in Applied Economics, Department of Applied Economic Analysis, University of Alicante, Spain.

Editor in Chief, *Journal of European Real Estate Research*

Chair of Real Estate Advisory Group – REM, United Nations Economic Commission of Europe (UNECE)

Director and Chair of Transformation of the Valencian Economic Model

Acknowledgements

We wish to thank all those who participated in our interviews. Special thanks to Aaron Block (Co-Founder and Managing Partner, MetaProp), Andrew Knight (RICS Global Data and Tech Lead), Ben Lerner (Managing Director, Lerner Associates PropTech M&A), Brad Greiwe (Co-Founder and Managing Partner at Fifth Wall), Bridget Wilkins (Head of Digital Citizen Engagement at the UK Department for Levelling Up, Housing and Communities), Dr Chlump Chatkupt (Founder and CEO, PLACEMAKE.IO), Jessica Williamson (PropTech Strategist at the UK Department for Levelling Up, Housing and Communities), John Fitzpatrick (Senior Managing Director and Chief Technology Officer at Blackstone), Michael Beckerman (CEO CREtech and CREtech Climate), Nick Moore (Founder and Executive Director, Lionpoint Group), Raj Singh (Managing Partner at JLL Spark), Sammy Pahal (Director General, UK PropTech Association), and Zach Aarons (Co-Founder and General Partner at Meta Prop).

We also thank several reviewers and colleagues who have helped us improve the quality of the book. Special thanks to Jonathan Dickins, Dr Jamiu Dauda (Leeds Beckett University), Professor Steven Hicks (University of Sheffield), and other academics at the Urban Studies and Planning Department at the University of Sheffield whose critical perspectives and thoughts shaped the output of this book.

We wish to acknowledge the amazing administrative and clerical support provided by Helen Clark and Imogen O'Riordan, and the graphical support provided by Anna-Maria Kotciuba. We also thank Ed Needle, Patrick Hetherington, Martha Luke, and other publishing professionals from Taylor and Francis for their remarkable help from the conception to the publication of this book.

Last but not least, our most important debt is to our families and friends, who have put up with us during the most intense days of this project and who have supported us throughout. Special thanks to Seyi Oladiran, Professor Johnson Oladiran, Desi Oladiran, Laolu Oladiran, Belinda Dickins, Richard Llyod, and Coot Lloyd. This book would not have been possible without their consistent support.

We have made every effort to minimise the errors and omissions. However, some errors and omissions may still have crept through. Regardless, we take responsibility for all errors. We would be grateful if readers kindly let us know of errors and omissions and suggested corrections as they engage with the book. We will address them in any future editions of this book.

Olayiwola Oladiran
Louisa Dickins

Glossary of Terms

3D City Models Digital representations of urban environments that allow for the examination of spatial relationships and simulation of urban processes using enhanced analytics and data.

Acquisitions The process of acquiring or buying something, often referring to businesses or assets.

Architecture, Engineering, and Construction (AEC) The sectors involved in designing, planning, and building structures.

Agency In the context of real estate, an agency refers to a company or organisation that acts as an intermediary between buyers, sellers, landlords, and tenants in property transactions.

Artificial Intelligence (AI) Science and engineering of intelligent machines or computer systems that simulate human intelligence.

Algorithm A set of rules or instructions for solving a problem or performing a task.

Analytics Capabilities Ability to collect, store, and analyse data to gain insights.

Angel Investors Individuals who provide capital for start-ups in exchange for ownership equity or convertible debt.

Application Programming Interface (API) Interfaces that allow different software applications to communicate and interact with each other.

Appraisal The process of assessing the value or worth of something, often used in the context of property valuation.

Augmented reality (AR) Technology that overlays digital information, such as images, text, or sounds, onto the real world, enhancing the user's perception of reality.

Asbestos A heat-resistant mineral once used extensively in building materials. It poses health risks when its fibres are inhaled, leading to lung diseases.

Asset Classes Categories of investments, such as stocks, bonds, and real estate.

Asset Management The strategic management of real estate properties to maximise their value, including tasks like maintenance, improvements, and financial planning.

Asset-backed Securitisation A financial process where loans or assets, such as real estate mortgages, are bundled together and turned into tradable securities, allowing investors to invest in these assets indirectly.

Automated Valuation Models (AVMs) Computer algorithms that estimate property values based on various data inputs.

Automation The use of technology to perform tasks or processes with minimal human intervention, often resulting in increased efficiency and reduced errors.

Business to Business (B2B) Refers to commerce transactions between businesses, where one business provides goods or services to another.

Big Data   Large and complex sets of data that are analysed to reveal patterns, trends, and associations, often requiring specialised tools and techniques.

Building Information Modelling (BIM)   A digital representation of a building's physical and functional characteristics, used for design, construction, and facility management.

Biodiversity   The variety of plant and animal species in a particular habitat or ecosystem.

Blockchain Technology   A digital system that securely records and verifies transactions across multiple computers, creating a transparent and unchangeable record.

Building Research Establishment Environmental Assessment Method (BREEAM)   A sustainability assessment method for buildings.

Brokerage   The business of arranging transactions between buyers and sellers.

Built to Rent (BTR)   Residential properties built with the intention of being rented out.

Built Environment   The human-made surroundings, including buildings, infrastructure, and landscapes.

Buzzword   A trendy or popular word or phrase that captures attention

Computer-Aided Design (CAD)   The use of computer software to assist in designing and drafting.

Cap Rates (Capitalisation Rates)   A measure used in real estate to indicate the potential return on an investment property, calculated as the ratio of net operating income to property value.

Carbon Emissions   The release of carbon dioxide and other greenhouse gases into the atmosphere, contributing to global warming and climate change.

Catalyst   Something that accelerates or triggers a process of change or transformation.

Central Business District (CBD)   The commercial and business center of a city.

Change   The process of something becoming different over time.

City Information Modelling (CIM)   An extension of Building Information Modelling (BIM) for urban planning and design, representing city-wide data and elements.

Cladding   The outer layer of a building, often made of materials like metal, wood, or stone.

Cloud Computing   Storing and accessing data and programs over the internet instead of a computer's hard drive.

Computerised Maintenance Management System (CMMS)   Software used to manage and track maintenance activities and resources.

Commercial real estate (CRE)   Properties used for business purposes, such as offices, retail stores, and industrial spaces.

Comparables   Comparable properties used to establish a baseline for property valuation based on recent market transactions.

Compliance   Adherence to rules, regulations, and legal requirements.

Computer Science   The study of computers, their technology, and their applications in various fields

Construction   The physical building of structures and properties.

ConTech (Construction Technology)   Application of digital technology and innovation to construction processes to enhance efficiency.

Correlation   Correlation is a statistical measure that quantifies the degree and direction of the relationship between two variables.

Customer Relationship Management (CRM)   A management system that helps maintain relationships with clients, allowing for personalised communication, follow-ups, and targeted marketing efforts.

Crowdfunding   The practice of funding a project or venture by raising small amounts of money from a large number of people, typically via the internet.

Cryptocurrency   Digital currency that uses cryptography for secure transactions, often associated with blockchain technology.

Customisation   Tailoring products, services, or experiences to individual preferences and needs.

Digital Assets (DA)   Digital assets are digital files or information that hold value, ranging from cryptocurrencies and digital artwork to software code and online content.

Data Aggregation   The process of collecting and organising large volumes of data from different sources for analysis.

Data Analytics   The process of examining large data sets to identify trends, patterns, and insights that can inform decision-making.

Data Analysis   The process of examining data to uncover patterns, relationships, and insights.

Data Privacy   The protection of personal and sensitive information from unauthorised access or use.

Data Security and Integrity   Protecting data from unauthorised access and ensuring its accuracy and reliability.

Data Standards   Common guidelines and rules for organising and structuring data to ensure consistency and interoperability.

Data Visualisation   Presenting data in visual formats to aid understanding and communication.

Database   Organised collection of data stored and accessed electronically.

Decarbonisation   The process of reducing or eliminating carbon emissions and transitioning to cleaner energy sources.

Degradation   Gradual damage or weakening of materials over time.

Digitisation   The process of converting information, data, or processes into digital formats for easier management, analysis, and accessibility.

Digital Asset Management   The management of digital assets, such as property rights and ownership records, using blockchain technology.

Digital Currencies   Virtual or digital forms of money, often associated with blockchain technology, such as cryptocurrencies.

Digital Intelligence   A set of skills and competencies related to understanding and navigating digital technologies effectively.

Digital Marketing   The use of digital channels to promote and advertise properties.

Digital Technology   Technologies that involve digital processing, storage, and communication of data, often referring to computers, software, and the internet.

Digital Technology Era   The period characterised by advancements in digital technology, influencing various aspects of industries, including real estate.

Digital Tokens   Units of value representing ownership or access rights to a digital asset, often stored on a blockchain.

Digital Tools   Software applications and technologies designed to enhance or optimise various processes or tasks.

Digital Transformation   The process of using digital technologies to change business operations and models fundamentally.

Digital Twins   Digital replicas of physical assets, processes, or systems that provide real-time data and insights, enabling monitoring, analysis, and optimisation.

Disintermediation   Disintermediation refers to the cutting out of middlemen/agents from real estate transactions. In this case, potential renters will contact the landlord directly.

Due Diligence   The process of thoroughly researching and investigating a subject, such as a property or investment, to gather relevant information before making a decision.

E-commerce   Buying and selling of goods and services over the internet.

Equality, Diversity, and Inclusion (EDI)   Efforts to promote fairness, equal opportunities, and diverse representation.

E-signature Software   Software for digitally signing documents.

Ecosystem   A complex network or interconnected system of elements, entities, or components that interact and influence each other.

Efficiency   The ability to accomplish a task with minimum wasted effort or resources.

Eminent Domain   The legal right of a government to acquire private property for public use, often with compensation to the owner.

Environmental, Social, Economic, Physical (ESEP) Efficiency   A framework for analysing how a technology or solution improves efficiency, addresses social aspects, impacts economics, and influences physical or spatial factors.

Environmental, Social, Governance (ESG)   A framework for assessing the environmental, social, and governance impacts of an investment or business, often used to evaluate sustainability practices.

Equity finance   Funding method involving personal savings or assets for real estate projects, including private equity investment and joint ventures.

Feasibility   The likelihood of a project or plan being successful and achievable.

Fintech   A portmanteau of "financial technology," referring to innovative technology solutions that enhance or automate financial services.

Fractional Ownership   Ownership of a portion of an asset, such as real estate, enabling smaller investors to participate.

Full Repairing and Insuring Lease (FRI Lease)   A lease where the tenant is responsible for all repairs and insurance costs.

For Sale By Owner (FSBO:)   Sellers list properties themselves without agents.

Funding Events   Instances where companies secure financial investments to support their operations and growth.

General Data Protection Regulation (GDPR)   A European Union regulation that addresses data protection and privacy rights for individuals.

Gearing   The extent to which a company's operations are funded by debt rather than equity.

Geographic Information System (GIS)   A system designed to capture, store, manipulate, analyse, manage, and present spatial or geographic data.

Graphics Processing Unit (GPU)   A specialised electronic circuit that accelerates the creation of images for output to a display.

Green Building   A structure designed and built using environmentally sustainable practices and materials.

Greenfield Development   Building on undeveloped land, typically in suburban or rural areas.

Hardware   Physical components of a computer or electronic device.

Hard Landscaping   The non-plant elements of outdoor spaces, such as pathways, walls, and paving.

Hedonic Pricing Models   Models that estimate the impact of specific property features on its value, considering both tangible and intangible factors.

Help-to-Buy Equity Loan   A UK government initiative allowing homebuyers to acquire properties with a five-year interest-free loan.

Heterogeneity   Diversity or variation within a particular group or context.

Homogeneity   The similarity or uniformity of properties within a certain category.

HVAC   Heating, Ventilation, and Air Conditioning systems.

HTML   A standard technology for structuring and presenting content on the web.

Initial Coin Offering (ICO)   A fundraising method for new cryptocurrency projects, similar to an IPO (Initial Public Offering) in traditional finance.

Illiquidity   The state of an asset or security that cannot be easily converted into cash without a significant loss of value.

Inertial Measurement Units (IMUs)   Electronic devices that measure and report a body's specific force, angular rate, and orientation.

Inclusivity   Ensuring that all individuals and groups are treated fairly, without discrimination.

Innovation   The introduction of new ideas, methods, or products that create value.

Instance-Based and Rule-Based Algorithms   Mathematical procedures that use specific cases or predefined rules to solve problems or make decisions.

Insulation   Material used to reduce heat transfer, keeping buildings warmer in winter and cooler in summer

InsurTech   The use of technology to enhance and streamline processes within the insurance industry.

International Cost Management Standards   Standards for managing and controlling costs in construction and real estate projects.

International Property Measurement Standards   Standards for measuring and assessing property dimensions and space.

International Land Measurement Standards   Standards for measuring land area and boundaries.

International Valuation Standards   Standards for valuing real estate properties.

International Valuation Standards Council (IVSC)   An organisation that publishes international valuation standards.

Interoperability   The ability of different systems, applications, or devices to work together and exchange information seamlessly.

Investment Portfolio   A collection of assets, including real estate, held by an investor to achieve specific financial goals.

Internet of Things (IoT)   A network of interconnected devices and objects that can communicate and exchange data.

Initial Public Offering (IPO)   The first sale of a company's shares to the public, making them available for trading on a stock exchange.

Information Technology (IT)   Refers to computer systems and networks.

Joint Venture   A business arrangement where two or more parties collaborate to achieve a common goal or project.

Key Performance Indicators (KPIs)   Measurable metrics used to evaluate the success or performance of an organisation or project.

Know Your Customer (KYC)   A process of verifying the identity and background of clients to prevent fraud and money laundering.

Land Administration Systems   Systems that record and manage information about land rights, ownership, and transactions.

Land Management   Activities related to the use and development of land resources.

Land Registry   A public record of property ownership and related transactions maintained by a government agency.

Leasing   Renting out a property to a tenant for a specified period.

LEED Systems   Reference to well-known building sustainability assessment methods, (Leadership in Energy and Environmental Design).

Lifecycle Stages   Different phases that a property asset goes through, from planning to disposal.

Liquidity   The ease with which an asset can be converted into cash without significantly affecting its price.

Loan to Value (LTV)   The ratio of a loan amount to the appraised value of a property.

Machine Learning   Technologies that enable computers to learn from data and perform tasks without explicit programming, often referred to as artificial intelligence.

Mainframe Computing   Early large-scale computing systems that formed the basis for more advanced digital applications.

Market Maturity   The level of development and stability within a market or industry over time.

Market Transparency   The availability of accurate and accessible information about market conditions, prices, and trends.

Market Value   The estimated amount for which a property should exchange between willing buyers and sellers in an arm's length transaction.

Mortgage   A loan taken to buy a property, where the property itself serves as collateral.

Megacities   Highly populous urban areas with populations exceeding several million people.

Metaverse   A collective virtual shared space, merging physical and virtual reality, where users can interact with each other and digital objects.

Mechanical, Electrical, and Plumbing (MEP)   Refers to the systems in a building responsible for mechanical, electrical, and plumbing functions.

Mezzanine Financing   A hybrid financing method combining debt and equity elements for real estate projects Mixed-use Development: A real estate project that combines different functions, such as residential, commercial, and recreational spaces.

Multiple Listing Services (MLS)   Database for sharing property information among brokers.

Mobile Technology   Technology used for cellular communication, including mobile phones and other devices.

Non-Fungible Tokens (NFTs)   Digital tokens that represent ownership of a unique item or asset using blockchain technology.

NoSQL Databases   Databases that store and manage unstructured or semi-structured data.

Object Storage   A scalable storage solution for unstructured data.

Operational Real Estate (ORE)   an emerging real estate asset class that contains specialist assets that are primarily utilised businesses operations and direct income generation income, e.g., hotels, pubs, purpose-built-student accommodation, care homes, filling station.

Peer-to-Peer Lending (P2P)   A decentralised method of borrowing and lending directly between individuals or entities, often facilitated through online platforms.

Purpose-Built Student Accommodations (PBSAs)   Housing specifically designed for students.

Pedagogical Approaches   Various methods and strategies used in teaching and education.

PropTech Education Integration Framework (PEIF)   A structured approach to integrating PropTech into real estate education.

PlanTech   The integration of digital technology with urban planning and design to enhance the efficiency and effectiveness of these processes.

Pay-Per-Click (PCP)   Advertising model where advertisers pay for clicks on their ads.

Predictive Maintenance   Using data analysis and technology to predict when equipment or systems might fail, allowing for timely maintenance to prevent breakdowns.

PropTech (Property Technology)   The application of digital tools and technologies to enhance the environmental, social, economic, or physical efficiency of real estate use, operations, and services.

Public-Private Partnerships   Collaborative efforts between public and private sectors to achieve common goals.

Research and Development (R&D)   Activities undertaken to create new products, services, or knowledge through systematic investigation.

Real Estate   Property, including land and buildings, used for residential, commercial, or other purposes.

Real Estate Investment Trusts (REITs)   Companies that own, operate, or finance income-generating real estate, allowing investors to access real estate assets without direct ownership.

Regeneration   Revitalisation or redevelopment of an area to improve its economic, social, and environmental conditions.

Regulatory Technology (RegTech)   Technology solutions that help organisations comply with regulations and manage risks more efficiently.

Remote Working   The practice of working from a location other than a traditional office, often facilitated by digital communication tools.

Residual Valuation   A technique to determine the value of a property's land by subtracting development costs from potential future value.

Retrofitting   The process of modifying or upgrading an existing structure or system to improve its performance, energy efficiency, or functionality.

Royal Institution of Chartered Surveyors (RICS)   A professional body for qualifications and standards in land, property, infrastructure, and construction.

Return on Investment (ROI)   Measure of profitability.

Software as a Service (SaaS)   A software distribution model where applications are hosted by a third-party provider and made available to customers over the internet.

Scaling   The process of expanding operations and increasing resources to accommodate growth and meet higher demand.

Shared Economy   An economic system where individuals share resources, products, or services with each other, often through digital platforms.

Smart Buildings   Buildings integrated with technology to optimise energy usage, comfort, and efficiency, often involving sensors and automation.

Smart Cities   Urban areas that use technology to improve efficiency, sustainability, and the quality of life for residents.

Smart Contracts   Self-executing contracts with the terms of the agreement directly written into code.

Smart Homes   Residences equipped with IoT devices and technology to enhance automation and convenience.

Soft Skills   Non-technical skills that are important for effective communication and collaboration.

Solid-State Storage   Storage technology using memory chips to store data.

Spatial   Pertaining to space, especially in relation to its dimensions, arrangements, and usage.

Space-time Commoditisation   The transformation of physical and temporal spaces into valuable commodities that can be traded in the global marketplace.

Structured Query Language (SQL)   Language used to manage and manipulate databases.

Stakeholders   Individuals, groups, or entities with an interest or involvement in a particular project or industry.

Start-ups   Newly established companies, often with innovative business models or technologies.

Security Token Offering (STO)   A fundraising method similar to ICO, where tokens represent ownership in a real-world asset or security.

Structured Data   Information organised in a specific format for easy processing.

Sustainability   Practices that aim to meet present needs without compromising the ability of future generations to meet their own needs, often focusing on environmental, social, and economic aspects

Tokenisation   The process of converting rights to an asset into a digital token on a blockchain.

Syndication   A method of raising capital for real estate projects by pooling funds from multiple investors.

Tokenisation   The process of representing assets or ownership rights as digital tokens on a blockchain.

Transparency   Openness and honesty in the disclosure of information and decision-making processes.

UK PropTech Association (UKPA)   A non-profit membership organisation focused on driving digital transformation in the property sector in the UK.

Unicorns   Privately held startups or companies with a valuation of over $1 billion.

Unstructured Data   Information that lacks a specific format or organisation.

Unique Property Reference Number (UPRN)   A unique identifier for properties in the UK.

Urban Design   The process of designing and shaping the physical layout and organisation of cities, towns, and other urban areas.

Urban Development   The planning, growth, and management of cities and urban areas.

Urban Planning   The process of designing and shaping cities and urban areas to achieve sustainable development and liveability.

Urbanisation   The process of an increasing population moving from rural to urban areas.

User Interface   The visual elements and controls through which a user interacts with a computer system, software, or hardware.

Vacancy Rates   The percentage of unoccupied rental properties or spaces in a given area.

Value-Add Investments   Real estate investment strategy that involves making improvements to increase a property's value, often focusing on management and operational enhancements.

Valuation   The process of estimating the financial value or worth of a real estate asset.

Value Chain   The sequence of activities or processes that contribute to creating and delivering a product or service to customers.

Venture Capital   Investment provided to early-stage and high-potential companies in exchange for ownership or equity.

Velocity   Speed or rate of motion.

Viability   The ability of something to be done effectively or sustainably.

Virtual Reality (VR)   A technology that creates a simulated environment for users to interact with using specialised equipment.

Web Scraping   Extracting data from websites.

Workflow Automation   Using technology to automate and streamline manual tasks and processes in a business workflow.

Yield   The income generated by a property as a percentage of its value.

Zoning   The division of a city or municipality into different zones for specific land uses and development regulations.

Section One

Introduction

Change is a fundamental element of life and human existence. The Greek philosopher, Heraclitus, neatly summed it up: "change is the only constant in life". People change, systems change, societies change, cities change, procedures change, and certainly the environment is changing. Indeed, there is hardly a "constant" in life. The property industry, however, has a reputation for being traditional, rigid, and slow to evolve. As wave after wave of technological advancement transformed sectors such as communication, media, finance, trade, and hospitality, the property sector seemed to evolve at a grindingly slow pace. In the last decade, however, technology-enabled innovations have transformed the operational and practice areas across the real estate value chain. The buzzword "PropTech" has therefore emerged amid this rapid transformation.[1]

Over this decade, PropTech has swiftly advanced accompanied by debates and discussions about the imminent digitisation of most real estate operations and practice areas. This recent accelerated growth has created enormous opportunities for investors, entrepreneurs, professionals, and industry leaders. There are currently over 8,000 PropTech firms globally with a multi-billion-dollar capital base. However, despite the huge capital in the PropTech space and the increasing global interest and attention from industry leaders, investors, and the academic community, the knowledge base is fragmented; so, conceptualising PropTech has been a challenge for students, academics, and right across the industry. This challenge is exacerbated by the limited scholarly contribution. Real estate students and educators, therefore, still lack a comprehensive and integrated knowledge base that they can draw on to conceptualise PropTech and real estate innovation.

PropTech may be complex but the use of simplified concepts, examples, and illustrations can untangle these complexities. This section, therefore, looks to demystify the concepts and ideas around real estate innovation and digital technology, thereby laying the foundation for the knowledge development of real estate students, educators, professionals, and members of the PropTech community. Readers can then understand the PropTech phenomenon from both conceptual and practical angles.

Chapter 1 begins with a description of some of the core concepts of PropTech and goes on to synthesise its various definitions and descriptions. It further proposes a simplified and easy-to-comprehend PropTech definition that captures its core elements, while enabling students and new enthusiasts to appreciate and understand the phenomenon. The chapter concludes by identifying inefficiencies associated with real estate use, operations, and services and proposes a conceptual framework that captures the core impact areas of innovation and digitisation.

DOI: 10.1201/9781003262725-1

Chapter 2 provides an overview of various digital tools and technologies that have transformed the real estate space. This provides a baseline description of these technologies so that readers can understand their mechanisms and applications with a focus on how they have been applied to real estate.

Chapter 3 concludes this introductory section with a review of the PropTech evolution and its associated investment capital over time, focusing on some of the major companies that have driven recent real estate digitisation and innovation. This will let readers gain an understanding of the major landmarks and the real estate innovation process.

Learning Outcomes (LO)

At the end of this section, the reader should be able to:

1 Understand how the blend of digital technology with real estate has enhanced the efficiency of real estate use, operations, and services.
2 Understand the basic principles and concepts of PropTech and real estate innovation and describe them in concise terms.
3 Describe the core digital technologies that have transformed real estate operations and practice.
4 Understand the global real estate digitisation process and global PropTech capital trends.

Note

1 PropTech combines the first four alphabets of two words: "Property" and "Technology".

Chapter 1

Background & Introduction to PropTech and Innovation in Real Estate

1.1   PropTech: What Is It?

PropTech is typically perceived as a sophisticated realm of real estate and often comes across as being too complex to understand. Simplifying it can lay the foundation for more advanced knowledge and comprehension. The illustrations below can help demystify and simplify the perceived complexities associated with PropTech. As illustrated in Figure 1.1, PropTech was derived by combining the first four letters of "property" and "technology".

As seen in Figure 1.2, our real estate-related needs have not changed per se, yet the approach to satisfying these needs has changed significantly in the last few decades. The four illustrations show some of the areas where traditional real estate operations and practices have been replaced using innovative and contemporary approaches. These advancements have been improved and enabled through the adoption of various digital tools and platforms. Real estate design, planning, construction, leasing, management, funding, finance, marketing, transaction, disposal, mortgage, valuation, and appraisal are core areas that have been transformed by automation and digital technology. It is these digital and automated real estate operations, practices, and products that have been wrapped up in the buzzword "PropTech". PropTech encapsulates the fundamental change occurring in the real estate sector which is often referred to as the digital transformation of the built environment.

Although one of the largest asset classes and a core element of the world economy (Saull et al., 2020), real estate has been conventionally considered to be a conservative and slow-moving industry (Braesemann and Baum, 2020). This sector is traditionally believed to be shaped by the importance of personal connection and the slow adoption of novel digital technologies (Porter et al., 2019). PropTech, then, is the result of recent accelerated innovations that have emerged through an intersection of various elements including data, analytics, advancement in digital technology, and innovative approaches to real estate use, operations, and services.

Innovation is one of the core underpinnings of PropTech (Block and Aarons, 2019), and it is best described as the development of new ideas, methods, or products – often following imagination, deep reasoning, and thinking unconventionally – to break the barriers associated with traditional systems of use, operations, or practice. Contemporary real estate innovations are now dependent on digital technology; this advance has been transformative to real estate operations. Associated improvements in data availability and analytics have

DOI: 10.1201/9781003262725-2

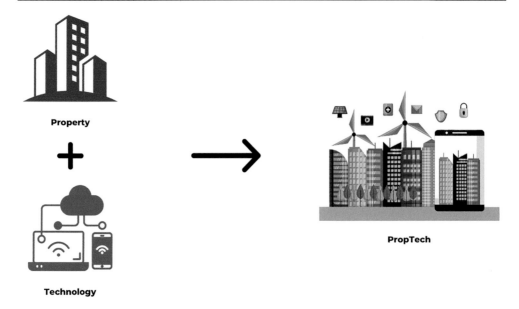

Property

Technology

PropTech

Figure 1.1 Illustration of the formation of PropTech.

also significantly supported this contemporary recasting. Although some of these innovations are not typically classified as PropTech (such as shared spaces, co-working, flexible working, and leases), they are inherently driven and sustained by digital technological tools.

Regardless of the level of variation in the adoption and acceptance of digital technologies in real estate, there is no argument that the real estate sector is riddled with some major inefficiencies: high transaction costs, illiquidity, heterogeneity, etc. Innovation and digital technology have, therefore, provided alternative approaches that can enhance the sector's efficiency through improved velocity, lower transaction cost, and improved liquidity and commoditisation (University of Oxford Research, 2017). Consequently, PropTech and other contemporary real estate innovations generally target deficiencies and inefficiencies associated with traditional real estate operations and practice, either directly or indirectly.

The PropTech phenomenon has led to new methods, approaches to practice, tools, and products in the real estate ecosystem, and this is increasingly attracting a significant volume of financial capital. For instance, data from Unissu (2022) reveals that in 2010, PropTech attracted an investment capital of approximately $900 million, but by 2021, the investment had multiplied twentyfold to around $19 billion. This is mirrored in the volume of PropTech firms set up. Approximately 1932 firms were set up between 1970 and 2010, and this exploded threefold to over 6,000 firms being founded in the following ten years alone. Additionally, PropTech has attracted a diversified pool of talent principally in terms of gender, age, ethnicity, nationality, and education (University of Oxford Research, 2017). As PropTech continues to attract a high volume of financial capital and an astronomical increase in human and entrepreneurial capital, it has unsurprisingly spawned several real estate practices and operational digital tools, platforms, apps, and new products.

It is difficult to define a boundary for PropTech, and debates continue as to its extent and scope. To illustrate, what should be classified as PropTech and what should not? Again, seeing that technology itself dates back several centuries, what generation of technology can be viewed as the starting point in the PropTech evolution? The second question is particularly critical as there are several postulations that digital tools have been part of real estate operations and practice for several decades. That old office staple, the Excel spreadsheet, for example, has been used to input and analyse data for property management and valuation spreadsheets way before more advanced software such as Argus.

Jan 1970 Helen (Manchester, UK)

Jan 2023 Helen (Manchester, UK)

SCENARIO	
Redevelopment project entailing the planning and design of a new residential area	
NEED	
Prepare and design planning	
APPROACH	
Physical site visit and inspection, evaluation of the urban plan and sketches, citizen engagement, drawing draft design, compiling reports.	Intelligent and automated planning system drawing on data from an urban area and combining this with algorithms from planning policy and law. Automated design such that corrections can be made more efficiently and simulation of various scenarios can be done.
INEFFICIENCIES AND SOLUTIONS	
Environmental inefficiency: carbon emission from search-related transportation. **Economic inefficiency:** lower competition; higher borrowing costs, longer search time. **Social inefficiency:** frustration, anxiety, fewer options; the difficulty of comparison and selection; the higher likelihood of getting a loan with unsuitable terms.	**Environmental solution:** less search-related travel and lower transportation-related carbon emission. **Economic solution:** higher competition; more products; lower borrowing cost; shorter search time. **Social solution:** variety; more options, easier choice process, identification and selection of mortgage products with higher utility; access to integrated information system for better decision making.

*The scenarios painted in these illustrations may play out in varied ways for different individuals. They are simply typical real estate-based scenarios and they do not represent any real-life events. The approaches, inefficiencies and solutions are not exhaustive and readers are encouraged to create similar tables of comparison.

Figure 1.2 Further illustrations with additional context.

Jan 1970 Ahmed (Dubai, UAE)

Jan 2023 Ahmed (Dubai, UAE)

SCENARIO	
Property and Facilities Manager of a high-rise building	
NEED	
Ensure that all the building systems function optimally and efficiently	
APPROACH	
Each system had to be monitored individually and it often required physical checks and maintenance.	Use of BIM, AI, sensors VR, AR, IoT, drones etc to monitor, manage and react to management issues remotely and using automated approaches.
INEFFICIENCIES AND SOLUTIONS	
Environmental inefficiency: high carbon emission from extensive use of equipment and facilities; energy wastage from the unmonitored use of facilities and systems. **Economic inefficiency:** high level of energy wastage leading to higher running costs; higher maintenance costs; longer operational time. **Social inefficiency:** frustration, anxiety, system breakdown; no pre-emptive maintenance, difficulty in managing various systems, personnel etc. **Physical/ spatial inefficiency:** space underutilisation; poor and inefficient use of space.	**Environmental solution:** lower energy use and less wastage; lower carbon emission. **Economic solution:** reduced energy consumption and wastage; lower running cost; lower maintenance cost; shorter operational time. **Social solution:** access to integrated information system for better decision and operation; pre-emptive maintenance minimising system breakdown, integrated management of personnel and facilities; better comfort of tenants; ease of use of facilities. **Physical/spatial solution:** optimal space utilisation.

*The scenarios painted in these illustrations may play out in varied ways for different individuals. They are simply typical real estate-based scenarios and they do not represent any real-life events. The approaches, inefficiencies and solutions are not exhaustive and readers are encouraged to create similar tables of comparison.

Figure 1.2 Continued.

Jan 1970 Bola (Lagos, Nigeria)

Jan 2023 Bola (Lagos, Nigeria)

SCENARIO
Relocating to London; does not know anyone

NEED
Temporary/rental accommodation

APPROACH	
Physical search hotels (temporary accommodation); and rental accommodation (medium term); involved visits to various local agents and several property inspections; difficulty in assessing and comparing various factors such as distance, amenities, and neighbourhood quality.	Online search (hotel.com, Zoopla, Rightmove, Airbnb); use of filters to identify most suitable properties; better comparison of properties based on distance, amenities, transportation routes etc; virtual inspection.

INEFFICIENCIES AND SOLUTIONS	
Environmental inefficiency: carbon emission from search-related transportation. **Economic inefficiency:** search cost, high cost of temporary accommodation, longer search time. **Social inefficiency:** frustration, anxiety, fewer options; the difficulty of comparison and selection; a higher likelihood of the selection of an unsuitable property.	**Environmental solution:** less search-related travel and lower transportation-related carbon emission. **Economic solution:** faster and cheaper search; lower cost and need for temporary accommodation. **Social solution:** variety; more options, easier choice process, identification and selection of property with higher utility; access to integrated information system for better decision making.

*The scenarios painted in these illustrations may play out in varied ways for different individuals. They are simply typical real estate-based scenarios and they do not represent any real-life events. The approaches, inefficiencies and solutions are not exhaustive and readers are encouraged to create similar tables of comparison.

Figure 1.2 **Continued.**

Jan 1970 Anish (New York, USA)

Jan 2023 Anish (New York, USA)

SCENARIO	
Recently saved up to buy his first home in Manchester	
NEED	
Residential mortgage	
APPROACH	
Physical visits to financial institutions to make applications.	Online mortgage search targeted at either traditional financial institutions (banks), FinTechs or peer-to-peer lending platforms; online or physical search for mortgage brokers.
INEFFICIENCIES AND SOLUTIONS	
Environmental inefficiency: carbon emission from search-related transportation. **Economic inefficiency:** lower competition; higher borrowing costs, longer search time. **Social inefficiency:** frustration, anxiety, fewer options; the difficulty of comparison and selection; the higher likelihood of getting a loan with unsuitable terms.	**Environmental solution:** less search-related travel and lower transportation-related carbon emission. **Economic solution:** higher competition; more products; lower borrowing cost; shorter search time. **Social solution:** variety; more options, easier choice process, identification and selection of mortgage products with higher utility; access to integrated information system for better decision making.

*The scenarios painted in these illustrations may play out in varied ways for different individuals. They are simply typical real estate-based scenarios and they do not represent any real-life events. The approaches, inefficiencies and solutions are not exhaustive and readers are encouraged to create similar tables of comparison.

Figure 1.2 **Continued.**

Would it be appropriate to classify this as PropTech simply because it aided certain real estate operations or practices? Although this textbook does not aim to provide definitive answers, it alludes to and acknowledges the broadness, blurred boundaries, and controversy around PropTech and its scope.

The next sub-chapter (1.2) specifically attempts to address some of these issues and examines the various PropTech delineations. It proposes a simple, yet all-encompassing PropTech definition.[1]

1.2   PropTech: Scope, Definitions, and Associated Issues

Scholars and industry experts have provided various definitions and descriptions to PropTech; nonetheless, there is still not a generally accepted definition. PropTech has become a buzzword and a descriptor for Property Technology (Chandra, 2019; Walker, 2019) following prior buzzwords such as "ReTech" (residential real estate technology) and "CreTech" (commercial real estate technology). It has also been referred to as real estate tech, RETech, or RealTech (Block and Aarons, 2019). With PropTech assuming various dimensions and applications, it is important to synthesise existing definitions and conceptual approaches to enable students and other stakeholders to gain a firm conceptual foundation that their understanding of what PropTech will be built upon.

Unissu (2020) provides some insight. The article analysed one of the first PropTech definitions provided by James Dearsley and Professor Andrew Baum:

> ….one small part of the wider digital transformation of the property industry. It describes a movement driving a mentality change with the real estate industry and its consumers regarding technology-driven innovation in the data assembly, transaction, and design of buildings and cities.

This definition is an essential part of the PropTech conceptualisation effort, and it captures two important aspects of current philosophies: the first element presents PropTech as a small part of the broader real estate ecosystem which is affecting (and has also been affected by) many and varied industries and their numerous innovations (University of Oxford Research, 2017); the second element captures the intangible or abstract realm of a change in mentality requiring knowledge, understanding, appreciation, and application of digital innovation – leading to a move from traditional practices. This definition, however, seems complicated and does not sufficiently clarify the classification of PropTech tools and its platforms; this definition, then, cannot be used to classify businesses or tools either as PropTech or not. Unissu (2020), however, builds on this interpretation and proposes a conceptualisation of PropTech tools through the lenses of physical (real) estate. According to Unissu (2020):

> … if a technology product applies to one or more of the lifecycle stages of any property asset, anywhere in the world, then that classifies it as PropTech.

This better captures the global nature of PropTech and introduces its "application" element. It also brings the asset lifecycle into focus. Yet, although this provides a useful framework to identify, analyse, compare, and contrast companies in the same stages of the asset lifecycle, the definition limits real estate to the building and management of the physical property. The real estate ecosystem transcends building development and management, covering other areas such as planning, finance, and valuation, which are not captured in this definition. PropTech overlaps various aspects of real estate from Fintech to the physical aspects of the buildings to hardware solutions like warehousing robotics. Furthermore, the proposition that any technology product that applies to a property asset is PropTech is also questionable: some digital tools, despite being applicable to real estate, are also applicable to other non-real estate sectors; they, therefore, may not necessarily be referred to as PropTechs.

Professional bodies such as the Royal Institution of Chartered Surveyors (RICS) have also attempted to propose a PropTech definition. In the RICS glossary page (RICS, 2022), PropTech is described as

> … the fusion of words 'Property' and 'Technology' and refers to all aspects of technology and how it impacts the built environment. This may include software, hardware, materials or manufacturing. PropTech is a general all-encompassing term but is often used to specifically refer to the small start-ups that are using technology to address market problems.

This definition provides a broader scope of PropTech contextualisation with the all-encompassing term of the built environment, specifying both hardware and software. It also acknowledges that PropTech is typically dominated by small start-ups that are using technology to address real estate market problems. This definition, though, appears to limit the scope of PropTech to small start-ups and excludes bigger players in the industry that are making a massive impact.

These preceding definitions appear complex, and this can make PropTech comprehension difficult for students and emerging PropTech enthusiasts with little or no prior exposure. In developing the PropTech Education Integration Framework (PEIF), Oladiran and Nanda (2021) propose a much simpler exposition in the glossary of terms. They define PropTech as the

> Application of digital and IT systems/tools to real estate operations and practice.

This definition is certainly simple and may perhaps be ideal for students learning. It emphasises the term "application". It also captures the broad scope of real estate and introduces the various areas of real estate to which PropTech is connected. However, it does not sufficiently capture real estate complexities. A further definition from Oladiran et al. (2022) builds on this. It describes PropTech as

> The deployment of digital and IT applications to the different elements of the value chain of real estate typically to enhance and maximise environmental, social, economic and physical efficiency.

This introduces "efficiency" into PropTech definitions, and it proposes an idealist's approach advocating that a PropTech product should enhance the environmental, social, economic, and physical (ESEP) efficiency of real estate. This further suggests that PropTech companies and products that do not satisfy all four of the listed efficiency frontiers should be excluded. However, PropTech tools have varying objectives and the solutions they provide may not necessarily address all four dimensions of inefficiency.

Following the critical analysis of the previous definitions, we propose the following definition:

> The deployment of digital and IT applications to the various sectors of the real estate ecosystem and value chain to enhance environmental, social, economic or physical efficiency of real estate use, operations and services.

This is simple yet encompassing. Although it appears to maintain the idealist approach and the ESEP efficiency dimensions mentioned in Oladiran et al. (2022), it presents them as a list of potential efficiency categories. In other words, a PropTech tool may not necessarily address all four of the efficiency categories, hence the replacement of the word "and" in the previous definition of Oladiran et al. (2022) with "or".

The keywords/phrases in this definition are expounded further below:

- "digital and IT applications": One fundamental element of PropTech is the use and adaptation of one or more digital tools and IT technologies in real estate. This suggests that for a product or platform to be categorised as PropTech, it must involve some digital and IT applications.
- "sectors and segments of the real estate ecosystem and value chain": This part of the definition captures the scope and dimensions of real estate that PropTech encapsulates. For a tool to be referred to as PropTech, it should relate to core real estate asset classes such as commercial real estate (office, retail, and industrial), residential real estate (rental, sales, and mortgage), operational real estate (such as purpose-built student accommodation, built to rent, hotels, healthcare), and other secondary sectors such as institutional real estate (e.g. schools, universities, police stations, and agricultural land). The value chain within these real estate sectors includes services and operations like planning, land management, urban design, development, construction, valuation, investment, finance, asset management, agency, brokerage, and conveyance.
- "enhance environmental, social, economic, or physical efficiency": This element of the definition captures the problem-solving and solution-driven elements of PropTech. It proposes that a PropTech tool should have the potential to address some of the areas of real estate inefficiency broadly categorised as environmental, social, economic, or physical. The definition in Oladiran et al. (2022) suggests that each PropTech tool or platform should fulfil all four dimensions; however, some PropTech tools may not necessarily address them all.

The four efficiency dimensions are further analysed in Section 1.3.

1.3   PropTech Environmental, Social, Economic, and Physical (ESEP) Efficiency Framework

As stated in Section 1.2, the real estate market is riddled with several layers of inefficiency and PropTech tools and products generally attempt to address these, serving as a bridge between the inefficiencies and their solutions. There is, therefore, the need to analyse the various dimensions of these shortfalls.

Real estate is one of the most heterogeneous sectors with a wide range of sub-sectors, assets, operations, and practice areas which, in many cases, overlap across the real estate ecosystem and value chain. The plethora of institutions and individuals in this sector makes planning, development, management, and associated transactions complex, difficult, and often inefficient. This is exacerbated by the lack of transparency in the markets, long delays, and the high cost of assets, transactions, and investment. Table 1.1 lists some of the real estate sector inefficiencies.

In a home purchase transaction in the UK, for example, there is the seller, the buyer, the estate agents (in some cases for the buyer and seller), the solicitors (for both buyer

Table 1.1 Examples of the Areas of Real Estate Inefficiencies

- High transaction cost
- High management and operational cost
- High capital involved
- Multiple parties involved in a single transaction
- Illiquidity
- Asset deterioration, wear and tear
- Information asymmetry
- Lack of market transparency
- High carbon emission
- Impact on wildlife and biodiversity
- Time intensive
- Various risks – economic, health and safety, etc.

Sources: Authors.

and seller), a mortgage broker, a mortgage institution, an insurance company, a building surveyor, etc. (as shown in Figure 1.3). The numerous parties involved will increase the level of information asymmetry, the transaction time, overall transaction cost bureaucracy, and conflicting considerations.

The kinds of issues highlighted in Figure 1.3 have underpinned the real estate and PropTech evolution. The ESEP efficiency framework, therefore, proposes four core dimensions of inefficiencies that PropTech tools should aim to address (Figure 1.4). This framework can either be applied as an *ex-ante* assessment tool (before the tool development) or an *ex-post* assessment tool (after the tool development). As an *ex-ante* assessment tool, the framework will assess the level at which an existing PropTech tool or product addresses the ESEP dimensions of real estate use, operation, or services that it seeks to improve. The *ex-post* application of the ESEP framework will examine the potential of a PropTech tool that is still being proposed or developed to enhance the ESEP efficiency dimensions. Each of these efficiency dimensions is discussed below and the inefficiencies refer to the traditional approach of performing an activity or task, generally around 1995 just before the dot.com boom.

1.3.1 The First "E" Dimension in the ESEP Efficiency Framework – Environmental

The first "E" in the ESEP framework represents the environmental efficiency dimension of the framework. Environmental consideration is now core to real estate use, operations, and service delivery. Real estate-related activities account for a significant proportion of global carbon emissions. More specifically, real estate construction, development, management, transactions, and other related activities are the core contributors to global carbon emissions. According to a United Nations Environment Programme (2020) report, 38% of total carbon emissions in 2019 were attributed to real estate-related activities: 10% from building construction, 17% from residential buildings, and 11% from non-residential buildings. Furthermore, 35% of global energy consumption that year came from real estate-related activity: 5% from buildings and construction, 22% from residential buildings, and 8% from non-residential buildings. Residential buildings, therefore, account for a significant proportion of both global energy and carbon emissions.

SELLER

BUYER

ESTATE AGENTS

SOLICITORS

MORTGAGE BROKER

MORTGAGE INSTITUTION

INSURANCE COMPANY

BUILDING SURVEYOR

Figure 1.3 Some of the parties that may be involved in a house purchase transaction.

Figure 1.4 The PropTech ESEP efficiency framework.

Building and construction-related carbon emissions arise from direct and indirect activities. Direct emission comes from building construction materials such as steel, glass, and cement, while indirect emission results from activities such as power generation for electricity and commercial heat. Building materials are typically sourced from trees and the quest to meet the growing demand for wood has led to major deforestation in several countries. Other real estate-related emissions can arise from the transportation of construction materials, transportation for site inspections, property viewings, and site meetings; emissions from fossil fuels are generated, for example, from the use of plant and machinery for construction and building maintenance and management. Furthermore, traditional professional reports, designs, floor plans, and other related tasks consume high volumes of paper. These are some of the unsustainable practices that have tainted the sector.

The real estate industry faces a huge challenge: decarbonising the existing building stock before 2050 (also known as the 2050 net zero target). A practical approach to achieving this target is to integrate this agenda into current building and construction designs. However, this alone may be insufficient, as 70%–90% of the buildings that will be in use by 2050 have already been constructed; 18% of commercial real estate in England and Wales do not meet the current minimum energy efficiency standards (Pi Labs, 2020).

Pi Labs (2020) also reveals that 35% of buildings in Europe are over 50 years old and an estimated 75% of these buildings are environmentally inefficient. With a projected increase in population and urbanisation in the global south and urban inhabitants projected to grow from 55% of the world's population to 68% (approximately 2.5 billion people), the report estimates that approximately £342 billion will be required to bring all homes up to the net zero 2050 targets in the UK residential sector alone. Commercial buildings are also energy intensive often running on fixed schedules without controls that are subject to detailed occupancy information. Furthermore, building elements such as lighting do not adjust with ambient daylight, and they are typically binary (switch on or off). These issues are, therefore, of notable importance with data showing that approximately 70% of building floor spaces that will be delivered by 2050 will be situated in areas with limited or no building energy codes in place (International Energy Agency, 2019).

Accordingly, it is becoming increasingly important for real estate stakeholders (space users), stakekeepers (built environment professionals and practitioners), and stakewatchers (professional bodies, regulators, and the public sector) to consider carefully the various channels through which innovation and digital technology can enhance environmental sustainability in the built environment.[2] In fairness, PropTech and contemporary real estate innovations in general make a considerable effort (directly or indirectly) to address the environmental inefficiencies associated with real estate. Smart buildings, for instance, have been designed to take energy-efficient decisions away from humans in order to minimise some of the environmental inefficiencies associated with building use. The first 'E' (environmental) dimension of the PropTech ESEP efficiency framework, therefore, examines the level of reduction in carbon footprint that PropTech and other real estate innovations provide, or can potentially provide, relative to traditional operations and practice. The environmental efficiency consideration should, therefore, be integrated into the design and planning of PropTech products and tools, targeting specific environmental issues that are associated with traditional practices and operations.

1.3.2   The "S" Dimension in the ESEP Efficiency Framework – Social

The "S" in the ESEP framework represents the social efficiency dimension with a focus on people and societies. This acknowledges the importance of the relationship between real estate and the physical and health conditions of real estate users, operators, and professionals. It connects to the abstract realm of real estate, so it is subjective and difficult to quantify or measure. It attempts to capture the social aspects and impact of PropTech and contemporary real estate innovations. The importance of these innovations is highlighted as mediums for the enhancement of human use of space. It stresses the utility, benefits (personal and collective), ease, comfort, health, security, and safety that these tools can attract.

The major question that the "S" investigates is: how does a real estate innovation or digital tool improve efficiency in the use, comfort, security, ease of use, health, and safety of the user, operator, professional, or other stakeholders? Other related questions include: how does the tool or product improve the ease of real estate transactions? How easy is it to maintain buildings, keep records, and manage them? Recent debates consider the relationship between real estate use and physical and mental health. It is, therefore, also important to consider how PropTech and real estate innovations improve the well-being of stakeholders using real estate and also stakewatchers that practice in and operate real estate assets and facilities.

There are other social dimensions that may appear to overlap with the economic. For instance, although housing (un)affordability is typically considered an economic issue, it also has its social aspects. Some of the outcomes of this unaffordability are inadequate housing, informal housing, slums, overcrowding, and homelessness. Consider this as a global issue: it is estimated that 1.8 billion people live in inadequate housing including slums, informal housing, and overcrowded settlement (Pi Labs, 2020). Looking at England, Scotland, and Wales, approximately 227,000 people were experiencing homelessness in 2021, projected to rise to 300,000 by 2023 (The Big Issue, 2023). Although affordability is not the only cause of homelessness, it is a significant factor. Lowering the cost of housing delivery may, therefore, help alleviate housing unaffordability.

Access to mortgage finance is also one of the key impediments to home ownership, but access to digital mortgage search, underwriting, and other elements of housing finance has now improved this process and may advance further through other Fintech mediums and peer-to-peer lending. Housing demand and supply mismatch is a critical factor as well: despite the high level of homelessness, there are still hundreds of thousands of empty homes globally. In England alone, The House of Commons Library (2020) reports that the number of empty homes as of October 2019 was approximately 648,000.[3] This mismatch may be addressed with digital solutions.

Contemporary real estate innovations and PropTech tools continue to provide new ways of developing and fostering social life and relationships: daylight-saving lighting technology provides easier control and aids in managing lighting remotely. Other tools include music, air quality monitoring, motion sensor cameras, advanced alarms, and security systems. Social improvement and market transparency can be seen through advanced techniques for capturing, storage, processing, management, and analytics of data.

Various online tools have made it possible for individuals to get a quick valuation for rental or sales purposes for their properties. The availability of information and data has improved the online search and selection of property for individuals. Remote viewing is gradually becoming the norm, particularly post-pandemic. Such tools can minimise anxiety and enable better search and selection experiences for individuals while also enhancing the jobs of real estate professionals and practitioners. Trust can also be improved through market transparency. Investors rely on data and multi-source information to make decisions. Real estate professionals also rely on data for reports and professional advice. Furthermore, the use of automated valuation models can make valuation a lot easier and more flexible, significantly reducing the time taken for mortgage and insurance decisions.

The subjectivity of the social dimensions of real estate innovations and digital technology makes it difficult to assess or measure the (potential) impacts of the tools or products. Cues can be taken from tools such as *Holmetrics* to measure factors such as engagement, job satisfaction, leadership, social capital, and burnout and how these relate to physical workspaces. This could involve, say, psychology, tone and sentiment analysis, agile management, and accountability. It will also be useful to have passive well-being platforms and tools that can give real-time insights into the well-being of tenants and occupants. One of the key benefits of this system is the enhanced insight and understanding of users' preferences and a viable feedback mechanism to rate users' experiences while occupying the space. This can further improve insight into the satisfaction, comfort, well-being, and ease of use for occupiers and, by extension, help to reduce vacancy rates in commercial properties and empty houses. This will further align with the current metamorphosis from the traditional "tenants" approach towards customisation.

The 'S' (social) dimension of the PropTech ESEP efficiency framework, therefore, examines the degree to which a PropTech tool or real estate innovation can improve the comfort, security, ease of use, health and safety of the user, operator, professional, relative to traditional operations and practice. The social efficiency consideration should, therefore, be integrated into the design and planning of PropTech products and tools, targeting specific social issues still associated with traditional practice and operations.

1.3.3 The Second "E" Dimension in the ES<u>E</u>P Efficiency Framework – Economic

The second "E" in the ESEP framework represents the economic efficiency of the framework. Economic consideration is an implicit core of real estate operations and practice. In most cases, real estate assets and associated activities have economic and financial implications. Real estate is capital-intensive and has intrinsic economic value; financial and economic considerations are, therefore, at the core of its use, operations, and transactions. The planning, construction/development, management, and transaction often require significant cash flow, and in some cases, these activities are supported by financial engineering tools that are critical for decisions to be made by financial institutions, investors, developers, etc. In many cases, investors, owners, and managers will seek to minimise costs and maximise profit. They will, therefore, do everything within their power and scope of influence to understand the various macro and micro risks, weaknesses, and threats of an asset or investment in order to mitigate them. In many cases, they also aim to identify the feasibility, viability, profitability, and such other elements that may allow a project to move forward. What's more, when that project becomes a reality, operators will typically seek to minimise financial expenditure while simultaneously maximising returns.

If time is money, then managing the ticking clock may be the biggest challenge of all. This can dominate development, investment, finance, and management. Delays in construction, for instance, can have massive implications, particularly to finance and costs. Consider construction equipment: it's sometimes borrowed for a fixed period and leases are time-critical. With real estate development and construction typically relying on gearing, delays in construction will affect the loan repayment and borrowing costs. Furthermore, because estate markets are fluid, these delays can be very costly. Understandably, professionals and other real estate practitioners tend to cost their activities and services on time taken. Managing and assessing risk is a central skill of the real estate practitioner: improvement in transparency and information availability can minimise risk, and by extension, cost and uncertainty offer clear financial benefits.

Real estate innovations and digital technologies are typically tailored towards the provision of economic solutions such as cost minimisation, profit maximisation, timesaving, risk management, and mitigation. The second 'E' (economic) dimension of the PropTech ESEP efficiency framework, therefore, examines the degree to which a PropTech tool or real estate innovation can minimise time and cost while at the same time maximising returns relative to traditional operations and practice.

The economic efficiency consideration, then, should be integrated into the design and planning of PropTech products and tools. These should target specific economic issues that are associated with traditional practice and operations and, of course, offer solutions.

1.3.4   The "P" Dimension in the ESEP Efficiency Framework – Physical

The "P" in the ESEP framework represents the physical efficiency part of the real estate, and it can also be termed "the spatial dimension". It accounts for issues relating to places, spaces, communities, and cities. The physical efficiency element captures the spatial dimension of PropTech and real estate innovation, and its impact – or potential impact – on space and places. This can range from micro-level spatial contexts (property-specific – like occupancy, workspace space utility, and optimisation) to macro-level aspects (such as neighbourhoods, cities, and regional economies). On a micro-level, the improvement of indoor space quality, such as space densities, retail layouts, and warehouse storage, can improve the efficiency of space within a building. On a macro-level, real estate activities often affect and define spatial and urban economies. Consider the rise of the student accommodation market in some cities such as Glasgow: this has significantly affected local housing and urban economies (The Herald, 2022). This dimension of the ESEP framework, therefore, captures the ways in which a PropTech tool or real estate innovation can improve both the immediate and extended environments of the physical attributes of the area where it is being deployed.

These tools can be used to enhance the effectiveness and efficiency of space-use planning in alignment with the highest and best use concept. This will ensure the reasonably probable and legal use of land, or an improved asset, to improve a site's financial viability.

Digitised urban design can also lead to the better integration of the features of the site in the design to achieve a higher level of effectiveness and also to preserve the natural features of a site. Furthermore, digitisation can ensure that building construction and change of use are carried out only when it is necessary. This can reduce the carbon footprint associated with construction activities. With better optimisation of outdoor spaces through more accurate predictive planning and spatial designs, better ordering of buildings and infrastructure, etc., the drive towards environmental efficiency can be improved on a wider scale.

One potential criticism of the physical dimension of the ESEP efficiency is the potential to overlap with the other dimensions. This is indeed a valid concern. For example, optimisation and maximising the utility of an office space may lead to lower vacancy rates and can be perceived as an economic efficiency dimension. However, we argue that failure to consider explicitly the spatial benefits of a PropTech or real estate innovative tool may create to a blind spot. In the example of office space optimisation, digital office space planners may reduce energy consumption through less space use and motion sensors (environmental); they may upweight circulation and comfort (social), and reduce space cost for the organisation (economic). Yet none of this captures the efficiency of the physical space itself. Because these digital tools enable organisations to estimate the actual space they need, those deploying the space are less likely to demand more than they need, and this, in turn, can reduce the wastage within the building. The efficient planning of office and retail space can promote optimal utilisation of the existing space. On a macro-level, space supply can become more predictable, and speculative supply can be minimised.

Secondly, the "P" dimension may have a limited scope of application, relative to the other dimensions, i.e., it may not apply to a wide range of real estate innovation and PropTech products. Where the others can be applied to the majority of PropTech tools, some may not necessarily offer physical or spatial solutions: mortgage finance digital tools and automated investment management may not tick the "P" box. Despite that, the physical dimension of the ESEP framework may be more prominent in the planning, construction, development, and management areas.

The 'P' (physical) dimension of the PropTech ESEP efficiency framework, therefore, examines the level of space maximisation and effective utilisation that a PropTech tool and other real estate innovations provide, or potentially provide, relative to traditional operations and practice. Where possible, this physical efficiency consideration should be integrated into the design and planning of PropTech products and tools, targeting specific spatial issues that are associated with traditional practice and operations.

This book, then, proposes the ESEP efficiency framework as a guide for real estate and PropTech stakeholders (users), stakewatchers (students, educators, professionals, practitioners, and enthusiasts), and stakekeepers (professional bodies, associations, and the public sector). This is expected to replace the erstwhile focus on economic and environmental factors with a more encompassing range of factors. Although already covering a broad range of issues, the ESEP efficiency framework can be expanded further and scholars, students, and enthusiasts are also encouraged to propose other dimensions that may not yet have been captured.

We said earlier that the framework can be applied either as a guide when developing a PropTech tool or product (*ex-ante*), or as an assessment tool for existing PropTech tools and platforms (*ex-post*), and the mechanism of use may vary. This book provides

Panel 1.5a: The four dimensions and efficiency scales

Efficiency scale	Environmental (En)	Social (So)	Economic (Ec)	Physical/spatial (Ph)
High (3*)	En3*	So3*	Ec3*	Ph3*
Medium (2*)	En2*	So2*	Ec2*	Ph2*
Low (1*)	En1*	So1*	Ec1*	Ph1*
None	En	So	Ec	Ph

Panel 1.5b: Illustration of Duke's ESEP Efficiency rating

En3*   So2*   Ec1*   Ph

Figure 1.5 Illustration of the PropTech efficiency ESEP framework application.

an illustration of one of the ways that this ESEP framework can be applied, (Figure 1.5). Figure 1.5a captures the ESEP dimensions and how they can be aligned to an efficiency scale: the vertical scale identifies the efficiency dimension; the horizontal scale measures the level of efficiency for each dimension. By mapping the level of efficiency of a PropTech tool or product from high within each of the ESEP dimensions, a graphical representation of the efficiency level of a tool can be made, reflecting the level at which a PropTech tool or product (for instance, a hypothetical product called "Duke" in Panel 1.5b) is efficient. In the case of Duke, Panel 1.5b can be interpreted to mean that Duke has a high En (environmental) efficiency, mid So (social) efficiency, and low Ec (economic) efficiency and does not have an identifiable Ph (physical) efficiency.

Although this illustration provides a somewhat crude and subjective scale and alignment, it can be developed further and its deployment can be automated using instance- and rule-based algorithms in machine learning and AI. This may also be applied in similar ways as BREAM and LEAD systems. Adopters of this framework may consider assigning a lower weighting to the P dimension due to its limited application across the real estate ecosystem.

This PropTech ESEP efficiency framework will underpin the core chapters in Section 2 and will serve as the conceptual underpinning for the analysis in the chapters and case studies.

1.4   Chapter Summary

This chapter has provided a background to PropTech and IT/digital real estate innovations and solutions, and reviewed different PropTech definitions. It has also proposed a simple, yet, all-encompassing definition to PropTech for easier comprehension and application. The chapter has further proposed the ESEP efficiency framework as a PropTech assessment and analytical tool. This section serves as the foundation for further exploration and in-depth analysis of the various areas of PropTech and real estate innovation. It also discusses the alignment of PropTech innovations with economic, environmental, social, and physical efficiency.

Notes

1 Some of the issues around the timeline and evolution of PropTech will be discussed further in Chapter 3.
2 The stakeholder classification used is based on the idea of the stakeholder classification in Fassin (2008).
3 This refers to the number of homes that had been empty for longer than six months.

References

Block, A., & Aarons, Z. (2019). *PropTech 101*. Advantage.

Braesemann, F., & Baum, A. (2020). PropTech: Turning real estate into a data-driven market? *SSRN Electronic Journal*. https://doi.org/10.2139/ssrn.3607238.

Chandra, L. (2019). *The PropTech guide: Everything you need to know about the future of real estate*. PropTech Asset Management LTD.

Fassin, Y. (2008). The stakeholder model refined. *Journal of Business Ethics, 84*(1), 113–135. https://doi.org/10.1007/s10551-008-9677-4.

International Energy Agency. (2019). *Perspectives for clean energy transition: The critical role of buildings*. International Energy Agency.

Oladiran, O., & Nanda, A. (2021). *PropTech education integration framework (PEIF)*. University College of Estate Management. Available online at: https://www.ucem.ac.uk/whats-happening/latest-publications/ (accessed 25 May 2023).

Oladiran, O., Sunmoni, A., Ajayi, S., Guo, J., & Abbas, M.A. (2022). What property attributes are important to UK university students in their online accommodation search? *Journal of European Real Estate Research*. https://doi.org/10.1108/JERER-03-2021-0019.

Pi Labs. (2020). *Transparency through technology*.

Porter, L., Fields, D., Landau-Ward, A., Rogers, D., Sadowski, J., Maalsen, S., Kitchin, R., Dawkins, O., Young, G. & Bates, L.K., 2019. Planning, land and housing in the digital data revolution/the politics of digital transformations of housing/digital innovations, PropTech and housing–the view from Melbourne/digital housing and renters: disrupting the Australian rental bond system and Tenant Advocacy/Prospects for an Intelligent Planning System/What are the Prospects for a Politically Intelligent Planning System?. *Planning Theory & Practice, 20*(4), pp. 575–603. https://doi.org/10.1080/14649357.2019.1651997.

RICS. (2022). *PropTech: Glossary*. Available at: https://www.ricsfirms.com/glossary/proptech/ (accessed 8 May 2022).

Saull, A., Baum, A., & Braesemann, F. (2020). Can digital technologies speed up real estate transactions? *Journal of Property Investment and Finance, 38*(4), 349–361. https://doi.org/10.1108/JPIF-09-2019-0131.

The Big Issue. (2023). *How many people are homeless in the UK? And what can you do about it?* (pp. 1–7).

The Herald. (2022). *Renting crisis: University of Glasgow students warned to not enrol amid lack of housing* (pp. 1–12).

The House of Commons Library. (2020). *Empty housing (England), number* (Vol. 3012).

Unissu. (2020). *What is PropTech?* Available at: https://www.unissu.com/proptech-resources/what-is-proptech (accessed 8 May 2022).

Unissu Data. (2022). Unissu Data. Accessed securely at: https://secure.holistics.io/users/sign_in (accessed in 5 August 2022).

United Nations Environment Programme. (2020). *2020 global alliance for buildings and construction: Towards a zero-emissions, efficient and resilient buildings and construction sector.*

University of Oxford Research. (2017). *PrepTech 3.0: The future of real estate.*

Walker, A. (2019). *Positivity or pessimism on PropTech?, Property.*

Chapter 2

Digital Technological Tools and Real Estate Applications

2.1 Data and Analytics

"Data is the new oil" is a popular catchphrase that highlights its increasing value. Albeit not totally accurate, this metaphor reflects the role that data has played in unlocking enormous potentials and opportunities in several sectors, and more specifically, in driving real estate and innovation, and digitisation. Data has different meanings, and it is often used in different contexts. For instance, an economist may describe data as a set of figures, factors or other information that can be analysed to increase the insight and understanding of a phenomenon, typically aimed at aiding the development of models and theories for prediction, forecast, and decision-making. Economists may collect micro-level data (e.g., individual, household, or firm level) or macro-level (pooling or aggregating data at the state, region, or country level) (Hill et al., 2008). From a business perspective, data is perceived as information that is collected, stored, and analysed to measure key performance indicators, identify trends and patterns, and then make decisions relating to products and services with the ultimate aim of enabling better decisions and improving organisational performance. In the context of business research, Collis and Hussey (2014) define data as known facts or things used as a basis for inference or reckoning.

From an IT or computing standpoint, data can be seen as the collection of raw facts, figures, or other types of information that can be processed, stored, and analysed using computers and digital systems. One of the early definitions of data in this context is the definition provided in the Institute of Electrical and Electronics Engineers (IEEE) Standard Glossary of Computer Science Terminology: "a representation of information in a formalised manner suitable for communication, interpretation or processing by humans or by automatic means" (IEEE 1990). More recent definitions of data include the Association for Computing Machinery (ACM) Computing Classification System where data is defined as "any digital representation of information, which can be stored, processed and transmitted by computers and computer systems" (ACM, 2012) and the definition by Provost and Fawcett (2013) where data is defined as "any facts, numbers or text that can be processed by a computer". In the context of IT and digital technology, data may take a wider variety of forms, including text, images, audio, video, etc. generated through a broad range of sources such as sensors, databases, social media, and user input.

For simplicity and reference purposes, this book will simply define data as various forms of facts and information that can be stored and analysed in several ways to enhance systems, operations, and services.

The four core elements of data are generation, aggregation, storage, and analysis.

DOI: 10.1201/9781003262725-3

2.1.1 Data Generation

Data generation takes various approaches, depending on the sourcing and collection techniques. Recent advances in data collection and sourcing have led to a significant increase in the volume now available globally. Data is available from different sources and in a variety of structured and unstructured formats. They can be generated and collected from traditional channels such as surveys and interviews and other more advanced systems such as transactional systems, social networks, social media, user interfaces, the internet, sensors, Internet of Things (IoT) dashboards, cameras, etc. Due to the vast amounts of data generated and collected on a daily basis (Han et al., 2012), and the variation of sources and the types generated, this era is often referred to as the "big data era". Before 2000, data was typically measured in numerical forms such as MB (megabytes) and GB (gigabytes); these are now being replaced with TB (terabytes), PB (petabytes), EB (exabytes), and ZB (zettabytes).[1] As shown in Figure 2.1, the volume of data and information that has been created, captured, copied, and consumed globally from 2010 to 2020 is estimated to be 233 ZB, and this is projected to almost triple by 2025. This substantial growth in data generation and use after 2019 is attributed to the significant increase in data use and demand following the COVID-19 pandemic, as remote working and other remote-based activities increased (Statista, 2021).

Big data generation is the result of our increasingly digitised lifestyle, i.e., our interaction with phones, computers, TVs, and other digital systems. We also generate it through our social media activities, financial transactions, and contactless banking. Furthermore, the GPS sensors on our mobile phones can also capture our commute and transportation patterns. The concept of big data is associated with certain elements that are known as the "Five V's of Big Data" framework which is used for conceptualising the characteristics and challenges associated with its use. They are volume, variety, velocity, veracity, and value (Yin and Kaynak, 2015):

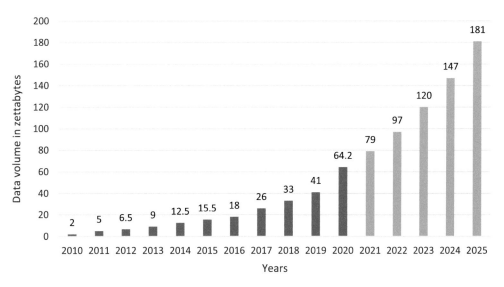

Figure 2.1 The volume of data created, captured, copied, and consumed globally from 2010 to 2020, and forecasts from 2021 to 2025.

- **Volume** refers to the amount of data being collected, processed, and stored. Big data typically involves extremely large datasets that traditional data processing systems will struggle to process.
- **Velocity** refers to the speed of data generation and processing. Big data is typically created and processed at a high rate, often in real time or near real time, and it requires a high and efficient processing system to keep up with the pace of data generation.
- **Variety** refers to the variation in the types that are included in datasets. Big data typically includes several forms of structured data (e.g., data in a database), unstructured data (e.g., text, images, videos, audio), and semi-structured data (e.g., JSON, XML).
- **Veracity** refers to accuracy and reliability. Big data is usually drawn from a variety of sources, and they may include errors, biases, and inconsistencies that will need to be identified and resolved.
- **Value** refers to the importance and usefulness for a particular purpose or application. Big data, when well analysed, can generate insights and actionable information for better decisions and outcomes.

Big data continues to revolutionise various aspects of life including politics, economics, culture, science, and business, with several benefits to services, management, advertising, operations, etc. It also has its challenges, particularly in relation to computer science and information technology – like data management, storage, processing, analysis, security, and visualisation.

Data access has also played a major role in transforming the real estate industry and is expected to remain valuable for future innovation and digitisation in real estate (Oladiran and Nanda, 2021).

2.1.2 Data Storage

Data storage is a key component of our data-driven world, evident in the growing demand for reliable, secure, and cost-effective data storage facilities. As shown in Figure 2.1, the volume of global data is projected to increase significantly over the next few years; this, therefore, suggests that data storage will become increasingly valuable and relevant. It enables firms and organisations to store large volumes in a centralised location to support their data management and analysis processes. Data loss and poor management can have devastating effects such as revenue loss, damage to reputation, and other legal and ethical problems. Furthermore, it can minimise the value that firms and organisations can draw from a dataset, regardless of the value that is embedded within the dataset. Reliable data storage solutions, therefore, support individuals, firms, and organisations to easily access their data and utilise them appropriately.

There are various types of data storage solutions. They include the following:

- **Cloud storage**: offers scalable, on-demand storage that can be accessed remotely from multiple locations typically using internet service. It can also be used for backup and achieving purposes. These storage systems include Dropbox, Google Drive, OneDrive Amazon Web Services (AWS) S3, Microsoft Azure, iCloud, etc.
- **Solid-state storage**: offers quicker access times and lower power consumption advantages, albeit, relatively expensive. Examples are USB flash drives, memory cards, hybrid drives, embedded multi-media cards (eMMCs), solid-state drivers (SSDs), etc.

- **Object storage**: provides a scalable low-cost storage solution for unstructured data often used for media content and big data analytics. They include IMB cloud object storage, digital ocean spaces, Amazon Simple Storage Service (S3), Scality RING, Ceph, OpenStack Swift, etc.
- **Magnetic storage**: offers high-capacity storage at low cost, thus making it one of the most popular storage technologies for years. Examples are hard disk drives (HDDs), magnetic tape storage, floppy diskettes, zip discs, magnetic stripe cards, magnetic bubble memory, etc.

Due to the overwhelming nature of big data, new and innovative storage solutions have also been developed. These systems are generally targeted at data from sources such as social media, customer interactions, sensors, etc. that can enable organisations to hold data for longer periods of time and improve analytics, forecasting, and projections leading to more informed decisions.

Some of the big data storage solutions are:

- **In-memory databases**: provide fast and scalable storage solutions for real-time analysis. They store data in memory and enable faster access and real-time processing. Examples are Microsoft SQL Server Hekaton, Apache Ignite, MemSQL, IBM solidDB, Oracle TimesTen, SAP HANA, etc.
- **Distributed file systems**: provide a scalable and fault-tolerant storage solution for big data. These systems distribute data across multiple servers, allowing for parallel processing and high availability. Examples are Hadoop Distributed File System (HDFS), GlusterFS, Lustre, Google File System (GFS), Amazon Elastic File System (EFS), IBM Spectrum Scale, etc.
- **NoSQL databases**: provide a flexible and scalable storage solution for unstructured data. These databases can store and process in various formats such as videos, images, text, etc. Examples are MongoDB, Cassandra, Couchbase, Redis, Amazon DynamoDB, etc.

Data storage has several challenges. One of the most remarkable challenges is security and integrity. Data incidents (such as breaches and cyberattacks) often occur when organisations or firms do not have appropriate technical or organisational measures to protect the data that they have in their possession. Data from the Information Commissioner's Office (2023) shows that 32,541 incidents were reported in the UK between 2019 and 2022 and cyberattacks accounted for 20% of these incidents. The UK General Data Protection Regulation (GDPR), through the principles of integrity and confidentiality mandates organisations, to have appropriate technical and organisational measures to protect the personal data they hold. Businesses and organisations (in the UK and beyond) are, therefore, always seeking robust security mechanisms to protect their data using encryption, access control, intrusion detection systems, data masking, etc. Scalability remains another challenge, particularly big data: as data volumes rise, firms and organisations are facing the challenge of scaling their storage solutions with speed and efficiency; these storage solution systems should ideally have the capacity to be increased when needed.

2.1.3   Data Aggregation

Data aggregation refers to the process of gathering, summarising, reformatting, and presenting data for further analysis. For data to be useful and meaningful, and for trends and patterns to be explored, they must be collated in a meaningful and useful way. Data aggregation minimises the volume of data that needs to be analysed. By using efficient aggregation techniques, multiple data sources can be collated into a dataset which will allow researchers or professionals to focus their analysis on the most important information, saving time and resources. Aggregation also allows for a more comprehensive analysis of complex datasets through the combination of data from multiple sources. This allows researchers to identify trends and patterns that might have been difficult and, in some cases, almost impossible using individual data points, and this can lead to new insights and improved decision-making processes. Furthermore, it provides unique insight that can be derived through the comparison of data from various sources. For instance, data from various firms can be aggregated to aid the comparison of firms, their input, output, and other factors; these can then be further used to identify efficient approaches to productivity.

As data continues to increase in volume, velocity, variety, veracity, and value, an important issue is how vast amounts of data are not just collected but also summarised and aggregated from disparate sources into a dataset. Analysing large datasets is increasingly important as the amount of data being generated continues to grow at an unprecedented rate (as highlighted in Section 2.1.1). This ability to aggregate data allows a more comprehensive analysis of complex datasets to enable a researcher or analyst to identify deep and hidden trends and insights that would previously be impossible to identify using traditional methods.

Despite the merits and importance of data aggregation, it has its challenges too. A major challenge is the mechanism of aggregating data from multiple sources and the associated risk that the data may be compromised with potential errors and inconsistencies. Furthermore, data indicators and units of measurement may be incommensurate with the data in other datasets (Braganza et al., 2017). It is, therefore, essential that the data are carefully validated and cleaned before they are aggregated to ensure the accuracy and reliability of the results and ahead of any inferences being drawn.

Consider data privacy and security. Depending on how it is done, the process of data aggregation introduces the potential risk that sensitive information may be inadvertently exposed; it is thus important to ensure that data privacy and security are maintained during the aggregation, and this can be achieved using access controls and encryption. Furthermore, aggregation can also affect data interpretation, potentially affecting the application of the results.

Regardless of the challenges, data aggregation – particularly of big data – remains a potent tool and pathway for analysing large and complex datasets and digitisation in several sectors. By aggregating relevant data from multiple sources to identify market trends and patterns, real estate investors, for instance, can make better and more informed investment decisions. Data aggregation can also significantly improve insight through research for better measurement, decisions, and policy guidance (Eisfeldt and Muir, 2016). Thus, with a careful process in place for data validation, cleaning protection, privacy, and security, and considering the context and its limitations, researchers and analysts can use the aggregation process to inform decision-making and continue to drive innovation in real estate.

2.1.4   Data Analysis

After data has been collected and, in some cases, aggregated, it can then be processed for further analysis. This means processing, examining, and interpreting to extract valuable information and gain insight. This can involve the use of various statistical and computational techniques applied to datasets. Data analysis typically involves cleaning, exploration, modelling, and visualisation.

- **Data cleaning**: involves checking for accuracy, completeness, consistency, etc. It is important that errors and inconsistencies be identified and dealt with early to ensure that the information is accurate and reliable. Errors should also be prevented from being carried forward and used for further analysis.
- **Data exploration**: involves examining the data for patterns or relationships between variables or factors of interest. This can enable a better understanding of the characteristics of a dataset, identifying data quality issues and generating hypotheses for further analysis.
- **Data modelling**: involves statistical and computational techniques to create a simplified representation of a complex phenomenon or system for hypothesis testing, forecasting, and predictions.
- **Data visualisation**: involves the presentation of the results for visualisation using tools and formats such as graphs, charts, images, etc. to make them easier to comprehend. This allows analysts to identify patterns and potential relationships in the data quickly and easily. It also means insights and findings can be communicated to stakeholders more effectively.

Data analysis is used in a wide array of sectors and fields of study: social sciences, economics, business, medicine, healthcare, engineering, real estate, etc. The increase in data generation and variation in data and data sources has made analysis and analytics more valuable. Data analysis has evolved from the use of statistical tools such as Microsoft Excel to other tools such as SPSS, Stata, R, Minitab, SAS, etc. Other more advanced programming tools include Python and SQL and machines such as neural networks and decision trees are increasingly used in data analysis to generate predictions and identify patterns in large datasets. The use of computer-based statistical analysis allows for more time to emphasise the interpretation of the results of the computations without having to expand enormous time and effort in the actual computations (Ott and Longnecker, 2016).

Data analysis and advanced analytics are enabling organisations to make more informed and evidence-based decisions to improve their productivity, effectiveness, and efficiency through the trends and patterns that they have identified. For instance, itemset mining is a well-known exploratory data mining technique used to discover interesting correlations hidden in a data collection (Apiletti et al., 2017). The data analysis can enable firms to identify the products or services that are most popular among their customers, allowing businesses to allocate their resources more efficiently. To stay ahead of the curve, firms and organisations need to continuously understand their customers and adapt to their needs through product improvement (Werder et al., 2020). The visual dimension of data analysis and analytics has also experienced significant development. For instance, it is now common to see geospatial data represented using maps in place of standard charts and graphs.

Advanced visualisations like 3D graphics, bubble charts, and three-dimensional plots are now used for high-dimensional data using MATLAB, Octave, Maple, and other tools.

Data analysis and advanced analytics are becoming increasingly important in the real estate sector, and this is significantly improving market insights and transparency. They are also improving property valuation and transactions. Real estate professionals and PropTech practitioners are leveraging the advancement in data analytics to gain a competitive edge in their market and professional domain through the provision of innovative tools, products, services, and platforms that can provide better services to clients, space users, space providers, managers, and other real estate stakeholders. PropTech product designers are establishing processes and systems to design products and maximise the value generation potential of the information at their disposal. Data and analytics-oriented real estate innovators are expected to continue to benefit from the lessons of these data-oriented innovative practices. Beyond that real estate researchers and academics are also benefiting from the advancement in data analysis and advanced analytical techniques. For instance, Heinig and Nanda (2021) used supervised machine learning algorithms to extract market sentiments from UK commercial real estate market data.

2.2   Computing, Internet, Mobile, and Cloud Technology

Advancements in computing, internet, mobile, and cloud technology in recent years have significantly transformed human activities, systems, and societies. The changes range from our work patterns, living conditions, learning, commute, communication, etc. These technologies have enabled tasks to be completed more efficiently, enabling better access to information and better connectivity with people. The approaches to various aspects of real estate use, operations, and services have seen significant changes resulting from the integration of these technologies. Each of these, though interlinked, will be discussed in sub-sections.

2.2.1   Computing

Computing has evolved dramatically since its inception several decades ago. The earliest computing devices were mechanical calculators such as the abaci that were used for arithmetic. This led to the development of early computing machines such as Charles Babbage's Difference Engine and Analytical Engine which were designed to perform more complex mathematical operations. The invention of electronic computers in the 1940s, however, served as a major "game changer" in the history of computing. The first electronic computer – ENIAC (Electronic Numerical Integrator and Computer) – was developed by John Mauchly and J. Presper Eckert at the University of Pennsylvania. Decades after, technological advancement led to the development of smaller but more efficient computers; the introduction of the microprocessor in the 1970s was a major milestone enabling the development of personal computers and other devices. Computing technology has directly and indirectly improved digital technologies such as mobile, internet, etc. Core components of modern computer systems are hardware, software, microprocessor, networks, and applications (Schneidewind, 2012).

Computing technology has significantly changed the real estate industry (Kontrimas and Verikas, 2011; Munawar et al., 2020; Ullah and Al-Turjman, 2021). For instance, advanced software is now enabling real estate agents to manage their listings, track their clients' preferences, and conduct market analyses with a higher level of efficiency and effectiveness, and CRM (Customer Relationship Management) tools enable property agents to store client data and use them for bespoke marketing campaigns. Furthermore, advancements in computing technology have enabled real estate professionals to access information about properties with better speed and efficiency. VR (virtual reality) and AR (augmented reality) are also transforming property viewings so that clients can now view properties in 3D, meaning potential space users can make better selection decisions with economic, social, and environmental benefits. Valuation has also been transformed: valuers and appraisers can now use a variety of approaches and tools ranging from traditional Excel spreadsheets to advanced algorithms and predictive analytics, to estimate property values and general market trends. Some property and asset managers use specialised software to track rental payments, maintenance, and facilities management in real time, allowing a faster and more efficient response. Blockchain technology is also being used for lease con-tracts, etc. These remarkable improvements in real estate operations and practice through computing have significantly improved the type and quality of services that real estate professionals can provide to their clients and space users.

Computing remains a goldmine with much untapped potential in various sectors. It is, therefore, important that researchers, professionals, policymakers, and other stakeholders continue to explore several channels to harness and use its extraordinary power.

2.2.2   Internet

The internet refers to a universal system of interconnected computer networks that use an Internet Protocol (IP) suite to communicate between devices and networks, typically linked through a range of wireless, electronic, and other optical networking technologies. There is no centralised management of the internet; instead, it is a collection of several individual networks and systems (Gralla, 1998). The internet emerged in the late 1980s/ early 1990s, and it has since facilitated the mass storage, transfer, and analysis of data. It is estimated that approximately 5.25 billion people now have access to and use the internet (Broadband Search, 2023), implying that two-thirds of the world's population currently use the internet.

The internet has had a profound impact on human activities, society, culture, and the world in general, changing the way we learn, work, communicate, interact, etc. People are now able to communicate and connect with each other, regardless of geographical location, using emails, social media, video conferencing, etc. Information is also more accessible using online search engines and encyclopaedias. This has been facilitated by several services such as electronic mail, telephony, file sharing, hypertext documents, and the World Wide Web (www) applications. These have further led to a transformation in commerce and trade (e.g., e-commerce), banking (e.g., FinTech), networking and col-laboration (e.g., social media), and several other areas of society.

The real estate industry is one of the areas and sectors that the internet has transformed. For instance, property search and selection have become much quicker and more efficient using online listing sites such as Rightmove, Zoopla, Zillow, and Realtor.com do not only enable buyers to search for properties based on their preferences but also provide images, videos, virtual tours, etc., minimising physical contact. Furthermore, access to data on property values, crime rates, school districts, and other infrastructure is enabling buyers to make better decisions about their property selection and investment.

The internet is also transforming property management and agency. Landlords use on-line platforms to advertise their properties and connect with potential tenants making the rental process more efficient. Real estate agents are also increasingly using social media to expand their reach to a wider range of clients and customers. Online management tools have made it easier for landlords and property managers to better handle tasks such as rent collection, maintenance requests, lease renewals, and other services more effectively and efficiently while reducing administrative costs and improving tenant satisfaction. These and several other internet-led transformations have increased the quality of real estate service delivery and reduced costs. They have also led to an increase in innovation, new business models, and improved market transparency in the sector, optimising the value of real estate use, services, and operations.

2.2.3 Mobile Technology

Mobile technology refers to the technology used for cellular communication. It has evolved from the first generation of two-way pagers to more advanced gadgets: mobile phones, GPS navigation devices, wristwatches, tablets, etc. The mobile technology evolution began with the use of wireless radio communication in the 1890s, and it currently involves more complex wireless technology, particularly internet connectivity and wireless devices. 1G was introduced as voice-only communication via "brick phones" in the early 1980s, and this was further developed into 2G for SMS in the early 1990s. By the late 1990s, 3G was introduced for faster data transmission speed to enhance video calls and internet facilities. 4G was introduced about one decade later, and the most recent generation of mobile device networks is 5G which reduces latency and cost. It also enhances energy saving and increases system capacity and large-scale device connectivity. The growth of mobile technology is linked to the seamless integration of information and communication technology, enhanced by high-coverage mobile communication networks (3G, 4G, and now 5G), high-speed wireless networks, various mobile information terminals, etc.

Mobile technology has also grown with the integration of apps (mobile applications). An app is a computer software application designed to operate on mobile devices. The first generation of apps was generally intended for productivity assistance such as email, calendar, contacts, etc.; however, these have now expanded to other areas such as leisure, gaming, e commerce, social media, religion, education, etc. Mobile technology has become a core enhancer of contemporary life, and it is increasingly contributing to the digitisation of systems and the society in general. Several digital activities now take place on mobile devices, and this trend is growing rapidly. In 2009, approximately 0.7% of internet traffic worldwide was from mobile devices; however, 54.4% of internet traffic as of 2023 is through mobile phone devices (Broadband Search, 2023). This is, therefore, further driving transformations in e-commerce, FinTech, social media, and several other sectors.

Mobile technology is also improving the experiences of real estate users, operators, and service providers. Several web-based platforms and real estate digital solutions have mobile phone app versions to enhance user experience and ease of navigation. These apps and general mobile devices have helped with property search and selection, property management, transactions, video and conference calls, and the IoT such as cameras, doorbells, light and energy saving, etc.[2] People can send and receive emails on the move, search for properties, visit websites, communicate and, in more recent times, hold meetings and conference calls, conduct virtual inspections, sign documents, and complete agreements. Furthermore, people can use mobile devices to access mortgage calculators and access other important information. Real estate professionals and service providers use mobile apps to target specific demographics, create personalised marketing campaigns, and analyse the effectiveness of their marketing efforts. These have driven improved service delivery and quality.

2.2.4   Cloud Computing

Cloud computing refers to the on-demand availability of computer systems resources/servers, especially data storage and computing power without direct active management by the user. It advocates a service-oriented computing paradigm where various types and forms of resources are organised virtually (Li et al., 2009). It offers scalable, on-demand storage that can be accessed remotely from multiple locations typically using internet service; it can also be used for backup and archiving purposes. Cloud computing also provides IT-supporting infrastructure with scalability, large-scale storage, and high performance as an effective way of implementing parallel data processing and data mining algorithms (Xu et al., 2009). It involves the grouping of networked elements such as infrastructure (compute, block storage, network), platform (identity, object storage, runtime, queue, database), and application (monitoring, content, collaboration, communication, finance); these may be exchanged to and from laptops, mobile devices, desktops, and servers. It can be provided through various forms such as public/utility computing, Xaas, MSP, or other hardware that can be taken as services to set up scalable virtual machines (Cheng Zeng et al., 2009).

The concept of cloud computing can be traced to the 1960s when the idea of time-sharing was introduced via RJE (Remote Job Entry). Full-time-sharing solutions became available in the 1970s, and telecommunication companies that hitherto offered primarily dedicated point-to-point data circuits began offering VPN (virtual private network) services in the 1990s with comparable quality of service. This gave the companies the ability to switch traffic based on the dynamics of the need to balance server use, making it possible to improve the overall bandwidth efficiency and effectiveness. As computers became increasingly decentralised, technologists explored the channels through which large-scale computing power could become better accessible to a wider range of users through time-sharing. This led to the use of algorithms to optimise infrastructure, platforms, and applications to prioritise tasks executed by computer processing units (CPUs) and improve the overall efficiency of the users. This evolved further in the first decade of the century (2000–2010) with the likes of Amazon (AWS in 2002) and Google (Google App Engine in 2008) which enabled developers to offer innovative applications independently and for users to easily deploy such applications and scale them to demand.

By 2010, Microsoft released Microsoft Azure, and this was followed by the open-source cloud software initiative known as OpenStack jointly launched by Rackspace Hosting and NASA. Microsoft Azure OS (Linux) and Amazon's AWS (and Google and IBM) are some of the latest fully managed cloud services that extend infrastructure, services, and tools to almost any customer data centre, co-location space, or on-premises facility for a truly consistent hybrid experience. A driving factor in the evolution of cloud computing is the need for risk mitigation, particularly internal power loss and the complexities associated with hosting network and computing hardware on-site, as well as the need to share information with workers located in diverse areas in real time to enable teams to work more efficiently and effectively regardless of location. The COVID-19 pandemic has made cloud computing even more popular following the high level of flexibility and security needed to facilitate remote working.

Cloud computing generally attempts to address quality of service and reliability-related challenges. Some core advantages include cost reduction, device location independence, improved agility, maintenance, multi-tenancy, performance monitoring, productivity, scalability, elasticity, and security. It does this using a group of applications running on computers or servers that are accessed over the internet and then offered as a service. Because of the distribution of functions of larger clouds over multiple locations, data centres continue to increase in prominence, and this, in turn, is expanding that segment of the operational real estate market.

Cloud computing has significantly contributed to the transformation of the real estate sector, providing tools and services to users, service providers, and operators. Real estate agents can store and access real-time property information, client data, and other important documents remotely, enabling better collaboration on documents in real time and from various locations. Property managers now use cloud-based platforms to manage properties, communicate with clients, and access important information regardless of location. They also use cloud-based software to streamline their management tasks from maintenance requests to rent collection. In turn, customers use this technology to access property listings, virtual tours, and other important information. Big data is particularly making cloud computing more valuable, and cloud-based platforms are enabling real estate professionals to store and analyse large amounts of information, providing insights into market trends, buyer behaviour, and other important factors. Property valuers use cloud-based platforms to access real-time data on market conditions, promoting more accurate valuations, predictions, and forecasts. Additionally, property investors can benefit from advanced analytics of market trends, increasing their ability to identify investment opportunities and manage their portfolios more efficiently.

2.3   Internet of Things (IoT)

The IoT is the interconnected network of physical devices and objects that exchange data over the internet. It refers to a physical object or conglomerate of sensors, software, and other digital systems that exchange and link data with other systems using the internet and other contemporary communication networks. This technology is embodied in a variety of networked sensors, systems, and products that utilise computer power, electronic miniaturisation, and network interconnections to offer new capabilities not previously possible (The Internet Society, 2015). IoT combines multiple technologies such as computing, commodity sensors, machine learning, and powerful embedded systems, and it is enabled by traditional fields of wireless sensor networks, control systems, embedded systems, and automation. The concept of a network of smart devices gained prominence in the early 1980s; however, the catchphrase "Internet of Things" was first used in 1999 by British technology pioneer Kevin Ashton to describe a system in which objects in the physical world could be connected to the internet by sensors.

From an operational perspective, IoT devices connect and communicate through various technical communication models as outlined by The Internet Society (2015):

- **Device-to-device model** refers to two or more devices that directly connect and communicate with each other (e.g., through a wireless network) in contrast to communication through an intermediary application server. These devices could use protocols such as Bluetooth, Z-Wave, and or ZigBee.

- **Device-to-cloud model** relates to IoT devices connecting directly to an internet cloud service like an application service provider to exchange data and control message traffic.
- **Device-to-gateway model** refers to devices that connect through an application-to-layer gateway (ALG) service as a conduit to reach a cloud service. In this case, there is an application software operating on a local gateway device which acts as an intermediary between the device and the cloud service and provides security and other functionality such as data or protocol transmission.
- **Back-End data-sharing model** relates to a communication architecture that enables users to export and analyse smart object data from a cloud service in combination with data from other sources. This allows the data collected from single IoT device data streams to be aggregated and analysed.

The use of IoT in real estate has attracted several benefits, particularly in urban management, building construction, maintenance, and management, leading to concepts and practices such as smart cities, smart homes, etc. These have been designed to improve energy efficiency and enhance security, safety, and comfort while reducing waste and minimising maintenance costs. Previously, some IoT devices were exclusively for the wealthy. For instance, some celebrities had integrated systems of sound, fancy lighting, sports sections, television screens, etc. around their homes, all connected through integrated systems controlled through a remote control or similar device. It was also not uncommon to have CCTVs and audio door access controls in such houses. The arrival of technologies provided by Google, Amazon, and Apple caused major disruptions and made some of these accessible and affordable. It is no longer unusual to see homes with automated (mobile phone-controlled) central heating systems, video doorbells, door locks, fridges, kettles, music systems, power sockets, alarm systems, etc. The growth of APIs (application programming interface) and other contemporary digital technologies including better broadband, Wi-Fi, or 4/5G mobile and the internet has made these facilities more accessible, and through an increase in supply, their prices are now more affordable too.

According to Brown (2018), IoT has made it easier for residential real estate space users to improve their home automation and this has several benefits. For instance, the use of central heating systems can improve the efficiency of energy consumption and so minimise energy costs (economic), and lower carbon emissions (environmental). Some smart home systems are connected to weather information, acting as the basis for boiler and heating systems to be regulated in real time, even without an on-site resident. Minimum and maximum temperatures can also be set and programmed. These systems offer social benefits such as ease of use with remote programming and control.

Sensors and smart metres are refinements used to monitor and control heating, ventilation, air conditioning (HVAC), and lighting systems used to adjust the lighting in rooms based on occupancy, time of day, and daylight availability. They are also used to monitor and manage building systems, like elevators and security set-ups. The data collected from sensors are analysed in real time, allowing building managers to identify potential problems and take corrective action before the issues deteriorate. IoT-enabled security systems are also used to enhance building security through smart cameras and access control systems, allowing reliable monitoring and control of access to buildings and prevention of unauthorised access. Tenants can now use devices such as thermostats, lighting systems, and appliances that are connected to the internet and controlled remotely. This allows users to control their environment from anywhere, using a smartphone or tablet. Further, smart locks offer remote access control to tenants providing an added layer of security. With benefits such as improved efficiency of building management, reduced maintenance costs, enhanced security, improved tenant experience, better convenience, etc., some tenants are now willing to pay more for smart buildings. Smart homes are, therefore, increasingly becoming valuable. Houses with better EPC (Energy Performance Certificates) ratings would be more attractive to tenants because they will be warmer and have lower running costs (Brown, 2018). These facilities and the potentially higher demand that they attract can, therefore, increase returns on real estate investment.

It should, however, be noted that IoT, particularly in residential properties, has associated challenges. These include potential security breaches from hackers, privacy intrusion (if the system is compromised), viruses, etc., and high upfront cost for installation. There are also potential issues around the obsolescence of the devices and associated technologies. Science moves fast and some devices have become obsolete in the last two decades; users of these devices may, therefore, be sceptical about their lifespan or the technology that is being deployed. Consider the lack of interoperability and of unified systems which allow a variety of different protocols and standards. IoT application in real estate is still in its early stages. With a high cost of installation and lack of awareness, some building owners and investors are reluctant to invest in this technology. Regardless of these challenges, IoT technology is expected to continue to transform real estate operations and practices with its high potential for innovation and impact.

2.4   Artificial Intelligence and Machine Learning

Artificial intelligence (AI) refers to the science and engineering of intelligent machines or computer systems that exhibit the simulation of human intelligence. In simpler terms, it can be described as the study of how to program computer systems to enable them to do what humans can do (Boden, 1996). Machine learning neuron network models use natural language processing to build intelligent entities that exceed the understanding, prediction, and perception of concepts as obtained in traditional models. Product-assisted learning is an example of AI built with the ability to learn from the feedback of users of deployed products and to produce applications to meet customers' expectations (Werder et al., 2020). The use of AI for data analysis has increased significantly in the last few decades and is attributable to the volume and variation in data, the increasing computing power, and the rise of sophisticated algorithms. These are aiding the extraction of signals which can now accomplish much more than before. AI aims to achieve the optimum blend of human performance and rationality based on four approaches highlighted in Russell and Norvig (2010):

- **Thinking humanly**: involves the effort to make computers think like humans, i.e., machines with 'minds' and the automation of activities such as decision-making, problem-solving, etc. that are naturally associated with human thinking.
- **Acting humanly**: involves creating machines that perform functions that require a high level of intelligence that would normally be performed by people and to make computers do things that humans currently do better.
- **Thinking rationally**: involves mental faculties with computational models that make it possible to perceive, reason, and act.
- **Acting rationally**: involves the design of intelligent agents which, in turn, allows intelligent behaviour in objects.

Machine learning is a subfield of artificial intelligence that embodies a set of approaches to AI and algorithms to solve specific tasks. Machine learning is often described as the science and engineering of setting up to perform a task and teaching the machine to learn. It addresses the question of how to build computer systems and applications that can be trained to improve automatically through a combination of rule-based algorithms (specific rules and conditions) and instance-based algorithms (adapting to recorded trends, patterns, and experience). By training a machine to solve tasks based on experience and data without necessarily needing to program it, the machine's performance can be improved significantly. This mechanism, therefore, will need less instruction but will require more structures and parameters for its different functions. This can allow algorithms to learn from and execute actions, thereby significantly improving their performance.

Machine learning is becoming a popular method for software development for vision, speech recognition, robot control, and several other applications. It has progressed dramatically over the last couple of decades from being a "laboratory curiosity" to a practical technology now widely used across various industries and sectors and increasingly being deployed for data analysis, forecasting, and predictive models. The combination of computer science and statistics makes this method even more valuable for data analytics (Jordan and Mitchell, 2015). Developers of AI now generally acknowledge that it can be a lot easier to train a machine system and provide it with samples of desired input–output patterns, rather than merely programming it manually by anticipating a desired response for all possible inputs. By taking the machine through a training experience, its performance on task execution can be significantly improved. Data-intensive machine learning methods have, therefore, been applied to various areas of science, technology, commerce, health, education, marketing, finance, insurance, security, etc., strengthening evidence-based decision-making processes.

AI and machine learning are also significantly transforming real estate use, operations, and services. AI algorithms can now analyse large volumes of data, including property listings, transactions history, and market trends, and these can enable real estate investors and landlords to make informed decisions relating to buying, selling, or renting properties. AI can also aid investment analysis for due diligence, site selection, underwriting, origination and portfolio optimisation, asset positioning, building construction and development, property management, etc. These areas of real estate are becoming more data-driven, and as more becomes available, AI is a valuable tool that can enable continuous real-time generation, collection, aggregation, management, and analysis. These can eventually lead to more dynamic operations of buildings and more broadly, intelligent monitoring, management, and optimisation of the built environment. This technology is also being used for predictive maintenance and fault detection which can facilitate risk mitigation, risk management, and automation of quality testing. Furthermore, property managers can use AI-powered tools to automate, first, their property management functions – like rent collection, and maintenance requests and, second, for asset valuation, achieved by combining process control, data quality control and yield data, yield data, etc. AI-powered tools can also analyse various factors such as property features, location, and market trends to estimate property values.

The application of AI to various areas of real estate has the potential to streamline operations and service delivery. The deployment of AI tools and their resultant automation of these activities can save time and money, and increase security, safety, and comfort. It can build up environmental sustainability by reducing the carbon footprint of traditional real estate operations and services. In turn, it offers the potential to reduce human error and improve precision regardless of the data volume.

AI and machine learning, however, have some limitations and challenges. The core one is the potential to exacerbate existing inequalities in the industry. Algorithms can be biased (depending on the training they have received) and this can lead them to perpetuate and amplify existing inequalities and biases. For instance, if certain variables that capture demographics or socio-economic groups are omitted from the training of the machine, the results may not accurately reflect the needs and preferences of the omitted groups. Consider too that an increase in AI-enabled professional practice may lead to job displacement. Conversely, AI can also create new employment opportunities but only if professionals develop the requisite skills to tap its potential. Some of these potentials will be discussed further in Chapter 9.

2.5   Blockchain

Blockchain is a distributed database or ledger that is shared among the nodes of a computer network with the data stored electronically. Efanov and Roschin (2018) define it as a distributed database containing records of transactions that are shared among participating members with each transaction confined by the consensus of a majority of the members; this makes it difficult for fraudulent transactions to pass collective confirmation. One of the distinct features of the blockchain (in relation to a conventional database) is the structure of the data. It collects information in groups (blocks) that hold a set of information. Another distinct feature is that it guarantees the security and integrity of the data record, and it typically does not require third-party verification, thus providing a secure and transparent method of recording and verifying transactions. This should minimise the inefficiencies associated with the middlemen in typical transactions.

The typical blockchain process begins with the entry of a new transaction which is transmitted to a network of peer-to-peer computer systems in various locations. The algorithm in the network of these systems confirms the validity of the transaction to be legitimate and proceeds to cluster them into blocks. These blocks have limited capacities and when filled, they are closed and linked to previously filled blocks, creating a chronological string of data or loop of transactions. These chained blocks or loops of transaction become permanent, marking the completion of the transaction; it is this loop (chain) that is referred to as the blockchain (Figure 2.2).

One of the core goals of the blockchain is to allow digital information to be recorded and distributed, but unedited, laying the foundation for immutable ledgers and records of transactions that cannot be destroyed. The blockchain is typically decentralised to ensure that no single individual, entity, or group has control, rather, all users collectively retain control. As these decentralised blockchains are immutable and irreversible, transactions may be recorded and retained permanently. The fact that the data held is decentralised to several network nodes at different locations enhances the integrity of the stored data and makes it almost impossible for it to be tampered with because changing the data on one computer will fail when it is time to cross-reference it with other systems – this can make it easy to identify the node with the incorrect information and thus help to establish transparency.

Although various forms of data and information can be stored on a blockchain, the most common has been as a transactions ledger. This has led to the development of cryptocurrencies, decentralised finance applications, non-fungible tokens (NFTs), and smart contracts. Cryptocurrencies are some of the most common applications of blockchain technology; in fact, blockchain is synonymous with cryptocurrencies in many climes. Currencies such as Bitcoin and Ethereum have disrupted the finance sector through the

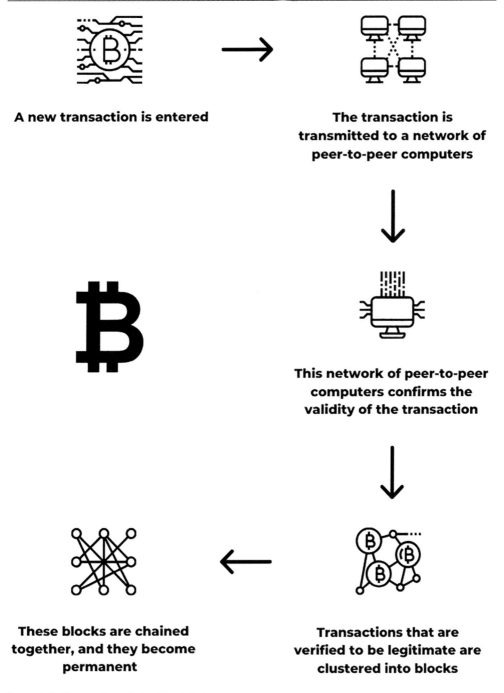

A new transaction is entered

The transaction is transmitted to a network of peer-to-peer computers

This network of peer-to-peer computers confirms the validity of the transaction

These blocks are chained together, and they become permanent

Transactions that are verified to be legitimate are clustered into blocks

Figure 2.2 Illustration of the blockchain process.

provision of secure and decentralised transactions and payment systems. From an academic viewpoint, it has cut across several disciplines including game theory, economic and monetary theory, cryptology, the internet, and computer sciences (University of Oxford Research, 2017). From a more practical standpoint, blockchain technology has been applied to various sectors such as banking, insurance, finance, sports, etc.

Like other sectors, blockchain technology is now transforming the real estate sector, and Brown (2018) provides useful insight into its role. Look at cryptocurrency-enabled payments. One of the first companies to facilitate the sale of property priced in Bitcoin is Cai Capital. Knox Group in Dubai announced in 2017 that they would accept payment for an off-plan apartment in Dubai in Bitcoin. This was followed by an announcement by Go Homes later that year that they had sold a home using cryptocurrency. A further announcement was made by a platform called "the collective" that they will receive their rent in Bitcoin.

If facilitating the tokenisation of real estate assets is an objective, blockchain can contribute. Some of the main challenges to investment are the high capital required and illiquidity. However, assets can now be divided into digital tokens, and these can then be traded on blockchain-based exchanges making it easier to buy and sell, and, therefore, accessible to a broader range of investors. Blockchain technology is also improving property trading and fund-raising. Brickblock, Habor, TrustMe, Dominium, and PropCoin are examples of property trading platforms that allow an asset-backed token, including property to be traded. Rental listing and management platforms include Rentberry in the US – a long-term rental listing and transaction platform built on blockchain technology.

Land administration has also been transformed through the provision of a tamper-proof method for recording property ownership. Traditional approaches to property ownership and records are prone to fraud and errors; however, blockchain technology allows enables the creation of a decentralised database that ensures that property ownership records are both transparent and accurate. Government parastatals are now adopting this technology at different levels. In the US, for instance, Cook County, Chicago partnered with a start-up called Velox RE to devise a pilot program in 2016 to record real estate transactions using blockchain for land registry records. This has spread to other US states which now have their land registry record tests. In the UK, HM Land Registry recently created a digital street to test blockchain functionality. Blockchain technology is now being explored for advancing transparent land administration in developing economies. According to Rodima-Taylor (2021), blockchain registries have contributed to the formalisation of property rights in the Global South.

In general, blockchain technology has sped up the process of information exchange process in real estate. The real estate market typically depends on intermediaries and these processes create multiple inefficiencies such as lengthy transaction time, higher costs, and risks. Blockchain has the capacity to complete several real estate processes with minimal friction. As discussed in Chapter 1, several parties are involved in a typical UK residential sales transaction (such as estate agents, building/structural surveyors, the Land Registry, conveyance lawyers, local authorities, and financial institutions). Although much of the data used by these parties is publicly available, information is typically stored in multiple databases which makes data access difficult. Furthermore, some of these roles overlap thereby directly, and indirectly adding to the overall transaction costs. This process can be significantly enhanced through a distributed ledger, referenced to the subject property, and made available on demand to all actors. Blockchain technology is also used for smart contracts for automated agreements and fund transfers. The emergence of eDocs and eSignatures to take the place of traditional paper-based contract execution is now common practice.

Blockchain technology, however, has some limitations and challenges. The fundamental theoretical principles and values of the technology have been criticised for not being realistic in practice. For instance, although the system is generally thought to exclude gatekeepers, it is difficult to maintain complete decentralisation as some entities and organisations that have more capital or resources may indirectly control the systems through their majority holding. Furthermore, although the goal is to minimise intermediaries, there are several platforms managing these operations such as wallets, exchanges platforms, Domain Name System (DNS), and cloud platforms which add a further intermediate layer. Additionally, although blockchain is theoretically expected to work based on the principles of self-regulation, trust-based interactions, distributed peer-to-peer collaboration, and decentralised governance, there have been reports of breaches of trust, scams, and unethical conduct. Some examples of high-profile cases of cryptocurrencies associated with scams, Ponzi schemes, or fraudulent activities include Bitconnect (BCC), OneCoin, Centra Tech (CTR), Prodeum (PDC), and BitKRX.[3] Blockchain has also been criticised for its advanced algorithm which requires high computing power with adverse environmental and economic implications. Nonetheless, international financial institutions, investment firms, and technology companies are continuing to invest significantly in research to evaluate the benefits and potentials of blockchain (University of Oxford Research, 2017).

2.5   Virtual Reality (VR) and Augmented Reality (AR)

VR is a simulated experience that is identical to the real world. Wohlgenannt et al. (2020) describe it as a virtual environment that immerses users to the extent that they have the feeling of being there. The term "virtual" refers to an element that is not physically in existence but is simulated to reflect or mimic reality; thus, though not physically present, computer software is used to produce a visual representation of reality. There are debates about the origin of VR. The works of Morton Heilig spanning the 1950s and 1960s are, however, perceived as one of the major landmarks of VR development. This began with the "Experience Theatre" which encompassed all the human senses effectively and evolved to the Sensorama – the prototype (mechanical device predating digital computing). Heilig also developed the "Telesphere Mask" which was referred to as a telescopic television apparatus for individual use, moving three-dimensional images with 100% peripheral vision, binaural sound, scent, and air breezes. This technology further evolved through the rest of the century, including the creation of the first head-mounted display for use in immersive simulation in 1968, and the Virtual Reality Modelling Language (VRML) introduction in 1994 which was intended for the development of virtual reality without dependency on headsets.

VR technology has advanced significantly in the twenty-first century with the advent of the SAS Cube (SAS3) developed by Z-A production, Barco and Clarte, and VRPack. The introduction of street maps by Google in 2007 was also a landmark development, providing panoramic views of several street positions, with other features such as stereoscopic 3D models introduced in 2010. More recent milestones include the Oculus Rift S and Oculus Quest (a standalone headset). Oculus Quest 2 was released in 2020 with new features including a sharper screen, reduced price, and increased performance.

AR is an offshoot of VR. It can be defined as a real-time direct or indirect view of a physical real-world environment that has been enhanced and augmented by adding virtual computer-generated information to it (Carmigniani & Furht, 2011). It is also known as "mixed reality" which uses computer technology to apply virtual information to the real world. In AR, virtual objects and the real environment are superimposed on the same space in real time, and in general, the information it provides differs from what humans can perceive. Google made one of the first attempts to develop AR through Google Glass and Apple also added AR functionality to its iOS 11 operating system in 2017 increasing users' access to AR. Although AR and VR emerged from 3D visualisation, they differ because AR overlays graphics onto the real world through a headset, smartphone, or other devices, while VR is a computer-based simulated, immersive reality which is typically accessed using a headset. With the headset on, the viewer can turn around and see things as if they are at that location.

These technologies are transforming several industries and activities such as education, gaming, entertainment, retail, business, etc.; real estate is no exception. Visualisation is one of the senses with the highest capabilities for high memory retention and the tangible (physical) aspect of real estate strongly connects to the visual senses; hence, this technology is increasingly applied and is particularly advanced in construction and development. VR-powered tools can now provide architects, developers, and other professionals with the ability to create and visualise designs in virtual settings in ways that enable them to make changes and improvements before construction commences. A 3D visualiser, for instance, takes 2D architectural CAD drawings and uses 3D software to create a model of a new real estate development that is accurate and to scale. Computer-Generated Images (CGIs) of development designs are presented in 3D before the construction commences, and this has further advanced from static images to 3D flythroughs or walk-throughs, where viewers can view the streets and proceed to view individual properties. For off-plan development, these tools can make interested buyers view the properties in advance, allay fears and, ideally, make a commitment to buy the property.

VR and AR are also transforming property viewings and visualisation. Real estate agents can now provide realistic 3D models allowing them to present their properties in innovative ways. These include immersive, 360-degree tours so that prospective purchasers may experience properties as if they were physically present. It applies to marketing as well; for example, a viewer can hold a smartphone or iPad over the brochure of an off-plan development and take a virtual tour of the site, further navigating into other specific properties of interest. In general, offering VR rather than merely providing photos can lead to higher conversion rates and by extension, higher take-up, and occupancy rates. This also effectively minimises the time that would hitherto have been spent with on-site visits and inspections for prospective tenants/buyers. This advanced preview system can shake out those not really interested in the property at the outset. The COVID-19 pandemic made virtual viewings more popular as more real estate agents increasingly offered virtual inspections, building on some pre-pandemic practices. These innovations significantly reduce property search time and costs while reducing commute-related carbon emissions.

2.6   Chapter Summary

This chapter has provided an overview of the core digital technologies and systems that have underpinned PropTech and real estate innovation. Various digital technologies have been discussed with highlights of their principles, mechanisms, and applications to real estate. It should be noted that there are several other digital tools that have been applied to real estate that have not been discussed in this chapter; Some of these will be looked at in Section 2.

Notes

1 1 terabyte is just over 1,000 gigabytes; 1 petabyte is just over 1,000 terabytes; 1 exabyte is just over 1,000 petabytes; and 1 zettabyte is just over 1,000 exabytes.
2 IoT will be discussed in more detail in Sub-section 2.3.
3 It should be noted that the authors do not have the ability to determine whether a particular cryptocurrency is a scam or not. The examples provided are simply based on reports of cryptocurrencies associated with scams or fraudulent activities.

References

ACM. (2012). Association for Computing Machinery (ACM) computing classification system. In *ACM computing classification system, 2012*. https://doi.org/10.1145/2366316.2366320.

Apiletti, D., Baralis, E., Cerquitelli, T., Garza, P., Pulvirenti, F., & Venturini, L. (2017). Frequent itemsets mining for big data: A comparative analysis. *Big Data Research, 9*, 67–83. https://doi.org/10.1016/j.bdr.2017.06.006.

Boden, M. (Ed.). (1996). *Artificial intelligence* (2nd ed.). Academic Press Inc. https://books.google.co.uk/books?hl=en&lr=&id=_ixmRlL9jcIC&oi=fnd&pg=PP1&dq=artificial+intelligence+&ots=JQQK3QqzPT&sig=2Jdisaj9Af040AuZKBYq278we6Y&redir_esc=y#v=onepage&q=artificial intelligence&f=false.

Braganza, A., Brooks, L., Nepelski, D., Ali, M., & Moro, R. (2017). Resource management in big data initiatives: Processes and dynamic capabilities. *Journal of Business Research, 70*, 328–337. https://doi.org/10.1016/j.jbusres.2016.08.006.

Broadband Search. (2023). *Key internet statistics in 2023 (including mobile)*. https://www.broadbandsearch.net/blog/internet-statistics.

Brown, R. (2018). *PropTech: A guide to how property technology is changing how we live, work and invest*. Casametro.

Carmigniani, J., & Furht, B. (2011). Augmented reality: An overview. *Handbook of Augmented Reality, 6*(4), 355–385. https://link.springer.com/chapter/10.1007/978-1-4614-0064-6_1.

Cheng Zeng, Guo, X., Ou, W., & Han, D. (2009). Cloud computing service composition and search based on semantic. In M. G. Jaatun, G. Zhao, & C. Rong (Eds.), *Cloud Computing, First International Conference, CloudCom 2009, Beijing, China* (pp. 290–300). Springer. https://link.springer.com/chapter/10.1007/978-3-642-10665-1_63.

Collis, J., & Hussey, R. (2014). *Business research: A practical guide for undergraduate and postgraduate students* (4th ed.). Palgrave Macmillan.

Efanov, D., & Roschin, P. (2018). The all-pervasiveness of the blockchain technology. *Procedia Computer Science. 8th Annual International Conference on Biologically Inspired Cognitive Architectures, BICA 2017, 123*, 116–121. https://doi.org/10.1016/j.procs.2018.01.019.

Eisfeldt, A. L., & Muir, T. (2016). Aggregate external financing and savings waves. *Journal of Monetary Economics, 84*, 116–133. https://doi.org/10.1016/j.jmoneco.2016.10.002.

Gralla, P. (1998). *How the internet works* (4th ed.). Que Publishing. https://books.google.co.uk/books?hl=en&lr=&id=iCMCwXLLdscC&oi=fnd&pg=PA1&dq=Gralla,+P.+(1998).+How+the+Internet+works.+Que+Publishing.&ots=sX7jxr6kwU&sig=rVZR-hMI5hZEnRxLjRQgiZhPQf4&redir_esc=y#v=onepage&q=Gralla%2C.

Han, J., Kamber, M., & Pei, J. (2012). Data mining: Data mining concepts and techniques. In *Proceedings - 2013 International Conference on Machine Intelligence Research and Advancement, ICMIRA 2013* (3rd ed.). Elsevier Inc. https://doi.org/10.1109/ICMIRA.2013.45.

Heinig, S., & Nanda, A. (2021). Supervised learning algorithms to extract market sentiment: An application using commercial real estate market. *Real Estate Finance, 38*(8), 147–160.

Hill, R. C., Griffiths, W. E., & Lim, G. C. (2008). *Principles of econometrics* (3rd ed.). John Wiley & Sons, Ltd.

IEEE. (1990). Institute of Electrical and Electronics Engineers. In *IEEE standard glossary of computer science terminology* (IEEE Std 6, pp. 1–84). https://doi.org/10.1109/IEEESTD.1990.101064.

Information Commissioner's Office. (2023). *Data security incident trends*. https://ico.org.uk/action-weve-taken/data-security-incident-trends/.

Jordan, M. I., & Mitchell, T. M. (2015). Machine learning: Trends, perspectives, and prospects. *Science, 349*(6245), 255–260.

Kontrimas, V., & Verikas, A. (2011). The mass appraisal of the real estate by computational intelligence. *Applied Soft Computing Journal, 11*(1), 443–448. https://doi.org/10.1016/j.asoc.2009.12.003.

Li, X., Zhang, H., & Zhang, Y. (2009). Deploying mobile computation in cloud service. In M. G. Jaatun, G. Zhao, & C. Rong (Eds.), *Cloud Computing, First International Conference,*

CloudCom 2009, Beijing, China (pp. 301–311). Springer. https://link.springer.com/chapter/ 10.1007/978-3-642-10665-1_63.

Munawar, H. S., Qayyum, S., Ullah, F., & Sepasgozar, S. (2020). Big data and its applications in smart real estate and the disaster management life cycle: A systematic analysis. *Big Data and Cognitive Computing, 4*(2), 1–53. https://doi.org/10.3390/bdcc4020004.

Oladiran, O., & Nanda, A. (2021). *PropTech education integration framework (PEIF)*. https://www. ucem.ac.uk/wp-content/uploads/2021/10/Harold-Samuel-Research-Prize-HSRP-20202021. pdf.

Ott, R. L., & Longnecker, M. (2016). *An introduction to statistical methods and data analysis* (7th ed.). Cengage Learning. https://books.google.co.uk/books?hl=en&lr=&id=VAuyBQAAQ BAJ&oi=fnd&pg=PP1&dq=an+introduction+to+statistical+methods+and+data+analysis+&ots=9 fzPKkcPHr&sig=VCKec-thquOFxvcMK57l599UJcw&redir_esc=y#v=onepage&q=an introduction to statistical methods and data an.

Provost, F., & Fawcett, T. (2013). *Data science for business: What you need to know about data mining and data-analytic thinking* (1st ed.). O'Reilly Media Inc.

Rodima-Taylor, D. (2021). Digitalizing land administration: The geographies and temporalities of infrastructural promise. *Geoforum, 122*(March), 140–151. https://doi.org/10.1016/j. geoforum.2021.04.003.

Russell, S., & Norvig, P. (2010). *Artificial intelligence: A modern approach* (3rd ed.). Pearson. https://archive.org/details/artificial-intelligence-modern-approach-3rd-ed.-russell-norvig

Schneidewind, N. F. (2012). *Computer, network, software, and hardware engineering with applications.* John Wiley & Sons, Ltd. https://books.google.co.uk/books?hl=en&lr=&id=4g7Kc0i_TT MC&oi=fnd&pg=PA1&dq=Schneidewind,+N.+F.+(2012).+Computer,+network,+software,+an d+hardware+engineering+with+applications.+John+Wiley+%26+Sons.&ots=rMbkQmejwo&sig= H-Kx2-jFcfQAECLn9GTv4V11LFI&redir_esc=.

Statista. (2021). *Volume of data/information created, captured, copied and consumed globally from 2010 to 2020, and forecasts from 2021 to 2025.* https://www.statista.com/statistics/871513/ worldwide-data-created/.

The Internet Society (ISOC). (2015). *The internet of things: An overview. Understanding the issues and challenges of a more connected world* (Vol. 80). https://d1wqtxts1xzle7.cloudfront. net/48790442/ISOC-IoT-Overview-20151014_0-libre.pdf?1473746977=&response-content-disposition=inline%3B+filename%3DThe_Internet_of_Things_An_Overview_Under.pdf&Expire s=1677767370&Signature=ABNs3ftJR5u4ovxcJsDvPLXJUATjBFVDP.

Ullah, F., & Al-Turjman, F. (2021). A conceptual framework for blockchain smart contract adoption to manage real estate deals in smart cities. *Neural Computing and Applications, 35*(7), 5033– 5054. https://doi.org/10.1007/s00521-021-05800-6.

University of Oxford Research. (2017). *PrepTech 3.0: The future of real estate.*

Werder, K., Seidel, S., Recker, J., Berente, N., Gibbs, J., Abboud, N., & Benzeghadi, Y. (2020). Data-driven, data-informed, data-augmented: How Ubisoft's Ghost Recon Wildlands live unit uses data for continuous product innovation. *California Management Review, 62*(3), 86–102. https://doi.org/10.1177/0008125620915290.

Wohlgenannt, I., Simons, A., & Stieglitz, S. (2020). Virtual reality. *Business and Information Systems Engineering, 62*(5), 455–461. https://doi.org/10.1007/s12599-020-00658-9.

Xu, M., Gao, D., Deng, C., Luo, Z., & Sun, S. (2009). Cloud computing boosts business intelligence of telecommunication industry. In M. G. Jaatun, G. Zhao, & C. Rong (Eds.), *Cloud Computing, First International Conference, CloudCom 2009, Beijing, China* (pp. 224–231). Springer. https://link.springer.com/chapter/10.1007/978-3-642-10665-1_63.

Yin, S., & Kaynak, O. (2015). Big data for modern industry: Challenges and trends (point of view). *Proceedings of the IEEE 103*(2), 143–146. https://ieeexplore.ieee.org/stamp/stamp. jsp?arnumber=7067026.

Chapter 3

Real Estate Digital Transformation and PropTech Investment and Finance

3.1 Real Estate Transformation and the PropTech Evolution

For several years, real estate use, operations, and service delivery remained rigid and conventional, although gradual changes began about half a century ago. In the last decade, however, there have been significant changes and innovations with PropTech driving the transformation. Although PropTech platforms, tools, and products are typically underpinned by digital advances, they are rooted in more traditional technologies. The evolution of PropTech takes its root in mainframe computing and the advancement in computing over the last five decades. This section will review changes to real estate use, operations, and service delivery using a two-generation approach: the first generation being the mainframe computing era and the second the digital technology era. The University of Oxford Research (2017) classifies the PropTech evolution using 1.0, 2.0, and 3.0 classifications; we, however, argue that PropTech 2.0 and 3.0 are underpinned by the same fundamentals, i.e., digital technology, and thus classify 2.0 and 3.0 as the "PropTech second generation".

3.1.1 PropTech First Generation (1970s–1999)

The first generation of PropTech refers to the era of computing and internet advancements. These laid the foundation for more advanced digital real estate applications. Due to the traditional nature and slow adaption in the real estate sector, advancements in computing technology in the mid-1900s and subsequent decades had minimal impact on the property sector. However, this began to change in line with the advancements in data and computing. The emergence of the mainframe and the subsequent introduction of personal computing in the late 1970s/early 1980s are considered major landmarks in real estate digitisation. These laid the foundation for the use of several applications, most notably, spreadsheets. The introduction of the Apple II and the twin floppy disc IBM PC XT enabled spreadsheet applications such as VisiCalc and Supercalc. This made way for Microsoft Excel which soon after became the industry standard tool for data analysis.

The increased efficiency and affordability of mainframe computing led to its high uptake rate which further led to the advancement of real estate service delivery in the mid- to late 1980s (University of Oxford Research, 2017). In real estate building and construction, for instance, Autodesk – a US-based software company[1] – was founded in 1982. The company developed software for architecture and construction which enabled the use of computer-aided design. Real estate investment and management services were also enhanced. Argus and Yardi, for instance, were first established in the mid-1980s to provide

DOI: 10.1201/9781003262725-4

commercial real estate investment and management analytics using software-aided tools. Similarly, CoStar was established in 1987 to provide information, analytics, and marketing services to the commercial real estate industry in the US, Canada, France, Germany, Spain, and the UK.

Real estate research also experienced significant transformation in the 1980s. Property Market Analysis (PMA) was established in the UK in 1982, laying the foundation for PC-enabled property research. The National Council of Real Estate Investment Fiduciaries (NCREIF) was also established in 1982 and used to develop a property index (NPI) for the US. This platform served the institutional real estate investment community as its data hub, providing a robust database of real estate assets from 1978. Furthermore, the Investment Property Databank (IPD) was established in 1985 to analyse data on the UK's commercial property performance, and Prudential in London and New York both established the first institutional property research teams using PCs.

The advancement of the internet and email technologies in the 1990s contributed importantly to the PropTech first generation as well. These technologies facilitated access, transfer, storage, and analysis of data. 1995–1996 is particularly referred to as the tipping point with the traffic on internet backbones[2] in the US increasing from 16.33 TB in 1994 to 1500 TB in 1996 during the dot-com boom, and this increased astronomically through the last part of the decade (Coffman and Odlyzko, 2002). This boom also supported the development of real estate online marketing. For instance, Craiglist in the US was established around this time, and Exchange and Mart in the UK also transitioned from print to web-based marketing around the same time.

These computer-based developments simultaneously occurred with the development of indirect real estate, resulting in various investment models such as debt and asset-backed securitisation, REITs, and derivatives markets. The advancement of these markets made software and other computerised systems even more valuable for quantitative data analysis, and more broadly, for real estate performance measurement, investment analysis, and research. This laid the foundation for the more advanced real estate use and services that emerged in the second generation of PropTech.

3.1.2 PropTech Second Generation (2000–2022)

The second generation of PropTech represents the era of significant integration of digital technology in real estate use, operations, and services. Various real estate sectors have been significantly transformed, and a wide range of digital tools and technologies have been introduced to various segments of the real estate value chain from the beginning of this millennium.

The residential real estate sector is tagged as the conduit for the transition of the second generation to the second with specific reference to brokerage and marketing. Multiple Listing Services (MLS) transformed real estate marketing, with websites being used to advertise rental properties, property search, and roommate search (through platforms such as apartments.com or Craigslist). Various regional MLS in the US used internet technology to centralise their real estate listings and include several property details and photographs. Regional firms and local MLS exchanges were incentivised to share some of their data from their databases to increase their business performance. Through this, they could gain better access to historic and current property listings with aggregated data from regional MLS. This was anchored by the National Association of Realtors (NAR). Examples of these platforms include Homefinder (launched in 1999), Trulia (launched in 2005), and Zillow (launched in 2006). Zillow was unique for providing property characteristics and market value look-up, and its subscription model provided sales leads or preferential placement by subscribed brokers.

Online listing and marketing platforms typically lead consumers to property agents, and prospective buyers or renters can contact them directly. Some of the online brokerage platforms were also targeted at for-sale-by-owner (FSBO) markets where sellers were property owners that listed their properties themselves and had buyers approach them directly without intermediation. These platforms include FSBO.com, HomesByOwner.net, etc. In the initial stages, the platforms were simple websites where the property owners only needed to input the property information and upload photos. However, only 7% of the recent home sales were from FSBO in 2018 (Saiz, 2020).

In the UK, Rightmove was launched in 2000 by four leading estate agents at the time (Sun Alliance, Halifax and Royal, Connells and Countrywide) and Zoopla was also launched in 2007. These platforms moved property search and listing (for sales and rental) from newspapers and specialist publications to online-based systems. The system where landlords traditionally targeted potential renters directly using newspaper adverts or specialised magazines was limited in terms of its scope. These platforms began to support people in their property search process and allowed agents to improve the reach of their properties to prospective customers. The online listing platform space has grown to become saturated and new platforms now have to deal with the challenge of highlighting their distinctiveness.

The shared economy also emerged in the mid-2000s. One of the first market participants of shortlets was Airbnb which was launched in 2008, and a similar platform for shared workspace – WeWork – was founded in 2010 to offer flexible workspaces. The IT 'revolution' also attracts opportunities for integrated CRE services companies (ICRESC), with multiple PropTech start-ups providing services in construction management, marketing, tenant representation, property management, smart buildings, asset management, underwriting, valuation, investments, legal, contracts, accounting, and other processes required by the CRE (commercial real estate) industry (Saiz, 2020).

There have also been new advancements in CRE transactions and data capturing. LootNet's holding company, CoStar, for instance, became a major repository of CRE data. Their platforms are used in both property transactions and commercial leases, and are thus attractive as listing platforms for agents of CRE interested in leases and sales. They enjoy a first-mover advantage as most brokers want to be listed there because potential customers and agents are already subscribed. The listing services were acquired in 2011 with their existing CRE data platform. Although other players such as 42 Floors, Real Massive, and Xceligent entered the market, CoStar has remained resilient.

The innovative tools and products that have emerged in this second generation of PropTech have been driven by unprecedented advancements in contemporary digital technology such as advanced analytics, IoT, AI, cloud computing, mobile communication, WiFi, and broadband. Although facilitated by these technological advancements, they have been complemented by e-commerce, social networking, open-source software, and the multi-platform world. Furthermore, open-source software and low-cost access to what used to be expensive technological tools, such as online payment services, have further facilitated real estate digitisation. Additionally, better connectivity between systems has transformed real estate. For instance, systems such as Argus and Yardi became open to collaboration; this, in turn, led to the development of compatible applications. Zoopla also launched an open application programming interface and API to allow developers to create applications using local data on their rental and sales listings using 15 years of data on sales prices.

Blockchain technology has also played a prominent role in the second generation of PropTech. Securrency, for instance, is an institutional-grade compliance-aware tokenisation, account management, and decentralised finance technology based on blockchain. It uses a process that makes the trade of value more efficient for a wide range of asset services such as leases. It introduced an elastic securitisation model that allows portfolios of assets to efficiently and securely expand and contract in alignment with market demand. This positively increases liquidity and access to both the dividend-producing asset and affordable funding capital for improvements and investments. Ethereum is a not-for-profit decentralised platform which operates smart contracts. These are applications that are operated based on programmed systems without downtime, censorship, fraud, or third-party interference. They enable developers to create markets, store registries of debts or promises, and move funds in line with instructions given in the past (e.g., wills or future contracts) without a middleman or counterparty risk. In the traditional server architectures, each application requires a unique server that runs unique codes in isolated silos, making data sharing difficult. The blockchain version allows any party to set up a node that replicates the necessary data for all the nodes to come to an agreement and this enables user data to remain private while apps remain decentralised.

In response to the growth and development of PropTech, and in line with the wide information pool it has created, several tracking platforms are in the space keeping tab of PropTech companies and the funds that they attract. These include Crunchbase, Unissu, Venture Scanner, etc. There are also several blogs and newsletters dedicated to the provision of PropTech news and the latest information such as the UKPA Newsletter, Real Estate Data (RED) Foundation Blog, CRE Tech Daily (US), and Prop Tech News (in the UK).

As Tagliaro et al. (2021) point out, PropTech is also associated with socio-cultural changes in the real estate sector. For instance, it has boosted property market diversity – attracting a wide range of talents with a strong female component and representation from different regions of the world. PropTech entrepreneurs are also drawn from diverse career and educational backgrounds.

3.1.3 The Impact of Covid-19 and Real Estate Innovation and PropTech

Due to the global outbreak of the Severe Acute Respiratory Syndrome Coronavirus 2 (commonly referred to as COVID-19), the World Health Organisation (WHO) declared a global health emergency on 30 January 2020. Because of the high mortality rate, high infection rate, and the inability of national health systems to contain its spread, the WHO declared a global pandemic on 11 March 2020. This led to a lockdown, i.e., the shutdown of national economies and severe restrictions on the movement of people. The lockdown disrupted several individual activities, institutional systems, and corporate operations. The COVID-19 pandemic is, therefore, classified as a black swan event[3] which has disrupted the various real estate sectors and various aspects of real estate use, operations, and service delivery; this ultimately resulted in an increasing prominence and usefulness of already existing real estate innovative and digital tools and platforms.

The office sector was one of the worst hit by the pandemic. Remote working became more prominent due to the lockdown. Working from home was already a well-established practice before the pandemic: 5% of the UK workforce (approximately 1.7 million workers) already worked from home as of 2019 (Office for National Statistics, 2020b) and 5.2% of the US workforce completely worked from home in 2017, while 43% worked from home occasionally (Hess, 2019); however, the pandemic led to a massive increase in remote working. By April 2020 (a few weeks into the lockdown), 46% of the UK labour force worked from home (Office for National Statistics, 2020a). Remote working also moved beyond occupational boundaries. Prior to the lockdown, remote working was most practiced in service occupations such as design, marketing, programming, media, and customer service (Saiz, 2020); however, this practice moved beyond professional and occupational borders. Also, before the lockdown, workplace apps and other digital tools were already being used to improve remote teams' meetings and collaboration as a way of facilitating flexibility and team cohesion (Jones Lang LaSalle, 2019). As a result of the pandemic, several digital tools became more prominent and used for meetings and corporate collaboration. These include Zoom, Microsoft Teams, Google Meet, etc.

As the pandemic lingered, workspace use and management, which hitherto had been relegated, began to receive some consideration. Corporate firms and institutions looked to develop strategies for business operations and office space utilisation. CEOs of global organisations with large space usage such as Google, Facebook, and Barclays are considering changes to office space use (Kalyan et al., 2020). These sentiments are reflected in a KPMG CEO Outlook Pulse Survey conducted in 2021 (KPMG, 2021) and Knight Frank's (Y) OUR SPACE (2021) survey data. There is growing interest in how hybrid workspaces (a combination of remote and office spaces) can be optimised to achieve a higher level of space efficiency and work productivity. Also, issues like office space quantity, quality, and ergonomics have become important (Oladiran et al., 2023). Technologies such as IoT and AI are now being used to collect and analyse data on office space use, and AR and VR tools are also used for viewings.

The retail and industrial real estate markets were also affected by the pandemic. Before, e-commerce was already a well-established sub-sector in retail (Dixon and Marston, 2002); this was reported to have led to a decline in demand further leading to increased vacancy rates and reduced the rate of sales in retail properties a few years before the pandemic (Zhang et al., 2016). Online shopping, however, intensified during the contagion. In the US, e-commerce sales increased by $244.2 billion (43%) in 2020 – the first year. This reflects an increase from $571.2 billion in 2019 to $815.4 billion in 2020 (United States Census Bureau, 2022). In the UK, 40% of UK shoppers in March 2020 reported that they had been shopping more online, compared to before the pandemic; this number, however, grew to 75% by February 2021 (Pasquali, 2023). As the demand for retail space decreases, the need for industrial space has also increased with the intensified demand for space for the storage and processing of goods ordered online.

The changes in the retail and industrial landscape have led to further innovation and digitisation in retail and industrial real estate. The use of smart buildings and advanced censors to optimise building performance and create more efficient, flexible, and responsive spaces is increasingly important. Data on occupancy, energy usage, climate control, etc. are being gripped to provide valuable insights into space users' needs and preferences to make real-time adjustments. AI too is increasingly used for predictive analysis and forecasting. AR and VR use in the retail real estate space also intensified. These technologies offer customers the opportunity to try on products, visualise their homes and businesses with new furnishings, and explore virtual showrooms from the comfort of their homes. As the retail sector continues to evolve post-pandemic, digital technological tools such as intelligent buildings, AI, AR, and VR will continue to support space users, operators, and service providers with more efficient approaches to using and managing their retail spaces.

Residential real estate has also seen significant changes following the global contagion. The lockdown confined individuals and households to their homes, and this created several new and highlighted existing issues too that previously received less attention. Individuals and households began to re-evaluate their housing needs and preferences in the context of confinement. Furthermore, work-life balance became more distorted as the boundaries between living, working, and leisure became blurred. These issues featured in major debates with recommendations and guidance notes issued for housing, planning, and homelessness by the Local Government Association, the Department for Levelling Up, Housing and Communities, and the Ministry of Housing, Communities, and Local Government in the UK. Emerging issues around well-being and inequality were increasingly debated: variation in housing quality, affordability, access to amenities, etc. across various communities, gender, and ethnic groups (Covid WIRED, 2021). These led to changing housing preferences in varying degrees. For instance, households in apartments and flats began to fancy housing units that had gardens and balconies. Furthermore, the need for office spaces at home became an important consideration. Residents also became a lot more conscious of their health and safety within their buildings. Residents in flats and apartments where central air conditioning systems were in place became increasingly aware of the issues surrounding their health and safety. These residential needs led to an increase in the need for IoT technology to monitor and regulate air and temperature control. There has also been an increased uptake of doorbells, cameras, etc. to minimise physical contact and improve safety and security.

Furthermore, with fewer people being able to go out to conduct house viewings, the use of virtual viewings during the pandemic increased significantly. For instance, Rightmove introduced a virtual viewing function during the pandemic, so prospective tenants or buyers can now view the interior of a property of interest without visiting. Online mortgage advisory and consultation also increased significantly during the pandemic and the rate of Automated Valuation Models (AVM) grew too. These models were largely used for mortgage valuations during the pandemic, particularly for newly built properties. The move from physical to more virtual services and operations in the residential sector during this time has attracted benefits including cost-saving, time-saving, improved comfort, lower carbon emission, etc.

The concluding section of this book will provide more information on the current PropTech generation, post-COVID-19 predictions, and the future of real estate innovation and PropTech.

3.2   PropTech Investment and Financial Capital[4]

PropTech has attracted a high volume of financial capital resulting from the enormous opportunities that it presents to investors and entrepreneurs. Finance and economic theory suggests that investment capital is driven by pull factors (i.e., factors that relate to the receiving countries) and push factors (i.e., factors relating to the sending source/sending countries). These pull and push factors may be macroeconomic (such as GDP, inflation, and interest rates), institutional (such as taxation, policy, and remittance), demographic (such as population, age dependency, and gender ratio), and geo-political (such as language, colonial history, contiguity, regional, and trading blocs). Insights on the impact of these have been extended to capital flow in real estate markets (Fadeyi et al., 2021, 2022; McAllister and Nanda, 2016). These studies suggest that a complex web of cross-country and global factors is responsible for real estate capital flow. However, little is known about PropTech investment and financial capital determinants. Furthermore, there is also limited knowledge regarding PropTech firms, the sectors they cover, and their growth pattern.

It is difficult to directly apply the insights from previous studies on real estate capital and service companies to the PropTech market. This is due to the real estate-related focus on physical assets that are bound by the fundamental principles of location, immovability, uniqueness, indestructibility, etc. PropTech tools, platforms, and products, although closely underpinned by real estate, have different characteristics and are heterogeneous. This may, therefore, require a blended approach where the real estate literature is blended with that of digital technology. This section provides insights on the growth of PropTech companies as a foundation for understanding PropTech investment capital. Historic and current trends are analysed using descriptive statistics: areas for further exploration are also highlighted.

3.2.1   PropTech Companies and Start-ups

A total of 8,201 PropTech firms have been established in the last five decades (1970–2021). Chart 3.1 provides the yearly breakdown of these numbers. It indicates that three PropTech companies were established in 1971 and the annual establishment of PropTech firms remained low for the next decade, slightly increasing the decade after. There was, however, a significant rise (double the number from the preceding year) in the number of new PropTech firms that were set up towards the end of the dot-com boom in 1999. This increase generally continued for the rest of the Noughties. In 2010, 292 PropTech firms were set up, bringing the total number of firms from 1970–2010 to 1940. The total number of PropTech launches in the final recorded decade (2011–2020) was 6,200. This represents a threefold increase in just one decade compared to the four preceding it. The sharp increase in the number of PropTech firms observed from 2008 occurred shortly after the launch of the iPhone. It is, therefore, argued that the functionality and utility that accompanied this innovation have supported the growth of PropTech tools and platforms. Conversely, the number of PropTech start-ups slowed down drastically in 2020 and 2021 (185 and 61 companies set up, respectively); this decline is likely to reflect the pandemic effects. It should be noted that not all these start-ups have survived, including 992 firms which have become inactive.

Location is a major factor in innovation. The data is, therefore, further explored for country-level trends. Table 3.1 shows the top 20 countries with the highest concentration of PropTech companies, and Figure 3.1 provides a global graphical view of the concentration of PropTech companies. The data reveals that there is a high concentration of PropTech companies in OECD countries.[5] This is not a surprise. It has been found that a few rich countries account for most of the world's creation of new technology (Keller, 2004); the reasons for this locational pattern are, however, unclear. Porter and Stern (2001) explored the influence of country-level factors on varying levels of innovation in OECD countries in comparison to emerging countries between 1975 and 2000. Their

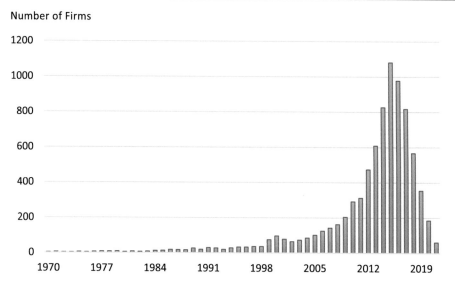

Number of Firms

Chart 3.1 Number of PropTech firms and years of establishment (1970–2021).

Table 3.1 Top 20 Countries with the Highest Number of PropTech Companies (as of 2021)

	Country	PropTech Firms
1	US	2,316
2	UK	1,095
3	France	572
4	India	473
5	Spain	440
6	Australia	439
7	Brazil	394
8	Germany	366
9	Netherlands	300
10	Canada	244
11	Switzerland	230
12	Sweden	205
13	Finland	194
14	China	181
15	Dem. Rep. Korea	171
16	Denmark	167
17	Norway	128
18	Turkey	122
19	Italy	115
20	Japan	115

Source: Authors' table using Unissu Data (2022).

study revealed that a relatively small number of factors relating to a country's business environment explain the massive variations. The work of Siedschlag et al. (2013) suggests that human capital and research and development (R&D) activity and expenditure are important. It is important to note that the major sources of this technological growth in

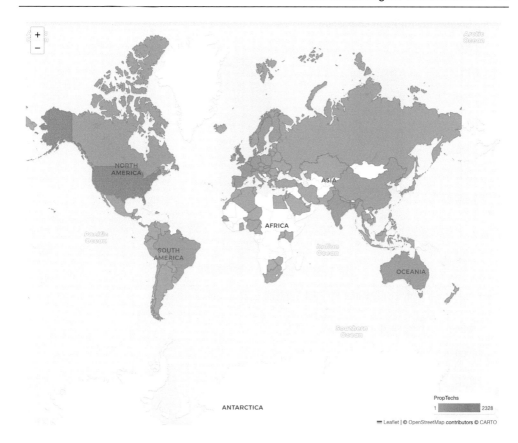

ANTARCTICA

PropTechs

1 [   ] 2328

Leaflet | © OpenStreetMap contributors © CARTO

Figure 3.1 Concentration of PropTech companies as of 2021.

OECD countries are cross-border, not domestic (Eaton & Kortum, 1999; Keller, 2002). From a real estate perspective, real estate market maturity and transparency may also influence PropTech capital flow, as markets that are more mature and have a higher level of transparency are more likely to have the data that can drive innovation and digitisation in real estate markets. This is supported by the JLL transparency index of 2022 (JLL, 2022) which shows that OECD countries are generally more transparent than other countries. As seen in Table 3.1 and Figure 3.1, the US has the highest concentration of firms (2,316) and this is followed by the UK (1,095). This data, particularly for the US, is consistent with the proposition of Keller (2001) that new ICT technologies have been developed at a faster pace in the US (relative to other countries). The key factors driving current PropTech companies' concentration should, therefore, be explored in further research.

Chart 3.2 also shows that PropTech companies are almost evenly split between residential and commercial real estate, although the firms have a slightly higher concentration in commercial (6,628) real estate than residential (6,180).[6] In terms of real estate activity, the post-construction stage of the real estate lifecycle has attracted the highest number of PropTech start-ups. Chart 3.3 shows that 1,561 firms provide tools and products for construction (building), 3,748 firms provide property management services, and a total of 4,411 firms provide brokerage/agency solutions.[7]

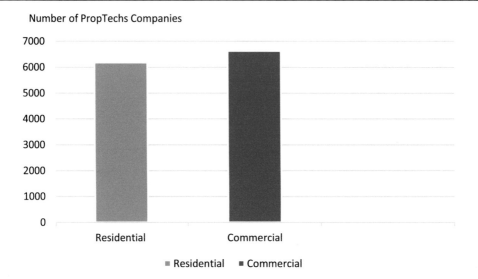

Chart 3.2 PropTech companies by real estate sectors (as of 2021).[8]

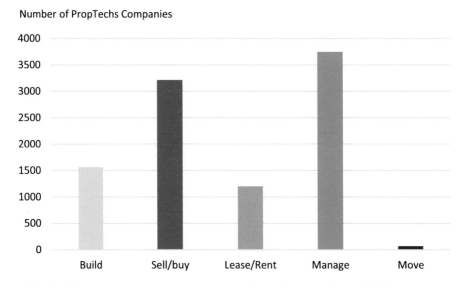

Chart 3.3 PropTech companies and the property lifecycle stages (as of 2021).

Chart 3.4 shows the distribution of PropTech companies based on job functions in the real estate sector. This is an indication of the job roles that PropTech firms support the most. The data shows that real estate design has the highest level of start-ups' interest with asset management following closely, while conveyancing is at the bottom of the scale. This is supported by Chart 3.5 which shows that firms providing solutions for construction and building maintenance have the highest concentration. Chart 3.6 further shows the PropTech companies by their business types. It reveals that the vast majority of PropTech firms are B2B (business-to-business), which suggests that their tools, products,

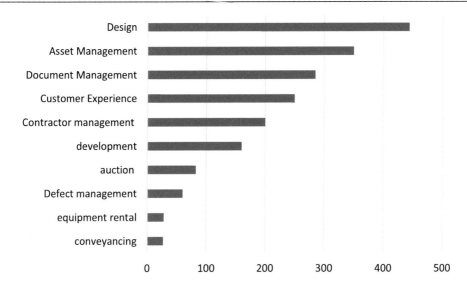

Chart 3.4 PropTech companies and the job functions that they support (as of 2021).

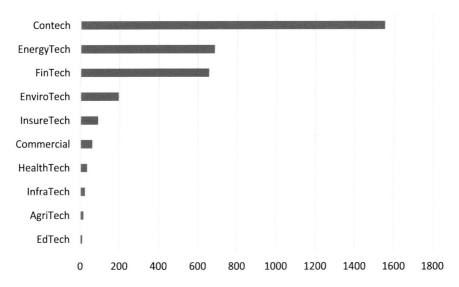

Chart 3.5 PropTech by sub-sector and industry (as of 2021).

and services are targeted at companies and organisations rather than individuals. As seen, some of the companies provide services and develop products to serve both corporate organisations and individuals.

As discussed in Chapter 2, several digital technologies have been applied to real estate in the last half-century. Chart 3.7 shows the breakdown of technologies that underpin the products and tools that PropTech companies have developed. It reveals that big data, AI, 3D modelling, and BIM are the most common contemporary technologies used by PropTech firms.

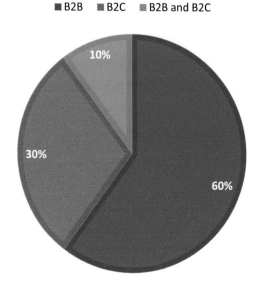

Chart 3.6 PropTech companies by business types (as at 2021).

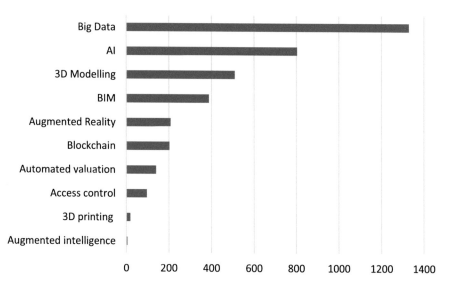

Chart 3.7 PropTech companies by core technology adopted (as of 2021).

3.2.2 PropTech Funding and Investment Capital

This sub-section analyses the financial capital that PropTech has attracted from 2010 to 2021. Chart 3.8 provides insight into the volume of investment capital in the PropTech space over the last decade. The data shows that approximately $870 million was invested in PropTech in 2010, and this increased to approximately $18 billion at the end of the

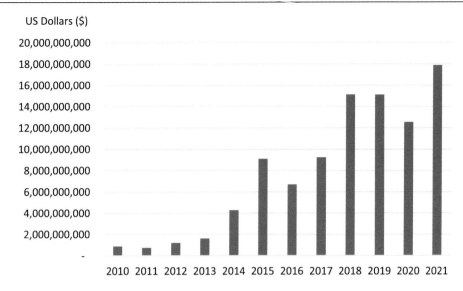

Chart 3.8 Total PropTech funding from 2010 to 2021.

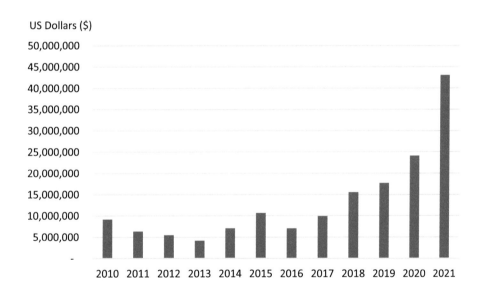

Chart 3.9 Average funding per year from 2010 to 2021.

decade. It is remarkable that PropTech investment capital began to see a significant increase in 2014 (rising from $1.6 billion in 2013 to $4.1 billion in 2014) and this peaked in 2021 ($17.8 billion). Furthermore, Chart 3.9 shows the average funding per year, derived based on the number of companies that received funding in those years and the total funds received in the year. As shown, the funding per capita has been on an upward trajectory since 2017 (apart from 2020 when it decreased mainly because of the COVID-19 pandemic).

As with every sector, the pandemic caused the PropTech investment market to slow down as investors became more cautious towards their portfolios and new deals. Start-ups typically depend on traffic in the built environment to demonstrate value, and the poor performance of several real estate asset classes during the pandemic did not help matters. The data in Chart 3.9 is particularly interesting, given that the number of funding events declined steadily in the same period (Chart 3.10). This may suggest a somewhat inverse relationship between these indicators. This may also be a hint that the PropTech investment market is maturing, and thus, the few funding events that have been organised have yielded positive results.

Like the distribution of PropTech companies, PropTech investment capital has also been unevenly distributed across the world. Figure 3.2a and b shows the total and average funding concentration globally. Figure 3.2a shows that a huge proportion of PropTech capital is concentrated in the US. However, the general outlook is slightly different from Figure 3.1 and Table 3.1. As seen in Figure 3.2b, China has the highest average funding, suggesting that the volume of funds per PropTech firm is higher than that in other countries. Table 3.2 provides further details. It shows that countries such as China, India, Argentina, and Israel, although not listed in the top 20 countries with PropTech companies, are part of the 20 countries that receive the highest PropTech capital. This also raises similar questions to those posed in Section 3.2.1 on the key factors driving PropTech capital flow. The literature on real estate (Fadeyi et al., 2021, 2022; McAllister & Nanda, 2016) and general technology capital flows (Cole et al., 2016) may provide insight into

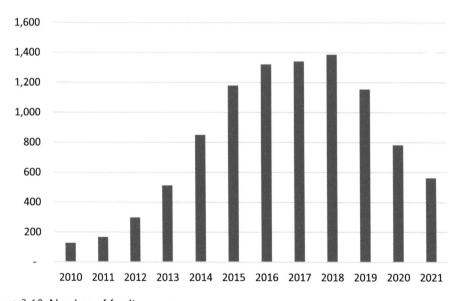

Chart 3.10 Number of funding events.

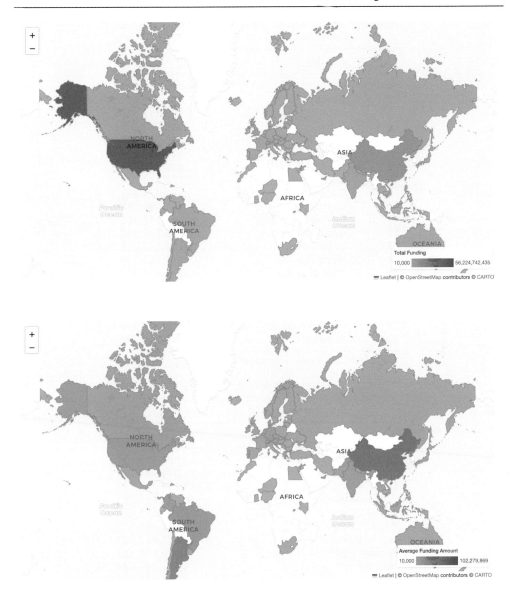

Figure 3.2 (a) Total PropTech funding amount from 2010 to 2021. (b) Average PropTech funding amount from 2010 to 2021.

the factors responsible for these trends from the property and technological perspectives. Yet, further empirical work would be valuable for demystifying this mechanism.

The breakdown of the funding by investment stage and the most funded PropTech vendors are provided in Tables 3.3 and 3.4. As seen in Table 3.4, the US once again dominates the list of countries with PropTech companies that have the highest funding.

Table 3.2 Top 20 Countries with the Highest PropTech
         Investment Capital

	Country	Total Funding ($)
1	US	54,262,385,724
2	China	12,066,132,841
3	UK	5,058,137,521
4	India	2,999,082,501
5	Dem. Rep. Korea	2,490,761,540
6	Brazil	2,100,407,214
7	Argentina	2,010,669,884
8	Singapore	1,773,230,776
9	Canada	1,597,769,263
10	Germany	1,593,580,548
11	France	1,442,735,390
12	Israel	1,286,221,696
13	Spain	1,109,637,012
14	Australia	566,563,250
15	Netherlands	330,052,421
16	Switzerland	329,008,962
17	Japan	273,317,735
18	Finland	266,187,762
19	United Arab Emirates	210,205,374
20	Sweden	194,881,424

Source: Authors' table using Unissu Data (2022).

Table 3.3 PropTech Investment Capital by Investment
         Stage as of 2021

Investment Stage	Total Funding
Series B	14,087,609,121
Series C	13,040,984,230
Series A	10,912,856,652
Venture – Series Unknown	10,342,681,723
Series D	9,902,677,658
Post-IPO Equity	7,655,469,035
Series E	7,473,637,999
Private Equity	6,516,800,154
Series F	4,567,115,784
Seed	3,304,045,086
Secondary Market	1,481,420,592
Series H	1,406,403,897
Series G	1,050,690,092
Convertible Note	852,079,705
Corporate Round	522,518,842
Undisclosed	423,563,049
Grant	197,574,159
Equity Crowdfunding	180,986,931
Angel	172,189,767
Pre Seed	167,662,265
Series I	95,920,000
Initial Coin Offering	42,802,792
Non-Equity Assistance	33,178,541
Product Crowdfunding	15,999,920

Source: Authors' table using Unissu Data (2022).

Table 3.4 Most Funded PropTech Companies (Cumulative) as of 2021

Vendor	Country	Total Funding
Airbnb	US	3,951,898,728
Ziroom	China	2,120,790,266
View Glass	US	2,044,985,500
Yanolja	Dem. Rep. Korea	1,948,500,000
Mercado Libre	Argentina	1,881,600,000
Opendoor	US	1,750,950,000
Katerra	US	1,648,210,844
Compass	US	1,528,000,000
ParkJockey	UK	1,501,900,000
Lianjia	China	1,405,714,728
Hippo Insurance	US	1,259,100,000
AvidXchange	US	1,128,171,523
ServiceTitan	US	1,099,817,129
Trax	Singapore	1,027,281,822
ThingWorx	US	1,000,000,000
ZhongAn	China	937,047,652
Samsara	US	930,100,000
Better Mortgage	US	905,612,733
Toast	US	899,000,000
Danke Apartment	China	874,773,503
Loft	Brazil	788,000,000
Postmates	US	763,050,000
QuintoAndar	Brazil	755,270,292
Tujia	China	755,000,000
Plume	US	722,357,341

Source: Authors' table using Unissu Data (2022).

3.3   Chapter Summary

This chapter has provided a review of the real estate transformation and highlighted major landmarks. It has also mapped various stages of the growth and development of PropTech and real estate innovation. Furthermore, this chapter has provided a review and analysis of global PropTech investment capital. PropTech tools, products, and companies will be discussed in more detail in subsequent chapters, and case studies will be provided to facilitate the understanding and appreciation of various real estate digital tools.

Notes

1 Now a multinational software corporation.
2 The internet backbone is the core of the internet where the largest and fastest networks are linked. They are typically owned and operated by commercial, educational, military, or government entities.
3 The theory of black swan events describes rare, unpredictable, surprising and outlier events which leave clear footprints and, in most cases, significantly change sector systems and operations (Runde, 2009).
4 The data in this section is mainly sourced from Unissu Data (2023); exceptions to this will be duly cited.
5 OECD (Organisation for Economic Co-operation and Development) is an international organisation comprised of 38 countries in America, Europe and Asia-Pacific and set up with the goal of shaping global policies for prosperity, opportunity and well-being. The full list of countries can be accessed on the OCEC website.

6 The dataset does not contain information on the operational real estate asset class.
7 The dataset does not contain information on other areas of the real estate value chain such as planning, design, and investment.
8 It should be noted that some PropTech companies and tools are applicable to more than one real estate sector, lifecycle stage, job functions, etc.

References

Coffman, K. G., & Odlyzko, A. M. (2002). Growth of the internet. In I. P. Kaminow, & T. Li (Eds.), *Optical fiber telecommunications IV-B: Systems and impairments* (pp. 17–56). Academic Press. https://doi.org/10.1016/b978-012395173-1/50002-5.

Cole, H. L., Greenwood, J., & Sanchez, J. M. (2016). Why doesn't technology flow from rich to poor countries? *Econometrica, 84*(4), 1477–1521. https://doi.org/10.3982/ecta11150.

Covid WIRED. (2021). *Wellbeing inequalities: HOUSING.* https://whatworkswellbeing.org/resources/covid-19-and-wellbeing-inequalities-housing/.

Dixon, T., & Marston, A. (2002). U.K. retail real estate and the effects of online shopping. *Journal of Urban Technology, 9*(3), 19–47. https://doi.org/10.1080/1063073022000044279.

Eaton, J., & Kortum, S. (1999). International technology diffusion: Theory and measurement. *International Economic Review, 40*(3). http://onlinelibrary.wiley.com/doi/10.1111/j.1468-2354.2009.00532.x/full.

Fadeyi, O., McGreal, S., McCord, M., & Berry, J. (2021). Capital flows and office markets in major global cities. *Journal of Property Investment and Finance, 39*(4), 298–322. https://doi.org/10.1108/JPIF-02-2020-0023.

Fadeyi, O., McGreal, S., McCord, M. J., Berry, J., & Haran, M. (2022). Influence of the global environment on capital flows in the London Office market. *Journal of European Real Estate Research.* https://doi.org/10.1108/JERER-05-2021-0029.

Hess, A. (2019). People who work from home earn more than those who commute—Here's why. *Cnbc,* 2019–2022.

JLL. (2022). *Global real estate transparency index, 2022 transparency in an age of uncertainty.* https://www.jll.de/content/dam/jll-com/documents/pdf/research/global/jll-global-real-estate-transparency-index-2022.pdf.

Jones Lang LaSalle. (2019). *How technology is fuelling the rise of flexible office space.* https://www.jll.co.uk/en/trends-and-insights/workplace/how technology-is-fuelling-the-rise-of-flexible-office-space.

Kalyan, S., Learner, H., & Moreira, R. (2020). *The future of office in the Covid-19 era.* CBRE (Issue July).

Keller, W. (2001). International technology difussion. In *National bureau of economic research* (Vol. NBERWorki). https://www-nber-org.sheffield.idm.oclc.org/system/files/working_papers/w8573/w8573.pdf.

Keller, W. (2002). Geographic localization of international technology diffusion. *American Economic Review, 92*(1), 120–142. https://www.aeaweb.org/articles?id=10.1257/000282802760015630.

Keller, W. (2004). International technology difussion. *Journal of Economic Literature, 42*(3), 752–782. https://doi.org/10.1257/0022051042177685.

KPMG. (2021). *Nearly half of global CEOs don't expect to see a return to 'normal' until 2022* (Issue March 2021). https://home.kpmg/xx/en/home/media/press-releases/2021/03/nearly-half-of-global-ceos-dont-expect-a-return-to-normal-until-2022-ceo-outlook-pulse.html.

McAllister, P., & Nanda, A. (2016). Does real estate defy gravity? An analysis of foreign real esate investment flows. *Review of International Economics, 24*(5), 924–948. https://onlinelibrary.wiley.com/doi/epdf/10.1111/roie.12228?saml_referrer.

Office for National Statistics. (2020a). Coronavirus and homeworking in the UK: April 2020. Homeworking patterns in the uk, broken down by sex, age, region and ethnicity. In *Special bulletin* (Issue July). https://www.ons.gov.uk/employmentandlabourmarket/peopleinwork/employmentandemployeetypes/bulletins/coronavirusandhomeworkingintheuk/april2020.

Office for National Statistics. (2020b). *Which jobs can be done from home? - Office for National Statistics. July 2020* (pp. 1–12). https://www.ons.gov.uk/employmentandlabourmarket/peopleinwork/employmentandemployeetypes/articles/whichjobscanbedonefromhome/2020-07-21.

Oladiran, O., Hallam, P., & Elliot, L. (2023). The Covid-19 pandemic and office space demand dynamics. *International Journal of Strategic Property Management, 27*(1), 35–49. https://journals.vilniustech.lt/index.php/IJSPM/article/view/18003.

Pasquali, M. (2023). *Changes in online buying among UK consumers since COVID-19 2020-2021 percentage change in online purchases due to the coronavirus (COVID-19) pandemic in the United Kingdom from March 2020 to February 2021*. https://www.statista.com/statistics/1230225/changes-in-online-buying-among-uk-consumers-since-covid-19/.

Porter, M. E., & Stern, S. (2001, July). Innovation : Location matters. *MIT Slogan Management Review*, 1–4. https://sloanreview.mit.edu/article/innovation-location-matters/.

Runde, J. (2009). Dissecting the black swan. *Critical Review, 21*(4), 491–505. https://doi.org/10.1080/08913810903441427.

Saiz, A. (2020). Bricks, mortar, and proptech: The economics of IT in brokerage, space utilization and commercial real estate finance. *Journal of Property Investment and Finance, 38*(4), 327–347. https://doi.org/10.1108/JPIF-10-2019-0139.

Siedschlag, I., Smith, D., Turcu, C., & Zhang, X. (2013). What determines the location choice of R&D activities by multinational firms? *Research Policy, 42*(8), 1420–1430. https://doi.org/10.1016/j.respol.2013.06.003.

Tagliaro, C., Bellintani, S., & Ciaramella, G. (2021). R.E. property meets technology: Cross-country comparison and general framework. *Journal of Property Investment and Finance, 39*(2), 125–143. https://doi.org/10.1108/JPIF-09-2019-0126.

Unissu Data. (2023). *PropTech market analysis data*. https://www.unissu.com/analytics.

United States Census Bureau. (2022). *Annual retail trade survey shows impact of online shopping on retail during COVID-19 pandemic*. https://www.census.gov/library/stories/2022/04/ecommerce-sales-surged-during-pandemic.html.

University of Oxford Research. (2017). *PrepTech 3.0: The future of real estate.*

Zhang, D., Zhu, P., & Ye, Y. (2016). The effects of E-commerce on the demand for commercial real estate. *Cities, 51*, 106–120. https://doi.org/10.1016/j.cities.2015.11.012.

Section Two

PropTech and Property Innovations across the Real Estate Value Chain

Section One introduced the principles and concepts that underpin PropTech and real estate innovations and provided a broad overview of the PropTech evolution. It demonstrated that PropTech spans various real estate sectors, sub-sectors, operations, and services. These operational areas and services are often referred to as the real estate value chain.[1] It is therefore important that real estate students, educators, professionals, and PropTech enthusiasts understand how digital technology and innovations have been integrated into various segments of the real estate value chain and across the core real estate sectors: commercial, residential, and operational.

Each chapter in this section focuses on a group of closely related real estate operations and services (illustrated in Figure 4.1). Figure 4.1 shows various groups of operations and services and illustrates the integration of digital technology in each of these areas. The outcomes of digital technology and real estate integration are captured within the oval PropTech sphere, with each of the orange segments showing the outcome of the integration of digital technology into each real estate section. The chapters in this section are based on the segments illustrated in Figure 4.1: Chapter 4 focuses on the integration of digital technology and innovation in planning, land management, and urban management; Chapter 5 covers digital technology and innovation in real estate construction, management, and maintenance; Chapter 6 covers innovation and digital technology in real estate investment, funding, and finance; Chapter 7 focuses on valuation and appraisal; and Chapter 8 covers innovation and digital technology in real estate agency, brokerage, marketing, and other allied service areas.

Each chapter begins with a review of the traditional operations and services in the real estate value chain segment being analysed, highlighting the core areas of inefficiency and challenges associated with traditional practices. This will be followed by an overview of some of the core digital technologies and applications within that real estate segment and their transforming effects within the value chain. This will include an analysis of emerging real estate digital functions and systems such as PlanTech in planning, building information modelling (BIM) in construction and building maintenance, and automated valuation models (AVMs) in valuation. Furthermore, the analysis will highlight the various ways through which these systems are addressing the inefficiencies highlighted. To provide further contextual and practical insight into specific value enhancement effects of PropTech, case studies are provided in each chapter.[2] The case studies are analysed using the environmental, social, economic, and physical (ESEP) efficiency framework introduced in Chapter 1 to enable the reader to understand some of the core environmental, social, economic, and physical solutions that they provide. A list and short descriptions of other

DOI: 10.1201/9781003262725-5

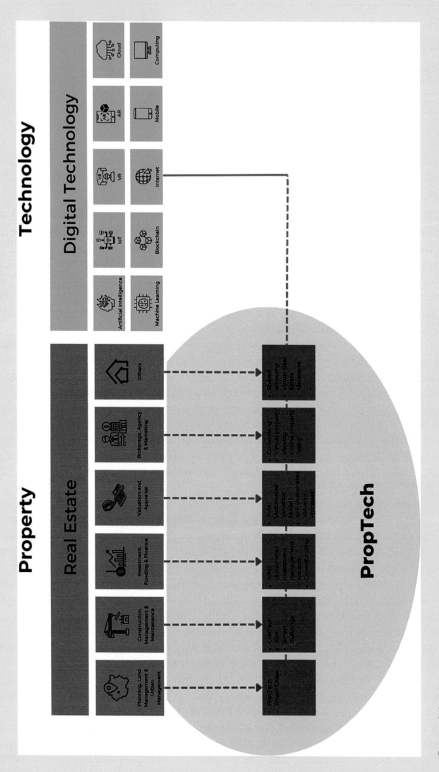

Figure 4.1

tools and platforms are provided, and readers are encouraged to explore and analyse them further using the ESEP efficiency framework.

Learning Outcomes (LO)

At the end of this section, the reader should be able to:

1 Describe the core inefficiencies associated with the various real estate uses, operations, and services.
2 Understand real estate innovation and digital applications, and their roles in addressing the various inefficiencies across the real estate value chain.
3 Understand the process of real estate innovation and, through the case study analysis, appreciate the process and impact of digitisation
4 Identify the current frontiers of real estate innovation and digitisation and develop ideas for further innovation and digital applications.

Notes

1 The segments of the real estate value chain include planning, land management, development, construction, investment, finance, portfolio management, property management, agency, and brokerage.
2 The information in the case studies was partly sourced directly from the companies and partly from their websites. This information is, however, objectively presented and analysed by the authors.

Chapter 4

Digital Technology and Innovation in Planning, Land Management, and Urban Management

4.1 Overview of Planning, Land Management, and Urban Management

Planning involves designing, mapping, and shaping places. It is intrinsically linked with land use, housing, transportation, job opportunities, social services, and the quality of the urban environment (Corburn, 2009). Planning may involve the development of land use plans, transportation, infrastructure, and other public facilities and services to promote social equity, economic growth, and sustainable development. Although the micro and macro conceptions of the politics of planning remain controversial among theorists, it is suggested that the planning practice ought to be considered an essential part of urban governance (Corburn, 2009; Fainstein, 2017; Yiftachel and Huxley, 2000). Traditional planning procedure revolves around sets of standards for sizes, locations, capacities, and other physical dimensions of highways, schools, houses, water systems, library buildings, etc., and it relies on long-range forecasts of demand, requirements, and, in many cases, professionals' judgements of future desired spatial conditions (Webber, 1965).

Planners are built environment professionals who engage in plan-making, development management, public engagement, and place-making (among other things). They are often tasked with ensuring that cities, towns, and urban areas grow and develop in an efficient and sustainable manner with adequate consideration for factors such as environmental protection, community needs, infrastructural needs, and population growth. They are also critically concerned with identifying animating and mediating between various interests in a development process, giving rise to the need for multiple skills (Tomaney and Fern, 2018). Developing and sustaining urban plans and policies requires a high level of collaboration with several stakeholders such as community citizens, community leaders, government institutions, and a wide range of built environment professionals.

Land administration systems are put in place to determine and record relationships between individuals and land regarding, the rights of use or ownership of land, and the process of making this information available to members of the public or community (Lemmen et al., 2017). It regulates relationships between spatial units, people, and the rights connecting them. These elements often entail the overlap of multiple entities, entitlements, individuals, spaces, etc. which can complicate the structure of the system and its wider coordination.

Various countries have planning and land administration systems, and these systems are fundamental to real estate markets, particularly the supply side of the market.[1] For instance, economic theory suggests that real estate supply challenges can lead to scarcity

DOI: 10.1201/9781003262725-6

of property in markets, and where demand forces are exceptionally strong, it can lead to acute shortages and significantly high rental values and prices. Furthermore, land use planning and management systems inform the allocation of land and the policy to support urban development. It is an activity that attempts to mediate market effects in land use and resource distribution (Porter et al., 2019). These aspects of the property value chain are important because they shape the places where we live, the type and quality of spaces provided in our living areas, types of jobs, housing affordability, access and capacity of amenities and infrastructure, quality of life, etc. Thus, policymakers, civil society groups, and community groups invest in and contribute to the development of planning systems and land administration.

4.2   Inefficiencies in Planning, Land Administration, and Urban Management

The traditional approaches to carrying out various planning, land administration, and urban management processes and tasks are riddled with inefficiencies. Some of the issues range from poor data gathering and analysis to inefficient decision-making processes, poor technical skills, and poor urban design and visualisation techniques. The manifestation of these issues may vary from country to country. In the UK, for instance, The Future of Planning in London Report highlights some issues (The London Assembly, 2022). They include the following:

• The planning system fails to improve home construction, particularly in areas where housing needs are worst.
• The planning system is too complicated and has lost public trust.
• Adopting a local plan takes a long period of time.
• The focus on design is minimal and there is little incentive for high-quality new homes and places.
• The process for negotiating developer contributions to affordable housing and infrastructure is complex, protracted, and unclear.

In many planning authorities and departments, planning and design are still done manually using archaic systems to review printed drawings, establish the conformity of developments to minimum standards, etc., and this makes it difficult to synthesise vast amounts of complex data on zoning, land use, and data from community engagement and environmental systems (Hartley, 2019; Webb, 2019). According to Webb (2019), local authorities have failed to use their websites to provide a clear representation of how their spaces are expected to change in the medium and long term, and planning committee meetings are difficult to understand due to the high technical language used.

Hartley (2019) further notes that the planning systems in some local authorities are inefficient as data collection on sites, evaluation of development proposals and citizens' engagement are rigorous, time-consuming, costly, and often generate large volumes of paperwork. Complex projects often require long technical reports compiled by separate consultants pieced together through a long-winded manual process, and it takes a long time to create a level of exclusivity as the documents are difficult to access and interpret. The often-voluminous reports contain recommendations and interpretations that are often made based on expert opinion which usually contribute minimally to the prediction of the impact of new projects based on the unique circumstances of a particular site. Furthermore, "expert opinion" may come with some level of bias and subjectivity which may be driven by cultural background, personal preferences, ideologies, and experience. These issues have social implications as citizens may continue to protest and resist developmental activities due to inefficient citizen engagement. There are also associated economic implications as planning officials may take too much time to complete their tasks, particularly when stakeholders are resistant to the proposed development. These issues can be compounded in the long term inhibiting economic and social development in cities, thereby increasing inequality in urban areas with additional spatial and environmental impact.

The traditional methods of land registration, management, and administration are also inefficient. Land registries typically utilise large volumes of data and are typically organised as centralised databases. Despite the huge financial resources allocated to data generation planning, development evaluation, control, and land management, these often end up being redundant due to poor system and data integration and communication. It is estimated that approximately 70% of the world's population lacks access to legally registered land titles (The World Bank, 2017). This problem of insecurity of land rights has implications for tax collection and infrastructure provision and limits the territorial authority of the state (Rodima-Taylor, 2021). Poorly managed land registration systems can also lead to incomplete registries that can fuel land disputes, corruption, and land disposition (Benbunan-Fich and Castellanos, 2018). Furthermore, secure land rights have the potential to reduce poverty and boost shared prosperity at the country, community, and family levels. These can improve social mobility and economic vitality. Land rights are also fundamental to stimulating investment in real estate and infrastructure and for the registration of property sales or mortgages. For instance, The World Bank (2017) revealed that launching an online service and boosting the capacity of the real estate cadastre in Macedonia significantly reduced the time taken to complete property sales and mortgage registration from two months to two days, and this further led to a growth of the mortgage market from EUR 450 million to EUR 3.4 billion.

These inefficiencies and challenges have been addressed using various innovative and digital technological systems and the next sub-section provides an analysis of some of these systems and services.

4.3   Innovative/Emergent Systems and Services in Planning, Land Administration, and Urban Management

4.3.1   Intelligent Planning Systems

A sub-set of planners and urban policymakers have long advocated the adoption of intelligent planning systems that draw data for urban areas and use them in combination with algorithms from planning policy and law to automate the development of optimal strategic planning and recommendations for planning decisions (Porter et al., 2019). Webber (1965) proposed the creation of intelligence centres that would interlink and collate data, formulate strategic plans, aid supply analysis, improve forecasting, aid incremental, multi-centred decision-making, and enact scientific principles in urban affairs. This proposition was aimed at minimising subjectivity, clientelism, and vested interest. The system was ultimately expected to learn from its actions to produce more effective and efficient outputs (what is now termed machine learning or AI).

Forrester (1969) further proposed a cybernetic approach to planning. This approach is based on the science of communication and control theory that is concerned with the comparative study of automated control systems. It positions the city as a structure of systems that could be subdivided into constituent parts and processes which can be modelled and simulated to capture its essence. In turn, these models can be used to plan and operate its function. In the 1970s, the systems perspective presented planning as an evidence-informed, structured, rational, applied science that could be performed using computational approaches. The development of the Geographic Information System (GIS) in the 1980s and 1990s as a platform for integrating and analysing spatial decision support and expert systems using a combination of planning rules and practices led to the further evolution of this system, particularly because of its decision-making abilities (Klosterman, 1997; Lee and Wiggins, 1990). Similar advancements in the 1990s saw the initial experiment with 3D urban and landscape models and VR technologies that improved the visualisation of the topography of existing and planned future environments (Doyle et al., 1998).

The first edition of The Future of Planning programme was conducted by Capital Enterprise in 2016, and this led to a study involving planners, property developers, architects, citizen groups, utility companies, and other stakeholders involved in the planning system in England. This study highlighted four areas of the English planning system that require transformation: the planning process (how plans are made), planning application system (the process for applying for planning permission), planning communication (how plans are communicated), and data use (the use of data within the system). According to Webb (2019), the use of data offers a significantly remarkable potential for development, particularly the development of products and services for the planning sector. Public ministries and departments tasked with planning are also acknowledging the role of digital technology in contemporary planning. The UK Government's Levelling Up and Regeneration Bill, for instance, acknowledges that the planning system is too complex and should now be based on contemporary technology (UK Ministry of Housing, 2020).

4.3.1.1   PlanTech

PlanTech refers to the integration of digital technology with urban planning and design to produce better living spaces. It provides innovative approaches to solve complex integrated planning and design challenges it offers new opportunities to collate and produce data, analyse them, and leverage them for insight. Although urban planning and design are still largely done manually, PlanTech is increasingly being used by planners and urban designers, and it has become a multi-billion-dollar industry. It is reported to be attracting more investment funding than BioTech and pharmaceuticals (Hartley, 2019).

Technologies such as 3D, GIS, CIM (City Information Modelling), and BIM are being explored for the utilisation of space intelligence for urban planning and design. BIM provides a platform for the full building cycle of a project to be viewed and analysed within a single integrated system and also provides a detailed, interactive 3D model beyond the traditional system of several 2D plans, sections, and elevations. Urban designers can therefore use this to dramatically update and recalculate scheduling and quantities of materials that will accompany changes to design and building specifications (Crotty, 2012).

Real estate developers and investors are core beneficiaries of PlanTech. With better access to planning data and improved transparency, developers and investors can carry out development and planning appraisals with better insight into the site and market; this can further enable them to make evidence-based decisions with better outcomes and benefits such as cost and timesaving. PlanTech also reduces the time it will take for planners and urban designers to complete their tasks, and it is now receiving increasing attention from professional bodies. The Royal Town Planning Institute (RTPI), for instance, featured PlanTech in one of its national events (Connected Places Catapult and RTPI Joint Online Event) in 2021. This increasing interest is based on the potential to attract benefits and opportunities such as better data management, support collaboration, development of more effective approaches to engagement with communities, better transparency, and efficient planning services. Hartley (2019) predicts that traditional approaches to urban planning will gradually give way to PlanTech as new tools and platforms continue to demonstrate higher levels of efficiency in comparison to traditional methods.

4.3.1.2   Challenges to Adopting Intelligent Planning Systems

Despite the enormous benefits of using intelligent urban planning and design systems, these systems are not without challenges. Porter et al. (2019) provide insight in this respect. Two core categories of the challenges will be briefly discussed: technical and institutional/political.

Technical Challenges: Developing technologies and incorporating them in mainstream professions often takes time – in many cases decades. The planning profession has been wary of a computational and technocratic approach to planning practice, and this makes the integration of intelligent planning systems problematic. For instance, GIS technology evolved from the 1960s and was not integrated into mainstream planning till the 1990s. There have also been other prototype 3D technologies for over three decades, however, these have only started becoming adopted in the planning profession in the last few years and in many cases, on a trial basis. CIM is also still at the initial stage of development and there are associated challenges which relate to software, requiring the use and integration of different platforms and packages that have their limitations and often necessitate the development of new workflows and the creation of bespoke code to bridge shortcomings. It is difficult to find a fully functioning open-source digital solution that can support the development of CIMs, and the game engine visualisation software for optimally displaying 3D environments is not generally configured to be used like a 3D GIS, for instance, which uses different coordinate systems and often lacks the required spatial precision. Furthermore, the integration of AI in planning is still generally in the infancy stage and requires substantial development to reach sufficient maturity and trust to support intelligent decision-making. Off-the-shelf platforms are rare, and this leaves users with the capacity to build and experiment with analytical predictive modelling and simulation tools.

There are also data-related issues. Detailed data reliable 3D models of landscapes are still relatively uncommon and there are associated challenges with urban data coverage, access, representativeness, quality, completeness, and metadata. These make it difficult to assemble relevant, timely, granular, high-quality, and interoperable data sets that are required for advanced analytical techniques. This places researchers in a position where they have to perform significant data cleaning and transformation to create workable and meaningful datasets (McArdle and Kitchin, 2016). These issues can lead to the presentation of low standard output (Porter et al., 2019), although this will depend on the level of advanced analytical skills and tools that the planner possesses.

Institutional and Political Challenges: Planning is highly influenced by political and institutional factors that cannot be narrowly reduced to data and standards. Cities and regions sometimes have competing vested interests and decision-making, which are underpinned by the art of negotiation and compromises. This can therefore make the decision-making process complicated, and bureaucratic hurdles can sometimes get in the way. The professional planning community and professional bodies' general concern for computational and technocratic approaches to planning can also compound these challenges. For these challenges to be conquered, there is a need for greater transparency, and a willingness to accept public opinion, debates, and contestation rather than enacting autonomous technocratic approaches. This can complement already existing epistemological and methodological approaches to planning.

In interviews with senior planners in Ireland about the potential use of CIMs in planning, Porter et al. (2019) reported that the interviewees expressed a number of doubts and concerns about its utility in aiding planning, although they acknowledged the potential benefits of such technologies. This suggests that a major shift in planning theory, ideology, and practice will need to occur for an intelligent planning system approach to become the standard practice, although it does not appear that there is currently a pathway to reaching this in the short-to-mid-term. Public institutions and professional bodies therefore need to proactively develop approaches to integrating technology within the planning system.

There are also regional and continental variations in the adaptation to intelligent planning systems. For instance, although GIS appears to be well entrenched as a supporting technology for urban and regional planning, and BIM is generally well entrenched in the Global North for the design and management of larger architectural, engineering, and construction projects, other contemporary technologies have only been partially implemented in the Global South and many remain in the experimental stage.

4.3.2   Digitised Land Administration

Land registries involve vast amounts of supporting data and are typically organised as centralised databases. Several digital tools are therefore increasingly integrated into the processes in a bid to address some of the inefficiencies and complexities. Digitisation of land registries is on the rise, and this is improving the transparency and efficiency of land administration. Blockchain technology is one of the most commonly adopted digital tools in land management in the last couple of decades. It is increasingly gaining prominence and is seen to be contributing to the formalisation of property rights in the Global South and enhancing the coordination of real estate markets in the Global North (Rodima-Taylor, 2021). It has garnered significant attention recently for building more reliable and transparent government record-keeping systems (De Filippi and Wright, 2018). Because records on a blockchain are distributed and verified by multiple nodes in a peer-to-peer digital network, new additions to the blockchain are cryptographically time-stamped, making tampering or accidental data loss unlikely in comparison to centralised digital registration systems. Blockchain-supported land registry records are therefore perceived to be more transparent and resilient.

Blockchain technology has several other potential land administration benefits: it improves public access to information and real-time verification of land ownership which can reduce manipulation of land records (Shang and Price, 2019). Furthermore, infrastructure for real estate transactions is slow, costly, and generally inefficient due to the number of intermediaries such as real estate brokers, solicitors, appraisers, and title companies. Blockchain technology has the potential to minimise these inefficiencies and to ensure that processing fees and deposits can be automated around smart contracts. It can also lead to the unbundling of property rights where owners or investors sell parts or shares in their property for micro-payments. This can enable the integration of local property records on national and international levels and, further combined with geospatial mapping and digital intelligent systems, could facilitate large-scale formalisation reforms.

Blockchain registries also have challenges. One of them is the legal and legislative framework that must be established before the system can be adopted. Another is the relative immutability of this technology; this can particularly hamper the operations of customary tenure norms in the Global South involving undocumented rights for vulnerable categories such as women and youths. Blockchain technology may be less effective in areas with very few existing land records because it works best in areas with strong existing records. This is a major problem because only about 30% of countries globally have some form of digital land registration system (Shang and Price, 2019). Additionally, the high cost of this technology has slowed down its adoption.

4.3.3   Smart Cities

The world is getting more urbanised, and more people are living in cities. Two hundred years ago, just 3% of the world's population lived in cities (Kalan, 2014). According to Brown (2018), 30% of the world's population lived in cities by the 1950s; that proportion has now increased to 50%, and it is projected to rise to 72% by 2050. Similarly, the number of urban residents is growing by approximately 60% annually (Kalan, 2014). One of the outcomes of urbanisation is an increase in megacities.[2] Back in 1950, New York was the only megacity in the world but the number of megacities grew to 24 in 2014 (Kalan, 2014). It is estimated that there will be 41 megacities with an estimated population of 730 million people by 2030 (Marlier et al., 2016). Urbanisation and the increase in the number of megacities are creating several challenges, particularly with infrastructural development not keeping the same pace. This creates challenges in planning and managing cities, particularly with the high level of urban dynamics being experienced in cities. The concept of smart cities has therefore arisen in a bid to improve city planning and management in the context of highly dynamic cities and to deal with the growth of megacities.

A smart city is an urban environment that utilises digital and other related technologies to enhance the performance and efficiency of regular city operations and the quality of services provided to its residents (Silva et al., 2018). Cities are becoming "smart" in terms of the ways in which routine functions, buildings, traffic, transportation, energy, and other infrastructure systems are planned, analysed, monitored, and controlled using automated systems to improve efficiency, equity, and quality of life for residents and visitors. New platforms, smart devices, and digital data are now governing relationships, capital flows, and behaviour in city planning (Fields, 2022; Porter et al., 2019). These entail the integration of big data, advanced analytics, visualisation, AI, IoT, etc., and it requires the use of various digital methods, e.g., sensors, inductive loop, and cameras, to collect it. Data can include information on individuals, devices, buildings, traffic, transportation, power plants, water supply, schools, crime, hospitals, community services, etc. These data can then be analysed and used by planners, city managers, and the government to plan, manage and govern the cities, and for the allocation of resources and amenities.

The activities in smart cities are usually coordinated in intelligence centres or control facilities that enable a holistic view and monitoring of city services and infrastructure. The Centro De Operacoes Prefeitura Do Rio in Rio de Janeiro, Brazil is an example of a facility that integrates administrative, statistical, and real-time data from 32 agencies and 12 private utility companies in order to manage day-to-day operations and plan the city (Luque-Ayala and Marvin, 2016). This concept can also be taken to a more micro-level. For instance, a firm known as Maalka has been working on district-level sustainability, developing expertise in creating eco-districts and sustainable districts, including the Pearl District in Portland and Kashiwa-no-ha in Chiba Prefecture of Japan.

CIM further extends the potential of the smart city model by creating 3D city models underpinned by the associated data and enhanced analytics which support the examination of spatial relationships and the simulation of urban processes under different conditions to facilitate informed decision-making concerning city management and planning (Thompson et al., 2016). The concept of digital twins which incorporates the digital representation of assets, processes, or systems in the built or national environment is also being promoted in several countries, particularly in the UK as a contemporary approach to the management of urban systems and infrastructure throughout their lifecycle (Bolton et al., 2018).

Huge financial resources are allocated to data generation in planning, development evaluation, and land management, and traditional systems, unfortunately, fail to put these data into optimal use; thus, the data often remain redundant due to poor systems of integration and communication. The cybernetic revolution with the growth of big data, IoT, and advanced computation including artificial intelligence has led to the improvement of infrastructure and facilities to increase operational efficiencies and inform infrastructure planning at a strategic level (Coletta and Kitchin, 2017). Open data initiatives have also made urban data more widely available, with urban dashboards that provide public tools for making sense of the data and several other applied urban data informatics and mobile apps. It should be noted that there are significant internal and inter-institutional politics and negotiation involved in creating and operating urban dashboards which inevitably shape the systems being created and what they convey (Kitchin, 2016).

The smart city model enables urban designers to dramatically update and recalculate the scheduling and quantities of materials that will accompany changes to design and building specifications (Crotty, 2012). Cities and now exploring ways of integrating their data such that they can be used multiple times to enhance economic, social, and spatial efficiency: financial resources can be more efficiently utilised and allocated, costs can be minimised, time can be saved, wastage can be reduced, and urban citizens can be better satisfied through data transparency and visualisation through the use of engagement platforms and apps. Carbon emission reduction potentials can also be identified, and environmentally efficient practices can be promoted and enhanced in urban areas. Authorities can also unify their data and systems to ensure that all stakeholders are working with the same information and assumptions, making it easier to build tools to access, interpret, and analyse the data. This effectively allows local authorities to maximise the value of the data that is generated as part of the planning and development control process, thereby reducing time spent on local planning, design, and control, as well as transparency and clarity for citizens and other stakeholders.

Barkham et al. (2018) suggest that digitising city planning and management has the potential to improve the quality of life and the productivity of cities in the long run. They further suggest that smart city technologies will reinforce the primacy of the most successful global metropolis at least for a decade or more. A few select metropolises in emerging countries may also leverage these technologies to advance the provision of local public services. In the long run, cities will adopt successful and cost-effective smart city interventions; nevertheless, smaller-scale interventions are likely to spring up in different areas in the short run. These systems generally have the potential to improve conditions of blighted or deprived neighbourhoods, which can generate gentrification and higher valuations. With increasing urbanisation being one of the global megatrends affecting our world, creating smart cities will help to make larger cities safer, sustainable, and more efficient (Brown, 2018).

4.4   Case Study Analysis

Following the review of core areas of digital transformation in planning, land administration, and urban management, this section analyses specific PropTech platforms and companies within this area of the built environment value chain. Each case study analysis begins with an overview and summary of key details of the platform/company; the solutions it provides are then analysed using the ESEP efficiency framework introduced in Chapter 1; the analysis concludes with a brief overview of the limitations and further prospects.

4.4.1   Case Study 1: Esri

4.4.1.1   Background, Key Details, and Technologies Utilised

Esri is the parent platform for a host of other platforms that take their use from ArcGIS.[3] It utilises GIS software, location intelligence, and mapping to improve the site selection process of buildings, facilities, and amenities for individuals and organisations. The company was founded by Jack and Laura Dangermond in 1969, and it has now grown to more than 5,000 employees located in 49 offices worldwide. It has 11 Research and Development Centres and remarkably invests 30% of its annual revenue into research and development. Esri has over 300,000 corporate users, including 90% of Fortune 100 companies, 30,000 cities, and local council governments, all 50 US states, and 12,000 universities. Its Headquarters is in Redlands, California, USA.

Esri's vision is that "technology, science and a geographic approach can be used by business and government leaders to make better decisions – building sustainable prosperity while safeguarding the planet…". At the heart of the platform is the drive to contribute to solving the housing shortage challenge in the UK through better access to land, and the tools on the platform are designed to address the issues of environmental sustainability of proposed buildings, ensuring that they are a good fit for the surrounding area. The company acknowledges that some of the greatest challenges that humankind faces – including climate change, sustainability, and social and economic inequality are interrelated and inherently tied to issues of geography and that a science-based, geographic approach can help to understand these interconnected problems holistically by integrating various forms of data. For instance, through the analysis of wind flow and speed, planners can determine the appropriate positioning of wind farms that can achieve an optimum level of renewable energy generation. Thus, they collect, analyse, and share a wide range of data to enable organisations to make smarter decisions about managing the world through collaborative digital geospatial systems.

The Esri technology cuts across various aspects of real estate. One of the most important aspects is the choice of location. Esri provides templates to create a robust data bank for developers of residential and commercial real estate in various areas. The platform enables a more informed choice of locations that are less prone to natural disasters. Furthermore, developers and agents can identify the best locations that suit their clients' needs and gain insight into the terrain type and land use choices. Agents also use the platforms to show potential clients what the surrounding areas look like in 3D, while also showing all the adjoining blocks, streets, parks, and other infrastructure. This enables the reduction in carbon emissions while saving cost, resources, and time. Agricultural property has also benefitted from Esri by way of FarmView which enables real estate brokers to find the best areas suited for certain agricultural activities. The style of visualisation and solution provision for each of these platforms set them apart from the traditional and less-efficient ways of getting these tasks done. Tools that function based on the Esri technology include ModelBuilder, GeoEnable, Skyward, FarmView, PRISM, Survey123, and Weather Ready Map.

4.4.1.2   Efficiency Analysis

Site search and selection for real estate development are time, cost, and labour-intensive, with very little certainty on the outcome. The solutions that Esri provides are itemised using the ESEP efficiency framework (Table 4.1).

Table 4.1 ESEP Efficiency Analysis of Esri

ESEP	Inefficiency with the Traditional Approach	How the Tool/Platform/Product Addresses the Inefficiency
Environmental	Property search and selection typically involves a high level of transportation and commute which increases CO_2 emissions.	A developer or prospective land buyer can view several sites and associated details without conducting site visits or inspections. Further information on the site and its surroundings can also be obtained using the platform; hence without the need to commute; CO_2 emissions are therefore significantly reduced in the site search process using this platform.
	Collaboration with stakeholders and other built environment professionals requires a high level of travel and commute.	Collaboration is done with reduced commuting; meetings can be held using remote meeting tools because various aspects of the site can be accessed remotely.

(Continued)

Table 4.1 (Continued)

ESEP	Inefficiency with the Traditional Approach	How the Tool/Platform/Product Addresses the Inefficiency
Social	Acute housing demand-supply mismatch due to information asymmetry. Although land is usually available for development, matching the type of demand with the available land is usually a challenge.	Information asymmetry in the site search process is significantly minimised, and the demand is better matched to supply.
	Site search and selection takes a lot of effort and labour, and in many cases, the site search and visits do not translate to getting suitable land.	The site and selection process is much easier and less labour-intensive. With better access to information, developers can more easily identify the site with the most feasible and viable potential for their development plans.
		Many more sites can be explored without having to visit any at the onset.
		There is quicker communication between the buyer and seller through the app reduces anxiety and stress associated with communication with the key parties in the transaction and getting relevant details about the site.
		It is easier to locate potential sites and follow the approval process in real time without the usual bottlenecks.
		The buyer can search using the app's filters to get the most suitable short list of properties.
		The use of 3D visualisation and "real-time walk-through" can give a first-hand feel of the space and surrounding areas.
		There are other details on soil quality, demographic distribution, climate, crime, violence, proximity to healthcare facilities, infrastructure, etc., on the platform and this will aid decision-making and reduce anxiety.
		ModelBuilder provides information on flood risk, proximity to historic monuments, existing local services, etc., which makes the site selection process more efficient.
		Farmview helps potential farm owners to study areas and know which areas would be best for their type of crop. This helps farmers to make more intelligent decisions to get farmland that would make their farming venture more profitable.

(Continued)

Table 4.1 (Continued)

ESEP	Inefficiency with the Traditional Approach	How the Tool/Platform/Product Addresses the Inefficiency
	Risk of site injuries and accidents.	There is a lower risk of site injuries and commute-related accidents due to reduced travel and site inspection.
	Collaboration with stakeholders and other built environment professionals can be cumbersome.	ArcGIS Online enables professionals to share asset maps with other professionals and users can use maps of development sites even down to roads, streets, buildings, and service lines while getting up-to-date project information. The PRISM app helps clients visualise high-resolution 3D images of areas, showing them a virtual 3D world of buildings, streets, and characteristics of that area. Clients can also virtually move around a space they like—using VR and AR. All of these improve collaboration.
	Limited choices of sites.	A wider variety of sites are available to the buyer to explore.
Economic	Site search and selection take a long time, and the land sales/purchase process typically takes several hours of inspection and communication.	Site owners can more easily find buyers with less effort as it is only sellers that are interested in the property (having fully explored the related details) that will contact them. The search, selection, and sales/purchase time is therefore significantly minimised. Transaction time and decision-making are also much quicker due to the availability of information and improved transparency of the sites. Project timelines can generally be shortened.
	High cost of site search and selection.	The platform minimises wastage and inefficiency in work done which can drive down project costs. Unexpected site challenges and inaccurate information is minimised, and by extension, unexpected costs during the project execution will also be minimised. Higher risk in real estate development translates to higher interest rates; improved transparency can reduce the risk in a project and thus minimise the associated costs such as lending and insurance costs. The cost of searching and selecting sites is significantly reduced as less time, labour, and commute are required.

(Continued)

Table 4.1 (Continued)

ESEP	Inefficiency with the Traditional Approach	How the Tool/Platform/Product Addresses the Inefficiency
	Inadequate information on a site and its surroundings leads to poor site selection or the land being underutilised; the potential value of a proposed project can also be underestimated.	Access to sufficient information can significantly improve the value maximisation potential of a site through the highest and best-use analysis. The developer or potential buyer will have access to vital information about the site and surroundings and this can influence the site plans and project feasibility and viability.
Physical/spatial	Inadequate information on a site can lead to space underutilisation.	The optimum use of a site can be identified and maximised as developers are looking to unlock its maximum value. The highest and best use can be identified through the information available to achieve spatial efficiency. The local economy of the area where the site is located can be enhanced through the maximisation of the site's potential. More accurate development plans and designs will create more functional spaces and improve the values of the spaces.
	Issues with spatial compatibility, and efficient utilisation of space (physical and external).	Better efficiency and use of space can be enhanced because of the spatial accuracy that the platform provides for both internal and external spaces. More accurate plans and designs will improve the functionality and the values of spaces.
	Inaccuracy of space can lead to poor planning and space value.	The high level of accuracy will minimise poor plans and use of spaces.

Source: Authors analysis.

4.4.1.3   *Further Prospects and Limitations*

Esri-powered platforms require high volumes of government-guarded data. Municipal data and council statistics are required to train artificial intelligence that makes the success of these apps. Without the necessary information and data shared by the custodian government bodies, these apps would struggle. A more generous and symbiotic relationship between the government bodies and the app administrators would help make the apps perform better with more accurate data that can help make the best decisions.

Because these apps are GIS-dependent, they need high-speed internet to function optimally. 5G or higher will be necessary to get these apps running smoothly as more users begin to take advantage of the apps, and this may limit the adoption in areas with lower internet coverage. Further development and adoption of AR and VR will also contribute to the development of Esri as these technologies would be the vehicles that will introduce real estate into the Metaverse which is increasingly gaining prominence. The world is getting more and more virtual, and people are beginning to dwell within the Metaverse for longer periods of time. Esri has the potential to adopt this technology to enable users to gain information and guidance on sites.

Further details on Pupil can be accessed from their website using this link: www.esri. com/en-us/home.

4.4.2   Case Study 2: LandTech

4.4.2.1   Background, Key Details, and Technologies Utilised

LandTech was initiated to address the complications in the search and selection of sites, particularly for self-builders. It was designed to connect potential buyers of off-market land to land owners by enabling both parties to more easily go through the process of securing the data and information concerning the parcel of land. This includes getting access to the approval template and exchange process and accessing funding for the development. The platform was inspired in 2014 by two individuals based on the need to support self-builders to access land more easily while minimising the long commute to different sites to source available sites with very little assurance of success. This inefficient land search process led them to develop solutions that could help real estate developers gain access to land more efficiently. The solutions were encapsulated in a digital platform aimed at getting rid of extensive travel and contact for land search. By 2015, the founders had their first clients who pre-ordered which enabled them to raise their first capital of £350,000.

LandTech[4] launched in 2016 by delivering three core services/products: *planning data* to help people find better sites, *land ownership data* to help potential buyers contact the right property owner and *market sales data* for the valuation of investment opportunities on the site. By 2017, they introduced *Sites Pipeline* to help people track potential deals, *planning applications alerts* to keep a live track of areas of the property market and *price-per-square-ft data* after a successful lobbying campaign to the government to get floor areas unblocked from EPC certificate – this helped people to target the right areas for their budget in site selection when multiple sites were being considered. They further combined land ownership data with Companies House data to show the family tree structure of landholding and ownership.

In 2018, *Policy Maps* were added helping people to see restrictions and potential red flags on sites. Further to this, the LandInsight *GO mobile app* was launched to help people save sites on their profiles even when on the move. In 2019, *LandEnhance* was launched, and the company was rebranded from LandInsight to LandTech, raising £2.5 million from JLL Spark. In 2020 *Ownership Alerts* were added to the platform to help people know when sites were added and sold/purchased. This served as a monitor for potential real estate opportunities. LandTech also added the *Shared Pipelines* function, making it easier for people to collaborate remotely. In 2021 *Substation Data* was added – helping people to see local power capacity and substation locations. *Drawing on maps* was also introduced to help people fully customise their maps and visualise their projects. A *Land Assembly* tool was also introduced to aid the combination of multiple sites into one project. *LandFund* was also launched to help platform subscribers complete development appraisals in the shortest possible time.

LandTech currently supports ten of the UK's largest housebuilders and thousands of SME property developers. As of 2021, they had supported the delivery of approximately 50,000 new homes in the UK. They estimated that their platform subscribers saved an estimated 430,000 hours that would have been spent searching for sites in 2021 – this includes 5 million planning applications accessed through 2.5 million comparable searches on the platform. They recently took a £42 million Series A round led by US-based investment firm, Updata Partners ("Updata"), followed by Flashpoint Secondary Fund with additional contributions from existing investors – JLL Spark and Pi Labs.

LandTech technology applies to the part of real estate concerned with the acquisition of land and the approval process timeline. It makes locating potential sites easier and makes it possible to follow the approval process timeline in real time; this enhances the prospects of securing a piece of land without the usual bottlenecks of acquisition and approval. The platform primarily targets developers looking to unlock the optimum value of land for various uses such as residential, commercial, and infrastructure. Their tools are also beneficial to other built environment professionals such as agents, lenders, and planners.

4.4.2.2   Efficiency Analysis

Site search and selection for real estate development are time, cost, and labour-intensive, with very little certainty on the outcome. The solutions that LandTech provides are itemised using the ESEP efficiency framework (Table 4.2).

Table 4.2 ESEP Efficiency Analysis of LandTech

ESEP	Inefficiency with the Traditional Approach	How the Tool/Platform/Product Addresses the Inefficiency
Environmental	Property search and selection typically involves a high level of transportation and commute which increases CO_2 emissions.	Using this platform, the site selection process will require less travel, as comparison and analysis can be done using information from the platform. The platform provides various levels of information that a developer requires without having to embark on journeys to check them out. CO_2 emissions are therefore significantly reduced in the site search process using this platform.
	Collaboration with stakeholders and other built environment professionals requires a high level of travel and commute.	Collaboration is done with reduced commuting; meetings can be held using remote meeting tools because various aspects of the site can be accessed remotely.
Social	Acute housing demand-supply mismatch due to information asymmetry. Although land is usually available for development, matching the type of demand with the available land is usually a challenge.	Information asymmetry in the site search process is significantly minimised and the demand is better matched to supply. This can support the drive to address the challenge of housing shortage in several countries.

(Continued)

Table 4.2 (Continurd)

ESEP	Inefficiency with the Traditional Approach	How the Tool/Platform/Product Addresses the Inefficiency
	Site search and selection takes a lot of effort and labour, and in many cases, the site search and visits do not translate to getting suitable land.	The site and selection process is much easier and less labour-intensive. With better access to information, developers can more easily identify the site with the most feasible and viable potential for their development plans. Many more sites can be explored without having to visit any at the onset. There is quicker communication between the buyer and seller through the app; reduces the anxiety and stress associated with communication with the key parties in the transaction and getting relevant details about the site. It is easier to locate potential sites and follow the approval process in real time without the usual bottlenecks. The buyer can search using the app's filters to get the most suitable short list of properties.
	The sale of land will typically require several hours of inspection and communication.	Site owners can more easily find buyers with less effort as it is only sellers that are interested in the property (having fully explored the related details) that will contact them.
	Limited choices of sites.	A wider variety of sites are available to the buyer to explore.
Economic	Site search and selection take a long time.	The platform significantly minimises the time that will be taken to search and select a site. Transaction time and decision-making are also quicker due to the availability of information and improved transparency of the sites. The team of booking consultants has micro-city knowledge so that they can advise and support potential buyers on the best part of the city to match their needs. Project timelines can generally be shortened.

(Continued)

Table 4.2 (Continurd)

ESEP	Inefficiency with the Traditional Approach	How the Tool/Platform/Product Addresses the Inefficiency
	High cost of site search and selection.	The platform minimises wastage and inefficiency in work done which can drive down the cost of the project. Unexpected site challenges and inaccurate information is minimised, and by extension, unexpected costs during the project execution will also be minimised. Higher risk in real estate development translates to higher interest rates; improved transparency can reduce the risk in a project and thus minimise the associated costs such as lending and insurance costs. The cost of searching and selecting sites is significantly reduced as less time, labour, and commute are required.
Physical/spatial	Inadequate information on a site can lead to space underutilisation.	The optimum use of a site can be identified and maximised as developers are looking to unlock its maximum value. The highest and best use can be identified through the information available to achieve spatial efficiency. The local economy of the area where the site is located can be enhanced through the maximisation of the site's potential. More accurate development plans and designs will improve the functionality and values of the spaces.

Source: Authors analysis.

4.4.2.3   *Further Prospects and Limitations*

LandTech aims to increase the value of the platform by embedding a more robust feedback system that will improve its efficiency and effectiveness. They further aim to support potential buyers through the site search to the stage where the site is ready for construction to begin. Expanding the service of the platform beyond England and Wales to Scotland has been explored; however, due to the bureaucratic bottlenecks with access to government-related data, this has been difficult to achieve.

Further details on LandTech can be accessed from their website using this link: www. land.tech.

4.4.3   *Case Study 3: Pupil*

4.4.3.1   *Background and Key Details*

Pupil is a spatial data company that has created an ecosystem to digitally map the built environment (internally and externally) to drive efficiency and maximise value. The platform is responsible for digitising buildings, and as stated on the website, "… what Google did for the outside world, Pupil is doing for interior space with precision, accuracy and reliability". The platform currently operates in Dubai, London, New York, and Singapore. Pupil was the subject of four years of extensive research and development. It has three core products: Spec, Stak, and Strat.

Spec (launched in 2019) provides digitised versions of residential real estate spaces. The tool provides a variety of digital services ranging from verified floor plans to photos, interactive 3D dollhouse experiences, and virtual tours. It also generates a digital brochure for residential properties which can then be shared using a web link.

Stak (launched in 2021) provides verified digital twins for commercial real estate. This is achieved through the combination of ground-breaking proprietary software with the latest SLAM technology to generate verified twins through the capturing and digitising of real estate with speed and accuracy.

Strat (launched in 2022) provides digitised registration services which include third-party survey services, area verification, leasing and title plans, professional area measurement to international standards, master community registration, onsite survey services, area demarcation, off-plan registration, and final property registration.

These products enhance various aspects of a building's lifecycle: acquisition, planning, development, construction, management, sales, and disposal. Through the power of AI, Pupil overlays its operating system to make spaces more connected, valuable, and sustainable. It achieves a high level of accuracy by combining the LiDAR scanning hardware (that scans up to 420,000 points of measurement every second) with its own pioneering AI and proprietary software. This enables scans that can be tracked on a smartphone in real time. The raw data is instantly uploaded to the cloud on completion of the capture from the field, where Pupil's AI engine turns these 3D models into millimetre-accurate floor plans in minutes. Pupil produces Spec Verified measurements with a 99% accuracy of a property's true size. This service is insured for accuracy up to £10 million, protecting the parties involved in transactions that are based on Spec Verified measurements. Pupil provides services to approximately 10% of London's real estate agents, and with its launch in the United Arab Emirates in 2021, and New York and Singapore in 2022, it is making a substantial impact in the property market. As of 2021, it had created 35,000 digital twins of residential properties (with a value of over $35 billion) and approximately 1 million virtual viewings of properties in London. This was particularly useful during the Covid-19-related restrictions on movement, saving approximately 11.6 million miles from being driven to physical property viewings.

4.4.3.2   Efficiency Analysis

Site and building planning and inspection require several site visits and related activities. These require time and money and increase CO_2 emissions and stress. The solutions that Pupil provides are itemised using the ESEP efficiency framework (Table 4.3).

4.4.3.3   Further Prospects and Limitations

Pupil's team of software developers and engineers is creating an AI that can programmatically audit a property and instantly identify all the objects within a home. It has already

Table 4.3 ESEP Efficiency Analysis of Pupil

ESEP	Inefficiency with the Traditional Approach	How the Tool/Platform/Product Addresses the Inefficiency
Environ-mental	Property viewings and inspection typically involve a high level of transportation and commute which increases CO_2 emissions.	Significant reduction in the number of property viewings and site inspection-related commutes which can lower carbon emissions. It is estimated that a driving distance of approximately 11.6 million miles to viewings and inspection was avoided due to as a result of using the app. Data can be captured from the first site visit and stored; this means that if additional assets are to be added or if modifications are required, there may be no need for another site visit. This will further reduce travel-related emissions. Site inspection, particularly in high-rise buildings may lead to high energy usage for power lifts, lighting, heating, etc. This process reduces these potential emissions.
	Collaboration with stakeholders and other built environment professionals requires a high level of travel and commute.	Collaboration is done with reduced commuting; meetings can be held using remote meeting tools because various aspects of the site and properties can be accessed remotely.
Social	Lack of trust and lack of transparency in the development and marketing process.	The use of graphical illustrations and pictures of the site and buildings increases trust and transparency. Being able to provide visualisation and other dimensional graphics further aids transparency. Having a sharable digital asset or site pack that contains photographs, accurate floor plans, site plans, and a virtual tour will improve transparency and access to information regarding a site. More information is made available and various visual perspectives are provided; this will increase transparency as stakeholders can have a better understanding of the city and the proposed changes.
	Inaccuracy and errors in measurement and record taking.	Because of the high level of accuracy in measurement when using these tools, errors, and inaccuracy are minimised and trust and transparency are enhanced.

(Continued)

Table 4.3 (Continued)

ESEP	Inefficiency with the Traditional Approach	How the Tool/Platform/Product Addresses the Inefficiency
	Face-to-face viewings and inspections take effort and labour, and in many cases, most viewings and inspections do not translate to deals being secured.	Being able to scan, track and assess a site or property using a smartphone eases the prospective client and the property owner in terms of physically inspecting the property and commuting. The data is stored in the cloud which makes remote access possible. Real estate agents using the platform can market their properties and enable them to reach many more potential clients. The virtual tour and pictures will be valuable for visual learners who relate better to images and graphics, and this is further balanced with text that read/write learners may prefer. This can further increase the level of information and trust.
Economic	Site visits, inspections, measurements, etc., take a lot of time to complete, and this time is duplicated with every new prospective client that wishes to view or visit the property.	Significantly shorter time will be spent on inspection, site visits and planning. The reduction in commute can also reduce commute-related costs.
	Cost that may arise from inaccurate measurement or information.	Due to the high level of accuracy and the digital footprint of each property, the potential costs that may arise from inaccuracy are minimised and this reduces overall project and building costs. The reduced risk can also lead to a decrease in insurance and lending costs. This is more so that the service it provides is insured for accuracy up to £10 million, thus protecting the parties involved in transactions where the parties involved rely on Spec Verified measurements.
Physical/spatial	Issues with spatial compatibility and efficient utilisation of space.	Better efficiency and use of space can be enhanced using this digital tool because of the spatial accuracy that it provides for both internal and external spaces. More accurate plans and designs will improve the functionality and value of the spaces.
	Inaccuracy in spatial information can lead to poor planning and space value.	The high level of accuracy will minimise poor plans and use of spaces.

Source: Authors analysis.

received awards from Innovate UK for its pioneering work in this field and further results have the potential to change the way that people around the world interact with buildings. The team of Digital Surveyors, who capture the properties that use their software now travel to each site capture using electric mopeds. This is a further step to enhance the environmental inefficiency in the process of carrying out their task. A further drive towards environmental sustainability is evidenced through their offset of 860 tonnes of CO_2 through the planting of 12,539 trees in partnership with Ecologi, their sustainability partner.

There are still other areas that could be developed further. The inclusivity and diversity of their user base can be further improved through better auditory facilities to meet the needs of auditory learners. Furthermore, although their practices generally have the tendency to improve the environmental efficiency of planning and urban design, the use of powerful computers and other associated technologies to access this platform is also leading to high carbon emissions from electricity generation for users, the platform managers, and other cloud-based storage systems.

Further details on Pupil can be accessed from their website using this link: www.pupil.co.

4.4.4   Cast Study 4: VU.CITY

4.4.4.1   Background and Key Details

VU.CITY is a smart city 3D interactive digital platform that brings together various built environment professionals and natural environmental disciplines under the same collaborative space to enable co-creation, co-design, and review of spaces, and urban development plans and projects. It enables all stakeholders on the project to make more informed decisions regarding spaces and improves collaboration and trust. The platform combines digital twins of cities with data to support its users in creating, refining, and sharing their journey of a city's transformation. It was formed in 2015 by GIA Surveyors – specialists in "rights of light," and Wagstaffs (now part of VU.CITY) – a PropTech agency that used Virtual Reality, Augmented Reality, and 3D technology for real-world application. The team has also grown from seven staff in 2017 to over 55 employees in 2022. The product started with modelling London, and by 2019, London (approximately 1,617 sq. km) was complete. By 2021 VU.CITY had over 3,000 sq. km of coverage, accurate to 15 cm, covering 20 UK and international cities. The platform provides solutions across residential and commercial buildings, ranging from regenerations to master plans. It is also comprehensively embedded across a diverse range of disciplines, including property development, spatial planning, and architecture within both the public and private sectors.

The "Smart Cities" function of the platform supports built environment professionals by providing them with future growth plans which enables them to understand the impact of the plans (or proposed plans) on the city. Used across the United Kingdom and Ireland's public and private sectors, this platform provides a shared visual perspective to understand the character of a city, facilitating a city-wide constructive narrative and inclusion with change. Through the use of 3D technology, design and planning functionality, and game engine technology, VU.CITY helps city changers make better design and planning decisions, faster. Ultimately, it creates a holistic space to test and discover how to maximise the potential benefits of a site and minimise its adverse effects.

VU.CITY covers over 16,000 trees accurate to height and canopy width and over 150 data layers from conservation areas, listed building points, flood risk zones, and much more. It also has over 3,000 major consented schemes visualised through the "cities" function which helps the users to understand how their designs/schemes work in the future context of the city. Nearly 90% of the London Boroughs use VU.CITY in their planning process: pre-application conversations, planning committees, and supporting local plan considerations. Other statutory and non-statutory consultation exercises include pre-application community consultations, pre-application discussions, and interdisciplinary team meetings on project progress. 30 of 32 local authorities in London have used the platform as part of their development management and forward planning approaches. A further 16 city and regional councils from across the UK and Ireland use the platform including Belfast City, Bristol City, Dublin, and Salford City Councils. The platform has over 200 licensees with their users primarily made up of architects, developers, planners, agents, and planning officers. VU.CITY is also being used by those in the architecture profession: 48% of AJ 100 (The Architect's Journal 100) use the platform for various functions including contextual design, place appraisal, options assessment, and design communication.

The company revealed that its revenue has grown 70% year on year since inception with £1 million of Innovate UK grant-funded projects. They offer licences to support academia and offer free individual student licences to students who want to explore the platform for their degrees. It has been used in various universities and institutions in student modules, lectures, and training programmes. This includes the Planning, Architecture, and Building Information Modelling course at Queen's University Belfast, and Architecture at both the University of Ulster and University College London (The Bartlett). They also have research collaborations with universities such as Queen's University Belfast.

4.4.4.2   Efficiency Analysis

As the population of cities continues to grow, natural resources (such as land) are increasingly being put under pressure and this is further mounting pressure on living standards. Planning systems also lack transparency and this often leads to a lack of trust and cohesion of built environment professionals throughout the process. The solutions that VU.CITY provides through its platform are summarised (in alignment with the ESEP efficiency framework) in Table 4.4.

4.4.4.3   Further Prospects and Limitations

VU.CITY aims to open up the platforms to international markets outside the UK and Ireland to support their planning processes. It also aims to expand the platform to other complementary industries e.g., asset management, facilities management, insurance, telecommunications, and security. Initially, users of the platform required a high-powered computer with a dedicated graphics card to run the software and this significantly limited its adoption. To mitigate this, VU.CITY introduced a cloud version that allows users to access the full platform from a web browser. Based on the feedback, the platform aims to deliver a lighter-weight subscription tier for users that do not require 3D visualisations.

Table 4.4 ESEP Efficiency Analysis of VU.CITY

ESEP	Inefficiency with the Traditional Approach	How the Tool/Platform/Product Addresses the Inefficiency
Environmental	A lot of paperwork is involved in planning.	Reduction in paperwork due to the advanced digital visualisation facilities of the platform. Using the traditional methods of paper and pen/ink, new drawings/printings will be required when corrections are made. VU.CITY allows corrections to be made and using the same visualisation tools, these may not need to be printed.
	Collaboration with stakeholders and other built environment professionals requires a high level of travel and commute.	Collaboration is done with reduced commuting; meetings can be held using remote meeting tools. The decentralised and remote access to the platform allows for various stakeholders and professionals to work on the platform from different locations.
	Planning plans and designs will typically be unable to achieve an optimum level of lower carbon emission and wastage reduction due to the inability to simulate various scenarios.	The tool allows planners, developers, built environment professionals, and other stakeholders to simulate and analyse various scenarios thereby improving the prospects of exploring the optimal potential for carbon reduction and waste minimisation.
Social	Uncertainty and lack of trust and transparency in the planning and development process.	The platform reduces the uncertainty in planning by providing rich layers of data and visualisation, thereby increasing the level of trust in the process. More information is made available and various visual perspectives are provided to increase transparency; stakeholders can have a better understanding of the city and the proposed changes. Crucial planning and design decisions can be made early in the planning of the project and the impact can be simulated and mitigated. Understanding the impact early on can lead to better decision-making. Better coverage and accuracy can improve micro and macro-level impact assessment. Proprietary data from the platform is useful as the platform curates all major schemes so that they can be visualised in 3D which traditional geometry did not support. Visualising the future context of a city provides insight for advocates of change and developers so that they are able to consider density, height and general design appropriateness.

(Continued)

Table 4.4 (Continued)

ESEP	Inefficiency with the Traditional Approach	How the Tool/Platform/Product Addresses the Inefficiency
	Inaccuracy and errors.	Due to the improvement in the planning and design process, inaccuracy and errors are minimised, and easily detected and corrected. The scale of the plans and designs can be enhanced, and this can be changed based on need and circumstances; thus, errors can be easily spotted.
	Lack of inclusivity in planning and development.	The platform enhances inclusivity through the unbiased and democratised framework for public and private sectors. The platform is accessible to people of various cultural and socio-economic backgrounds. The 3D element is valuable to visual learners who relate better to visuals; read/write learners can equally benefit from the textual information provided. This can further increase the level of information and trust in the planning process.
	Poor citizens' engagement leading to plans and designs that sometimes do not align with the citizens' views and perspectives.	This tool improves citizens' engagement through the opportunity for citizens to engage and explore proposed plans and development using various visualisation tools. Community engagement will become more robust as citizens have more information and can make more informed contributions and suggestions to enhance their communities and cities in general. Stakeholders can assess the real-world impacts of schemes at the earliest stage of concept design. Being able to show proposals in this way helps to de-risk the project for both developers and councils. Minimising the chance of poor designs and unpopular developments to the local community.
Economic	Slow planning and decision-making.	The time taken for planning and design of places and cities can be significantly reduced. Built environment professionals can make faster decisions on plans and proposed development based on the data provided, the visualisation aids and simulation facilities. By minimising protests and appeals, the time that would usually have been spent on appeals is reduced, further expediating project execution and completion.

(Continued)

Table 4.4 (Continued)

ESEP	Inefficiency with the Traditional Approach	How the Tool/Platform/Product Addresses the Inefficiency
	High cost of planning and development.	Having a viable plan with a higher level of certainty and transparency has the potential to reduce the cost of development. With more data and better visualisation, the risk perception of the project is minimised and this can increase the financial position of the project. This can lower interest rates and the cost of finance. Better information can improve forecasting and minimise the waste of financial resources.
Physical/spatial	Issues with spatial compatibility, planning, and design accuracy.	More efficient planning systems with better precision, scales and accuracy will lead to better precision in development and better quality of spaces. More accurate plans and designs will improve the functionality and values of spaces.

Source: Authors analysis.

There are still other areas that could be developed further. For instance, the inclusivity and diversity of their user base can be improved through better auditory facilities to meet the needs of auditory learners. Furthermore, although their practices generally have the tendency to improve the environmental efficiency of planning and urban design, the use of powerful computers and other associated technologies to access this platform can also potentially increase energy consumption generated from users, platform managers, and cloud-based storage systems.

Further details on VU.CITY can be accessed from their website using this link: www.vu.city.

4.4.5 Other Cases (Exercises for Students)

This section contains a short summary of case studies. Readers are encouraged to develop these case studies and analyse them using the ESEP efficiency framework (taking a similar approach to the previous section).[5]

4.4.5.1 Archistar

Archistar is a digital real estate development assessment and design platform which is aimed at making the process of development compliance smarter, easier, faster, and better coordinated and ultimately revealing the highest and best use. This platform has the added feature of generating simulations of potential designs for develop to conceptualise before committing to the purchase of a site. This makes the site selection and design process faster, cheaper, and more efficient and affective.

Further details on Archistar can be accessed from their website using this link: https://www.archistar.ai.

4.4.5.2 Built ID

Built ID enables decision-makers to deliver more socially impactful and sustainable places using award-winning digital community and occupier engagement platforms. It also enhances citizens' engagement with proposed developments.

Further details on Built ID can be accessed from their website using this link: www.built-id.com.

4.4.5.3 LandChecker

LandChecker provides their platform subscribers with an interactive map-based search experience from their desktop with access to reliable data for properties and sites.

Further details on LandChecker can be accessed from their website using this link: www.landchecker.com.au.

4.4.5.4 Land Insight

Land Insight is a platform that focuses on site appraisal as an integral element of the planning system. It aims to improve the process of site appraisal and aid developers in locating land suitable for development.

Further details on Land Insight can be accessed from their website using this link: www.landsec.com.

4.4.5.5 Neighbourlytics

Neighbourlytics is a social analytics platform that makes the human side of neighbourhoods easy to understand. It provides digital data with advanced analytics in urban life to minimise uncertainty in city-making and planning.

Further details on Neighbourlytics can be accessed from their website using this link: www.neighbourlytics.com.

4.4.5.6 Open Systems Labs

Open Systems Labs is a non-profit research and development lab that aims to redesign and automate the planning application process through the design and deployment of open-source tools and infrastructure that allows citizens, communities, governments, and businesses to operate in innovative ways to provide holistic solutions to planning issues.

Further details on Open Systems Lab can be accessed from their website using this link: www.opensystemslab.io.

4.4.5.7 SanpSendSolve

SanpSendSolve enables citizens to report issues in their communities to their local council and community leaders. The platform enables citizens to upload photos in real time for effective community feedback through real-time online tools.

Further details on SanpSendSolve can be accessed from their website using this link: www.snapsendsolve.com.

4.4.5.8   UrbanFootprint

Urban Footprint provides advanced visualisation of information which can improve data access and transparency for improved citizen engagement. It can be used to assess risk, understand markets, and make better decisions to enhance urban and community resilience.

Further details on Urban Footprint can be accessed from their website using this link: www.urbanfootprint.com.

4.4.5.9   Urbanly

Urbanly provides advanced visualisation of information which can improve data access and transparency for improved citizens' engagement. It empowers decision-makers by combining expert domain knowledge with model-based algorithms for more effective and efficient decision-making.

Further details on Urbanly can be accessed from their website using this link: www.urbanly.org.

4.5   Chapter Summary

This chapter has provided an analysis of various aspects of planning, land management, and urban design. It has also reviewed the core inefficiencies in traditional planning, land management, and plan management-related services. Furthermore, real estate innovations and digital systems in planning and land management have been analysed, and case studies have been provided to gain further insight into the impact of PropTech tools in planning and land management. Finally, other planning and urban management tools have been highlighted for readers to further explore and analyse.

Notes

1 There are different planning systems such as zoning and discretionary. Often times, these systems will adopt principles from other systems, some of which may be based on colonial link, cultural associations and several other factors.
2 Megacities in this context refer to cities with more than over 10 million residents (Li, 2003; Marlier et al., 2016).
3 ArcGIS is a technology-driven IT-based mapping and analytical software.
4 It was known as LandInsight at the time.
5 Readers are further encouraged to analyse other products, tools and platforms in this area using the ESEP efficiency framework.

References

Barkham, R., Bokhari, S., & Saiz, A. (2018). *Urban big data: City management and real estate markets, Urban Economics Lab MIT Center for Real Estate and DUSP* (p. 37). https://mitcre.mit. edu/wp-content/uploads/2018/01/URBAN-DATA-AND-REAL-ESTATE-JAN-2018-1.pdf.
Benbunan-Fich, R., & Castellanos, A. (2018). Digitalization of land records: From paper to blockchain. *Thirty Ninth International Conference on Information Systems 2018, ICIS 2018* (pp. 1–9). https:// d1wqtxts1xzle7.cloudfront.net/58965423/ICIS_Frompapertoblockchain-LandRegistry-Final20190419-73239-1vlu99f.pdf?1555717317=&response-content-disposition=inline%3B+fil ename%3DDigitalization_of_Land_Records_From_Pape.pdf&Expires=1681298449&Signature=.

Bolton, A., Enzer, M. and Schooling, J., 2018. The Gemini Principles: Guiding values for the national digital twin and information management framework. Centre for digital built Britain and digital framework task group.

Brown, R. (2018). *PropTech: A guide to how property technology is changing how we live, work and invest*. Casametro.

Coletta, C., & Kitchin, R. (2017). Algorhythmic governance: Regulating the 'heartbeat' of a city using the Internet of Things. *Big Data and Society*, *4*(2), 1–16. https://doi.org/10.1177/2053951717742418.

Corburn, J. (2009). *Toward the healthy city: People, places, and the politics of urban planning*. MIT Press.

Crotty, R. (2012). *The impact of building information Modelling: transforming construction* (1st ed.). Routledge.

De Filippi, P., & Wright, A. (2018). *Blockchain and the law: The rule of code*. Harvard University Press.

Doyle, S., Dodge, M., & Smith, A. (1998). The potential of web-based mapping and virtual reality technologies for modelling urban environments. *Computers, Environment and Urban Systems*, *22*(2), 137–155. https://doi.org/10.1016/S0198-9715(98)00014-3.

Fainstein, S. S. (2017). New directions in planning theory. *Foundations of the Planning Enterprise: Critical Essays in Planning Theory: Volume 1*, *35*(4), 139–166. https://doi.org/10.4324/9781315255101-17.

Fields, D. (2022). Automated landlord: Digital technologies and post-crisis financial accumulation. *Environment and Planning A*, *54*(1), 160–181. https://doi.org/10.1177/0308518X19846514.

Forrester, J. W. (1969). *Urban dynamics*. MIT.

Hartley, L. (2019). *Smart cities, how it works, big data: PlanTech explained*. Neighbourlytics. https://neighbourlytics.com/blog/plantech-explained.

Kalan, J. (2014). Think again: Megacities. *Foreign Policy*, 69–73, 12. https://www.proquest.com/openview/db0fdd0084fd0b0790cd206e8b5f1586/1?pq-origsite=gscholar&cbl=47510.

Kitchin, R. (2016). The ethics of smart cities and urban science. *Philosophical Transactions A*, *374*(2083). https://doi.org/10.1098/rsta.2016.0115.

Klosterman, R. E. (1997). Planning support systems: A new perspective on computer-aided planning. *Journal of Planning Education and Research*, *17*, 45–54.

Lee, Y.-C., & Wiggins, L. L. (1990). MEDIATOR: An expert system to facilitate environmental dispute resolution. In *Expert systems: Applications to urban planning* (pp. 197–221). Springer. https://link.springer.com/chapter/10.1007/978-1-4612-3348-0_13.

Lemmen, C., Vos, J., & Beentjes, B. (2017). Ongoing development of land administration standards. *European Property Law Journal*, *6*(3), 478–502. https://doi.org/10.1515/eplj-2017-0016.

Li, H. (2003). Management of coastal mega-cities - A new challenge in the 21st century. *Marine Policy*, *27*(4), 333–337. https://doi.org/10.1016/S0308-597X(03)00045-9.

Luque-Ayala, A., & Marvin, S. (2016). The maintenance of urban circulation: An operational logic of infrastructural control. *Environment and Planning D: Society and Space*, *34*(2), 191–208. https://doi.org/10.1177/0263775815611422.

Marlier, M. E., Jina, A. S., Kinney, P. L., & DeFries, R. S. (2016). Extreme air pollution in global megacities. *Current Climate Change Reports*, *2*(1), 15–27. https://doi.org/10.1007/s40641-016-0032-z.

McArdle, G., & Kitchin, R. (2016). Improving the veracity of open and real-time urban data. *Built Environment*, *42*(3). https://doi.org/10.2148/benv.42.3.457.

Porter, L., Fields, D., Landau-Ward, A., Rogers, D., Sadowski, J., Maalsen, S., Kitchin, R., Dawkins, O., Young, G., & Bates, L. K. (2019). Planning, land and housing in the digital data revolution/the politics of digital transformations of housing/digital innovations, PropTech and Housing – The view from Melbourne/digital housing and renters: Disrupting the Australian rental bond system and Te. *Planning Theory and Practice*, *20*(4), 575–603. https://doi.org/10.1080/14649357.2019.1651997.

Rodima-Taylor, D. (2021). Digitalizing land administration: The geographies and temporalities of infrastructural promise. *Geoforum*, *122*(March), 140–151. https://doi.org/10.1016/j.geoforum.2021.04.003.

Shang, Q., & Price, A. (2019). A blockchain-based land titling project in the republic of. *Innovations: Technology, Governance, Globalisation*, *12*(3–4), 72–78. https://direct.mit.edu/itgg/article/12/3-4/72/9852/A-Blockchain-Based-Land-Titling-Project-in-the.

Silva, B. N., Khan, M., & Han, K. (2018). Towards sustainable smart cities: A review of trends, architectures, components, and open challenges in smart cities. *Sustainable Cities and Society*, *38*(January), 697–713. https://doi.org/10.1016/j.scs.2018.01.053.

The London Assembly. (2022). *The future of planning in London: Planning and regeneration committee.* https://www.london.gov.uk/who-we-are/what-london-assembly-does/london-assembly-work/london-assembly-publications/future-planning-london-report.

The World Bank. (2017). *Why secure land rights matter*. News. https://www.worldbank.org/en/news/feature/2017/03/24/why-secure-land-rights-matter.

Thompson, E. M., Greenhalgh, P., Muldoon-Smith, K., Charlton, J., & Dolník, M. (2016). Planners in the future city: Using city information modelling to support planners as market actors. *Urban Planning*, *1*(1), 79–94. https://doi.org/10.17645/up.v1i1.556.

Tomaney, J., & Fern, J. (2018). Contexts and frameworks for contemporary planning practice. In J. Fern, & J. Tomaney (Eds.), *Planning and practice: Critical perspectives from the UK* (1st ed.). Routledge.

UK Ministry of Housing, C. and L. G. (2020). *Planning for the future: White paper August 2020* (Issue August). https://assets.publishing.service.gov.uk/government/uploads/system/uploads/attachment_data/file/958420/MHCLG-Planning-Consultation.pdf.

Webb, S. (2019). *PlanTech - A new market for digitla planning products and services.* Capital Enterprise. https://medium.com/capital-enterprise/plantech-a-new-market-for-digital-planning-products-and-services-885678f9de89.

Webber, M. M. (1965). The roles of intelligence systems in urban-systems planning. *Journal of the American Planning Association*, *31*(4), 289–296. https://doi.org/10.1080/01944366508978182.

Yiftachel, O., & Huxley, M. (2000). Debating dominence and relevance: notes on the 'communicative turn' in planning theory. *International Journal of Urban and Regional Research*, *24*(4), 907–913. https://doi.org/10.1111/1468-2427.00286.

Digital Technology and Innovation in Real Estate Development, Construction, Management, and Maintenance

5.1 Overview of Development, Management, and Maintenance of Real Estate

5.1.1 Building Construction

Building construction is the process of creating structures for various purposes such as residential, commercial, industrial, institutional, and recreational. It involves planning, designing, financing, securing permits, site preparation, material selection, construction, and project management. Building design and construction typically requires multidisciplinary expertise from architects, engineers, and other professionals across various phases of a project (Muthumanickam et al., 2023). The process of construction is an important stage of a building's lifecycle as it sets the foundation for further phases; this stage provides the best opportunity for a building's efficiency to be maximised. Building construction typically begins with architectural and engineering design, where floor and building plans and specifications are developed to highlight the layout, dimensions, materials, and other important details of the building. Important considerations in this stage include functionality, structural integrity, compliance with building codes and regulations, and aesthetics. The building design is generally carried out in response to the existing conditions on a site. These conditions may present varying degrees of challenges, and the architect is expected to develop a new set of conditions or solutions as part of the design process (Ching, 2023).

The design is followed by the construction which entails the site preparation, excavation, levelling of the land, and laying the foundation. After this, the building's formwork is erected, and this includes the construction of walls and roofs and the installation of structural elements such as columns and beams. This is typically followed by the installation of building elements such as doors, windows, insulation, exterior cladding, electrical systems, plumbing systems, ventilation, heating, and air conditioning (HVAC). After this, interior work such as painting, flooring, tiling, and installing fixtures is carried out to complete the construction.

The building construction process varies by size, type, and purpose of the building being constructed, and various built environment professionals need to collaborate effectively to ensure that the building is delivered to quality, standard, cost, and time. Project management is therefore an important aspect of building construction. It plays a crucial role in coordinating and overseeing the various activities involved in the construction process including scheduling tasks, ensuring quality control, managing the construction staff, and adhering to timelines and budgets.

DOI: 10.1201/9781003262725-7

5.1.2 Building Maintenance

After a building has been constructed, building maintenance is required. This refers to the activities and tasks carried out to keep a building in good working condition, preserve its value over time, and ensure its functionality. The process involves a range of routine, preventive, and corrective/reactive measures to address various aspects of a building's upkeep, including its physical, mechanical, electrical, and environmental systems.

Building maintenance may be preventive or reactive. Preventive maintenance entails conducting regular inspections, lubrication, and minor repairs to address potential building problems and prevent them from getting worse, while reactive maintenance details work done after a building or facilities' failure or weakness. Reactive maintenance includes the following:

a Regular inspections: This involves routine inspections with the aim of identifying maintenance needs, code violations, safety hazards, or other issues that require attention.
b Regular cleaning: This refers to the upkeep of a building's interior and exterior surfaces such as floors, windows, fixtures, and common areas.
c Structural maintenance: This entails the assessment and retrofit of the structural elements of a building such as walls, roofs, foundations, and floors to maintain their integrity and prevent damage.
d Assessing safety and security systems: This is done to ensure that fire protection systems, alarms, safety equipment, security systems, emergency exits, etc. are functional.
e Electrical system maintenance: This involves the inspection of electrical systems to ensure that they are functional as well as the replacement of faulty wiring, safety issues, and other electrical components.
f Plumbing maintenance: This entails checking for leaks, maintaining plumbing fixtures, clearing clogged drains, and repairing or replacing faulty plumbing components.
g HVAC (heating, ventilation, and air conditioning) system maintenance: This involves the inspection, cleaning, and servicing of HVAC systems to ensure air quality and optimal performance.
h Landscaping and exterior maintenance: This is the maintenance of the outdoor areas of the building, including landscaping, parking lots, walkways, signages, and lighting.
i Record-keeping: This involves the detailed recording of maintenance activities, repairs, warranties, and service schedules to facilitate efficient property management and for future reference.

Effective and efficient maintenance can potentially extend the lifespan of a structure, minimise operational disruption, protect a building's value, and promote occupancy safety and comfort.

5.1.3 Building and Facilities Management

Building and facilities management is also an important building function. It refers to the process of overseeing and coordinating the various operations and activities within a building or property to ensure its efficient and effective functioning. The buildings being managed can be residential (e.g., apartment complexes, houses, flats), commercial (e.g., offices, shopping centres, warehouses), and operational real estate assets (e.g., student accommodation and data centres). Building and facilities management ensures that a building, its systems, and facilities are operational, well-maintained, and upgraded to create a safe, comfortable, and productive environment for the users. Facilities management more specifically integrates organisational processes to maintain the agreed services that support and improve the effectiveness of its primary activities (Bilal et al., 2016). These services are essential for office buildings with rapidly increasing demand for high-level reliability and performance of facilities by office building users and corporations (Shin et al., 2018); it should, however, be noted that these services transcend office space use as residential assets (e.g., built-to-rent and luxury flats) and other operational real estate assets (e.g., hotels, purpose-built student accommodation, and data centres) are increasingly becoming complex and complicated in their operations.

Some of the core building and facilities management functions are summarised as follows:

a Facilities and equipment management/maintenance: Ongoing maintenance and repair of the building infrastructure and equipment (HVAC, electrical systems, plumbing, elevators, etc.), routine maintenance tasks, and coordination with external service providers for specialised repairs.
b Tenant and occupancy management: Handling lease agreements, tenant communication, rent collection, move-ins, and move-outs, enforcing building rules and regulations, and enhancing occupancy satisfaction and retention for buildings that are occupied.
c Budgeting and financial management: Supervising the financial aspects of building operations, developing budgets, monitoring expenses, ensuring efficient use of resources, negotiating contracts with service providers, analysing cost-saving opportunities, and reporting financial performance to property owners and stakeholders.
d Security and safety: Implementing security policy to protect the building, its occupants, and assets, installing security systems, access control, surveillance (cameras), fire detection, prevention, and suppression system management, developing and enforcing safety protocols, conducting regular inspections, and provision of emergency response plans.
e Compliance and regulation: Obtaining necessary permits and licences, conducting inspections, addressing compliance issues, and updating team members on local building codes, regulations, and industry standards to ensure compliance.
f Sustainability and energy efficiency: Implementing sustainable practices and energy-efficient solutions to reduce the environmental impact of buildings through initiatives such as energy-efficient lighting, waste management, water conservation, and renewable energy integration.
g Vendor and contractor management: Coordinating the activities of vendors, contractors, and service providers, sourcing and evaluating contractors, negotiating contracts, and ensuring quality service delivery.

A building and facilities manager should therefore possess a diverse skillset and knowledge base such as organisational and communication skills, financial intelligence, problem-solving, and technical knowledge of building systems.

5.1.4 Demolition

Building demolition refers to the process of intentionally dismantling or destroying a structure to make way for new construction or repurposing land. It may also be carried out because of obsolescence or safety concerns. The professional managing a demolition process needs to ensure a controlled and safe removal; thus, the process requires careful planning, strict adherence to safety regulations, and the use of specialist equipment. Building demolition therefore must be managed and supervised by a qualified and experienced professional who will adhere to professional standards and local regulations to ensure the safety of workers, the public, and the environment.

Key elements of demolition are given as follows:

a Environmental considerations: Demolition contractors attempt to minimise the environmental impact of the building demolition such as dust, noise, and emissions. They thus use water sprays (for dust control), employ noise barriers, and adhere to local regulations for air quality.
b Safety precautions: Demolition professionals assess the building's structural integrity, potential hazards (such as asbestos or lead-containing materials), and nearby infrastructure to determine appropriate safety measures. Safety considerations may include scaffolding, safety nets, protective barriers, dust control, etc., to ensure the safety of site workers and the surrounding environment.
c Demolition techniques: Various methods can be deployed to demolition works, depending on the site, building characteristics, surrounding environment, and project requirements. The common methods are mechanical demolition using cranes or excavators with attachments like hydraulic breakers or shears, implosion using controlled explosives for large structures, or deconstruction where the building is carefully dismantled to preserve reusable material.
d Planning and permits: Demolition requires thorough planning and obtaining the necessary permits from local authorities. Thus, the demolition plan is developed to include the sequence of operations, safety measures, waste management strategies, expected timeline for demolition, etc., and this is usually passed on to the local authorities for necessary permits.
e Utility disconnection: Demolition contractors plan to safely disconnect utilities such as water, electricity, gas, and telecommunications prior to demolition. This may require them to coordinate with utility companies so that the services are disconnected, and in some cases, for underground and overhead lines to also be disconnected.
f Salvage and waste management: Building demolition typically involves salvaging valuable materials and recycling or disposing of the generated waste. Items such as doors, windows, fittings, fixtures, and other reusable construction materials may be salvaged for resale, reuse, or donation. In such cases, demolition contractors need to adhere to waste management protocols of segregating materials for recycling and ensuring proper disposal of hazardous materials.
g Site remediation and preparation: This is aimed at removing the remaining debris, levelling the land, and preparing it for future land use or development. It entails backfilling excavated areas, grading the site, and ensuring proper drainage and soil stabilisation

5.2 Inefficiencies in Traditional Real Estate Construction, Management, and Maintenance

With the increasing global need to cut down on energy consumption and carbon emissions, the building and construction industry has been labelled as the sector with the highest potential for energy saving and carbon emission reduction (Hauashdh et al., 2022;

Lin et al., 2018). There are several challenges and inefficiencies associated with the various aspects of building and construction-related activities, spanning the lifecycle of property; these will be discussed based on the four major building-related activities: sourcing and transporting of materials, construction, building maintenance, and facilities management.

5.2.1   Sourcing and Transportation of Building Materials

One of the fundamental challenges of building construction is the sourcing of building materials and this has adverse environmental, social, and economic effects. Although building industry professionals are gaining awareness of environmental sustainability, sourcing materials with low carbon footprint, recycled content, and renewable resources is challenging, and the sourcing and transportation processes are also inefficient. The manufacture and transportation of some building materials have an exceptionally high environmental impact and a considerable amount of energy is spent in these processes (Venkatarama Reddy and Jagadish, 2003). For instance, the world's manufacture of cement contributes up to 10% of all industrial carbon dioxide emissions (Brown, 2018). It is therefore important to assess the environmental impact of building materials including energy consumption, waste generation, and emissions. There are also transportation and logistics challenges, particularly for large building materials. Building materials are often manufactured outside the construction site and they are typically conveyed to the construction site using various international and regional routes and transportation systems (Han et al., 2020). These activities have a significant environmental impact.

Global or regional supply chain shocks arising from natural disasters, political instability, trade disputes, or pandemics can also disrupt material production, transportation, or availability. This can affect shipping costs, transportation logistics, coordination with suppliers and contractors, handling of fragile or sensitive materials, etc. leading to a general increase in cost and time of delivery. This may result in changes to the cost of building materials and delays, making it difficult to accurately estimate the budget and time for construction projects and potentially leading to cost overruns or delays. These factors can also affect quality control and standards as contractors may source for alternative materials which might be less suitable and with lower quality to meet the project deadlines and minimise the cost associated with delays. Maintaining consistent building and construction quality standards is important and ensuring that building materials meet the prescribed standards is essential for constructing safe and durable buildings; the failure to ensure this can lead to compromising the structural integrity and longevity of the building.

Additionally, there are also ethical issues associated with the sourcing of materials such as child labour, unsafe working conditions, and unfair labour practices in the manufacturing or extraction of raw materials.

5.2.2   Building Construction Processes and Activities

There are several problems associated with the traditional methods of building construction and these often adversely impact the progress, cost, quality, safety, and overall success of a building construction project. One of the core issues in building construction is the design-construction interface problem, particularly in large buildings (Sha'ar et al., 2017). Closely related to this are the design and documentation errors. Incomplete or inaccurate design and documentation can lead to construction challenges. Design flaws, errors, omissions, conflicts, inconsistencies in plans and specifications, inadequate coordination among different disciplines, inaccurate measurements, improper structural calculations, insufficient details, etc., can cause confusion and delays during the construction and can also result in compromised safety and the need to rework certain parts of the structure.

Another problem that often arises during building construction is quality control and adherence to building codes and regulatory standards. For instance, despite the UK Government's revision of building regulations in 2002, 2006, and 2010 towards more stringent energy efficiency standards, Pan and Garmston (2012) reveal that the level of compliance with energy building regulations of new residential properties in England and Wales was poor. As stated in Section 5.2.1, challenges with sourcing and transporting building materials can lead building contractors to source for alternative materials and these can sometimes be sub-standard or low-quality materials with significant consequences during the construction. Inferior materials can further lead to compromised structural integrity, premature deterioration, and increased maintenance or repair needs in the future. Non-compliance with building codes, permits, and regulations can also result in significant issues during construction, and failure to meet regulatory requirements can result in fines, legal complications, construction delays, the need to carry out corrective measures, and, in some cases, the shutdown of her project. Non-compliance with building regulations may not necessarily be the result of systematic and purposeful disobedience. In some cases, this may be the result of the lack of skills and knowledge of the required standards on the part of the operatives and shortcomings in site management and toleration of sub-standard workmanship (Baiche et al., 2006).

Safety remains a major issue across the global construction industry (Zhou et al., 2012) and is sometimes compromised. Construction sites are inherently hazardous environments, inadequate safety protocols, poor communication, insufficient training, insufficient supervision, or non-compliance with safety regulations can occur if proper protocols are not observed. These can lead to accidents, injuries, and even fatalities. Data from Great Britain's Health and Safety Executive (2023) on construction-related deaths showed that there were 36 fatal injuries to construction workers, 21 fatal injuries to employees, and 15 fatal injuries to self-employed construction professionals from April to December 2022. The data further revealed that there were three fatalities to members of the public in the same period. This supports the statement by Chiponde et al. (2017) that the construction industry has been known to expose not just employees but also members of the public to hazards and accidents. Turner et al. (2017) further reveal that site-based construction workers suffer from a high prevalence of mental distress and in many countries, the levels of suicide are high when compared to other industries resulting from high levels of psychosocial factors, pain, and injuries. Ensuring a safe working environment is therefore essential in protecting the well-being of workers and members of the public, as well as mitigating potential legal and financial liabilities.

Human resource and skills-related challenges also affect the building construction process. The challenge of skills shortage in construction could get worse with the younger aged demographic group increasingly finding the sector less attractive. According to Autodesk and CIOB (2020), the construction industry is unattractive and needs stimulation and a reviewed image to attract talent. Inadequate construction skills, lack of attention to detail, shortcuts taken by the construction team etc. can result in poor workmanship, further leading to defects, structural weakness, improper installations, and reduced quality of the buildings after construction. There are also communication and coordination challenges among members of the project team. Due to the high number of professionals and artisans involved in building construction, there can often be misunderstandings, delays, conflicts of opinion, and reworking of parts of the building. This can negatively impact productivity. As the Autodesk and CIOB (2020) report shows, productivity in the construction industry has increased by only 1% in the last 20 years.

The highly fragmented and disjointed nature of the construction industry may also be one of the primary drivers of the delay and other inefficiencies. It is often challenging to connect all members of the construction team, supply chain, small-size organisations, consultants, contractors, and other key parties. Disputes may also arise between the project owners, contractors, sub-contractors, suppliers, or consultants. These disputes relate to payment disputes, scope changes, contract interpretation, or disagreement over project specifications. These issues will often lead to additional costs, strained relationships, and project delays. Project delays and budget overruns are recurring construction problems. The Climbing the Curve Global Construction Project Owner's Survey reveals that approximately 75% of projects in the preceding three years were completed more than 10% after their original deadline (KPMG, 2015). Autodesk and CIOB (2020) further reveal that some buildings are still being delivered 50% late and 50% over budget and there are still defects on site.

The environmental impact of building construction is also an important consideration. Reducing the energy intensity of a building needs to be considered during the building design stage (Lin et al., 2018). The construction process typically has a significant carbon footprint resulting from the materials being used, equipment, transportation, etc. According to Yan et al. (2010), building construction-related activities consume high quantities of energy and emit large amounts of greenhouse gases (GHG). There are therefore concerns about waste management, pollution prevention, and energy efficiency. Inadequate waste disposal practices, pollution from construction activities, non-compliance with environmental regulations, etc., can lead to environmental degradation with legal consequences and negative public perception.

There are several other factors that can arise in building construction. For instance, adverse weather conditions, labour shortages, equipment breakdowns, late delivery of building materials, design changes, etc., can disrupt the project schedule and cause delays. Furthermore, cost overruns can arise from unexpected site conditions, changes in project scope, price fluctuations of materials and labour, inadequate cost estimation, etc. Failure to anticipate and address these issues and poor cost management can result in financial challenges, delay in the project completion, and ultimately compromise the viability of a building project.

In general, the inefficiencies associated with building construction can have significant economic, social, environmental, and physical implications. For instance, fines, penalties, and remedial measures will incur additional costs and delays will affect the project time (economic); compromised quality of the building can lead to defects and safety hazards and disputes can lead to relationship breakdown (social), reworking and fixing defects will require further energy consumption, waste generation and carbon emissions (environmental) and poorly constructed buildings can affect both internal and external space utilisation (spatial/physical). The COVID-19 pandemic has further exacerbated these challenges (Sierra, 2022). Thus, ensuring that the building design and construction process is optimised, streamlined, and efficiently managed with effective communication, collaboration among stakeholders, and adherence to quality and standards (among others) can significantly enhance the efficiency of the building design and construction process (Dauda et al., 2022).

5.2.3 Building Maintenance and Facilities Management

It is estimated that 75%–80% of the overall costs of a building's lifecycle arise during the operation and maintenance phase for buildings and facilities (Madureira et al., 2017). Shin et al. (2018) provide a useful overview of facilities management functions. These processes and functions are complex and traditional approaches are generally inefficient. Building maintenance professionals and managers are often employed to manage these challenges; however, these tasks intrinsically have challenges that can affect the efficiency, functionality, and safety of the building (Cao et al., 2015; Hauashdh et al., 2021; Lin et al., 2018). Maintenance strategies include preventive, corrective, and condition-based maintenance (Lee and Scott, 2009). Preventive maintenance is technically different from condition-based maintenance. It is a hybrid of both preventive and corrective because is carried out at the onset of a failure, in contrast to preventive which is before, and corrective which is after failure.

One of the fundamental problems of building maintenance is poor maintenance planning (Barrelas et al., 2021). The lack of building maintenance planning often results in inefficient maintenance activities. This can further lead to delays, overlooked tasks, and incomplete maintenance works. It should be noted that in some cases, even when maintenance plans are in place; these plans are deferred, postponed, or neglected, and this can lead to the deterioration of building components and systems over time, resulting in increased repair costs, reduced building performance and a potential increase to safety risks. Lee and Scott (2009) identify several other issues: inadequate maintenance resources, insufficient budget, poor resource allocation, poor optimisation of operation cost, and wastage. The lack of a budget or a limited maintenance budget can particularly restrict the building manager's ability to comprehensively address all the maintenance needs. Poor maintenance funding can also lead to prioritising critical maintenance tasks while neglecting less urgent, albeit still important maintenance activities. The lack of efficient maintenance practices and systems often leads to high maintenance costs.

Several things can go wrong within a building. According to Roth et al. (2004), there is a widespread existence of building and equipment system faults. For instance, structural problems can arise due to foundation settlement, cracks in walls or floors, corrosion or structural elements, and damp or water damage. If these structural issues are not identified and addressed promptly, they can lead to further deterioration, reduce occupant safety, and undermine the structural integrity of the building. The efficient maintenance and management of buildings are also inhibited by equipment and system failure. A building's equipment and facilities are susceptible to unexpected breakdown, malfunction, and systems failure. For instance, HVAC and mechanical systems play a crucial role in maintaining occupant comfort; problems such as inadequate heating or cooling, malfunctioning controls, poor air quality or energy inefficiency can arise requiring maintenance or repair. Evidence suggests that various faults that occur in building and HVAC systems usually lead to higher energy consumption and inevitably to worse thermal comfort (Du et al., 2014).

There may also be plumbing issues such as leaks, pipe bursts, or drain problems causing water damage to the building, finishes, and other systems. Plumbing problems, if left unattended, can further lead to damage, mould growth, and potential health hazards. Electric system malfunctions such as power outages, faulty wiring, and overloaded circuits can also develop suddenly. There can also be a sudden breakdown of fire and life safety equipment such as fire alarm systems, sprinkler systems, emergency lighting, and exit signage. These can abruptly disrupt the building operations, and cause discomfort to occupants, requiring immediate attention and repair. This can particularly be challenging in situations where parts need to be ordered or where specialist attention is required for the facility or equipment to be fixed.

Bilal et al. (2016) also highlight three key challenges faced by the majority of facilities management systems. They include inefficient and time-consuming searching interfaces, lack of a unified interface for facilities management systems to exchange information, and inability to store and process large volumes of data generated by these systems. These issues call for the application of data and advanced analytics in the development of facilities management systems.

Fault detection and diagnostics are also an essential part of building maintenance and facilities management, and whole-building diagnostics has been the subject of extensive interest in the last decade. In some cases of equipment or facility breakdown, there may be an inaccurate, incorrect, or incomplete diagnosis of maintenance challenges or system failure. It is estimated that close to 30% of the energy consumed by commercial building HVAC and lighting systems results from inadequate control and sensing (Kim and Katipamula, 2018). Thus, accurately, and promptly identifying the immediate and remote causes of a system failure or equipment breakdown remains a major building maintenance challenge. Maintenance tasks will therefore need to be performed with proper skills, attention to detail, and adherence to standards to avoid poor workmanship. Inadequate knowledge about equipment systems, safety protocol, maintenance procedures, etc. can lead to inefficiency, sub-standard work, and increased safety risk and can potentially damage the building components or systems. This calls for the maintenance personnel to receive the appropriate level of training and to engage in continued professional development to update their knowledge and skill base.

The building and facilities manager also needs to ensure that there is effective communication among maintenance personnel, building users, and other stakeholders to minimise misunderstanding, delays, and effective maintenance activities. Incomplete or lack of documentation of maintenance activities such as inspections, repairs, and equipment services can make it difficult to track maintenance history, identify recurring issues, or demonstrate compliance with regulatory requirements.

In carrying out the building maintenance, several other challenges arise. For instance, the unavailability of spare parts is a key issue. For a building repair to be completed, parts of the equipment may need to be repaired or replaced. If spare parts are not readily available, it can cause delays in maintenance, and waiting for the procurement or delivery of required items can prolong the downtime of equipment or systems, impacting building operations. There are also potential safety hazards associated with maintenance activities. Building maintenance requires working at heights, handling hazardous substances, or working with electrical or mechanical systems. Failure to follow proper safety procedures, lack of personal protective equipment (PPE), or inadequate training can lead to accidents, injuries, or other occupational hazards. Additionally, the environmental impact is also important. Building operations and maintenance leave a significant carbon footprint and some problems arise from energy inefficiency, water waste, and inadequate waste management. Sustainable and effective strategies for building maintenance have the potential to reduce energy consumption and carbon dioxide emission (CO_2) (Hauashdh et al., 2022); non-compliance with environmental regulations can therefore impact the building's environmental footprint and increase operating costs (Figure 5.1).

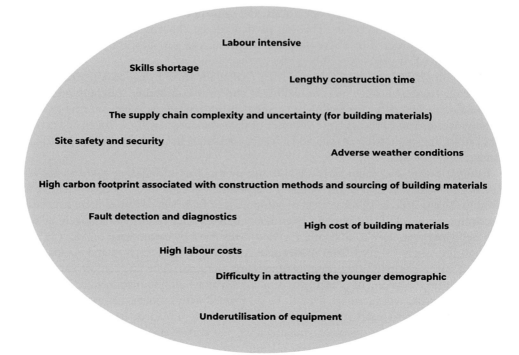

Figure 5.1 Provides a summary of the key inefficiencies associated with real estate construction, maintenance, and management.

As highlighted in this section, these inefficiencies have adverse environmental, social, economic, and physical impacts. Population growth, economic expansion, and urbanisation are expected to increase the global demand for construction output by 85% in 2030 (Autodesk and CIOB, 2020). This, therefore, underscores the importance of innovative solutions to enhance the environmental, social, economic, and physical efficiency of building construction, maintenance, facilities management, and other related services and operations. Section 5.3 provides a review of some of the core emergent systems and digital technological advancement in these segments of the real estate value chain.

5.3 Innovation/Emergent Systems and Services in Real Estate Construction, Building Maintenance, and Facilities Management

As building and construction become more complex, traditional methods and approaches are increasingly becoming insufficient and ineffective, and automated processes are gradually being adopted (Park and Kim, 2015). Several innovative practices and digital technologies have been applied to real estate construction and development to ensure that the construction process is more efficient and of higher quality. A range of innovative tools and digital products have demonstrated a range of opportunities and benefits to the construction industry. The solutions in this sector range from software to hardware and provide novel construction methods, new materials, site and project management tools, IoT, asset tracking, waste management, predictive maintenance, equipment utilisation, and workflow automation, among others.

Although the building construction industry is not adapting to digital technology at the same pace as other sub-sectors, progress is being made and this is changing the construction and development landscape with an increasing drive to implement new approaches to meet the ever-growing needs of improved efficiency, productivity, and sustainability. According to Witra (2022), $8 billion has been invested in contemporary construction technology alone, and a recent survey and report by Bimobjects (2022) reveal that 61% of architects, engineers, contractors, and owners/operators expect more digital tools to emerge post-COVID-19. This sub-section provides an analysis of innovative and emergent PropTech tools, platforms, products, and services that have emerged in construction, building maintenance, and facilities management in the last few decades.

5.3.1   ConTech and Digital Construction

Traditional building construction goes through the standard design, bid, and build method which typically requires the client/stakeholder to hire an architect to design the structure; this design is then passed on to the general contractor who would hire sub-contractors to execute the project. As illustrated in Section 5.2, this approach had several inefficiencies. Furthermore, this system made it impossible for the input of the general contractors, engineers, and sub-contractors to be integrated into the design process. To address this challenge, the integrated project delivery (IPD) method where the architect, general construction manager, and project manager worked together became normal practice. Using the IPD approach, all the stakeholders work together, which sometimes results in slow decision, thus leaving the manager at risk of delivering on time and within budget.

For any new technology to be accepted as a solution in construction, it must be perceived to be more efficient for contractors, owners, and designers in comparison to traditional methods (Kreiger et al., 2019). Digital construction and ConTech have emerged in the last few years to address some of the inefficiences associated with traditional buildings. Digital construction refers to the utilisation and application of digital technology to enhance the process of developing and operating the built environment. Similarly, ConTech (construction technology) can be defined as the application of digital technology and innovation to construction to improve efficiency, productivity, and safety. These relatively new terminologies attempt to encapsulate the drive to achieve improved efficiency in building construction through innovative methods and digital tools and ultimately to streamline processes, improve project management, and optimise resource utilisation. They involve the integration of digital tools, data, and workflows throughout the construction process from planning to designing and construction. They can be seen as the process of utilising digital tools to make the delivery and operation of the built environment more efficient, safer, and more collaborative which enables the owners, operators, and users to maximise the benefits of an asset at every stage of its lifecycle.

5.3.1.1   Core Digital Tools and Applications in ConTech and Digital Construction

The growth of innovation and digital technology in construction has taken different forms with a variety of digital tools and applications. These include data and analytics, Virtual Reality (VR), Augmented Reality (VR), Internet of Things (IoT), robotics, drones, cloud computing, etc.

5.3.1.1.1   DATA AND ANALYTICS

The construction industry has experienced a significant increase in data and analytics from diverse disciplines that contribute to the various elements of the building's lifecycle (Bilal et al., 2016). The growing demand for advanced technologies in the construction industry is driven by both private and public stakeholders who aim to promote sustainable, healthy, and financially viable buildings. Construction and development data provide valuable information that enables developers and building contractors to forecast the needs of a specific project, mitigating risks and achieving better efficiency of buildings and the construction process. Every construction project generates enormous data, and the ability to harness and analyse these data is highly valuable, providing an intelligent basis for decision-making and policy formulation. Furthermore, it can also influence the establishment of parameters for building design and the efficiency of construction and maintenance processes.

The design stage of a building construction is an essential aspect of ensuring building efficiency and data is a key factor in this stage. Buildings are expected to have aesthetic appeal and harmonise with the surrounding environment; however, they also need to adhere to building regulations and planning policies. Building contractors, therefore, need to consider the requirements of developers and clients while also ensuring that they meet these standards. This often leads architects, engineers, and other members of the design team to balance aesthetics with building performance and efficiency. As the demand for certifications such as BREEAM and LEED continues to grow, achieving building performance and efficiency is also becoming more important. Product information also becomes essential during this stage, requiring designers to thoroughly analyse data on product geometry and performance and select components that align with the building project. It is also important to assess factors such as u-value, corrosion resistance, ventilation rate, fire resistance, and the contribution of each component of the building to its Energy Use Intensity (EUI). Using predictive analysis, it is now possible to gain insight into the risks associated with each construction work and site and these predictive capabilities can lead to better decision and mitigation plans. Furthermore, data from sub-contractors and suppliers as well as the design plans and site information can be used to determine which risks may be influenced during the construction process. These capabilities combine to improve data collection, further analysis, and visualisation, and also the enhancement of collaboration, productivity, and decision-making in construction projects.

Due to the volume and velocity of the data generated from digital construction, advanced analytical techniques are being used to analyse the data and gain new insights and perspectives on construction projects. Bimobjects (2022), however, highlights the risks associated with big data, cautioning that it can become outdated, and siloed systems can lead to misinformation that can further lead to mishaps. Poor data management costs the global building industry approximately EUR1.57 trillion, and the lack of building data leads to a generation of 3.8 times more carbon emissions than their original design intended (Autodesk, 2020). Thus, improvement in data and advanced analytics can address challenges such as cost overruns, low productivity, and environmentally efficient design in the construction sector.

5.3.1.1.2   COMPUTING AND ASSOCIATED TECHNOLOGY

Computing, software, mobile, internet, cloud, and other associated technology are foundational to ConTech and digitisation in construction. Software development in construction has particularly witnessed significant advancement driven by advanced computing power.

Internet and mobile computing also play crucial roles in digitising construction processes and asset maintenance. According to Autodesk and CIOB (2020), computing has attracted benefits such as streamlined marketplaces for completing transactional tasks and leveraging crowdsourcing, where numerous project stakeholders collaborate online to solve problems. This enables contractors to efficiently access the most suitable skills quickly and with a greater level of ease. Furthermore, blockchain technology is enhancing transparency within project transactions and construction processes. Blockchain particularly ensures that contractors receive payment promptly while also ensuring that errors are minimised. Additionally, blockchain technology has the potential to revolutionise the leasing process of heavy equipment such as cranes, compressors, and excavators. This technology can also automate the management of the extensive paper documentation that needs to be signed, tracked, and recorded.

Cloud-based computing is also an integral element of digital construction. Cloud-based platforms enable real-time collaboration, data storage, and access to construction project information remotely. It also provides an opportunity to improve communication, streamline document management, and facilitate remote collaboration among project team members. Cloud computing is increasingly improving the ability of small and large-scale contractors to access unlimited processing power on demand facilitated through their connection to vast amounts of data and intelligent algorithms. Leveraging the infinite computing capabilities offered by cloud computing has the potential to further enhance the ability to solve highly intricate design challenges and provide optimal design solutions. A wide range of parameters can be simulated, including material selection, finance options, procurement strategies, physical dimensions, and labour considerations which can result in more effective designs that maximise returns and minimise costs. Furthermore, the availability of big data can significantly improve predictive analysis, enabling contractors to identify issues within the supply chain, optimise cash flow management, and identify root causes of projects going over time and budget.

InEight, a construction management software company also formed a strategic agreement with Felix Group Holdings, a global company operating a cloud-based enterprise SaaS procurement management platform and vendor marketplace to further enhance their project management capabilities for project owners and contractors across planning, construction, and turnover phases of the project lifecycle. This platform adds comprehensive procurement and vendor management capabilities and enables stakeholders to have a deeper supply chain understanding by leveraging up-to-date vendor insights during the estimating process.

5.3.1.1.3   INTERNET OF THINGS (IOT)

IoT is becoming more common in construction. It involves the connection of physical devices and sensors to a network, enabling the collection of real-time data from construction sites and relaying them to a centralised interface or dashboard. It entails real-time monitoring and collection of data on various aspects of a construction site such as temperature, humidity, air quality, noise levels, and equipment performance. This real-time data improves the monitoring of site conditions, safety, security, and equipment maintenance. Smart devices and wearables on the site (such as smart helmets, safety vests, and trackers) can provide real-time information on the workers' location, health and safety conditions, and equipment usage which can enhance safety and productivity on construction sites. Furthermore, it enables the tracking and management of construction assets, such as tools, equipment, and materials. By attaching IoT devices and sensors to these assets, their location, condition, and usage can be monitored, preventing loss, theft, or misplacement and optimising resource allocation.

Firms are increasingly leveraging IoT technology as a means to achieve strategic objectives (Jeyamohan and Clarke, 2022). One of the core features of IoT technology on construction sites is the ability to remotely monitor and control construction processes and equipment. This includes remote access to surveillance cameras, equipment control, and data analysis, enabling project managers and stakeholders to monitor progress, detect issues, and make informed decisions without being physically present on the site. IoT can also be used to monitor and optimise energy consumption in construction sites and buildings. By integrating these with energy systems, energy usage is now tracked, analysed, and optimised, and this is improving energy efficiency and saving costs. This is also valuable for predictive maintenance as the devices can gather data on equipment performance and conditions. By analysing these data, potential equipment failures or maintenance needs can be identified in advance, reducing downtime and improving overall equipment reliability.

5.3.1.1.4   ARTIFICIAL INTELLIGENCE (AI) AND MACHINE LEARNING

AI and machine learning are also core digital elements in ConTech and digital construction. They particularly enhance risk mitigation strategies in project management and construction. At the planning and design stage, AI algorithms are being used to analyse large volumes of data, including building codes, regulations, historical designs, and user preferences to generate optimised designs and assist in the planning process. This technology is aiding architects and engineers to create more efficient, sustainable, and cost-effective designs. AI algorithms are also used to generate multiple design options based on specific parameters and constraints, providing a wide range of possibilities and by extension, aiding the identification of innovative and optimised design solutions to achieve faster and more efficient decision-making. The construction process can also be better optimised through AI-enabled construction schedule analysis and resource allocation. AI algorithms can identify potential bottlenecks, schedule conflicts, and resource inefficiency, providing project managers and contractors with more efficient decision-making capabilities and productivity.

AI-supported construction attracts several other benefits such as quality control, predictive analytics, asset and facilities management, and safety monitoring. AI-powered computer vision systems, for instance, can analyse images and video footage captured on a construction site to identify defects, anomalies, or safety hazards. This automated analysis enhances quality control and streamlines inspections, minimises error, and ensures compliance with safety standards. AI algorithms are also used to analyse historic data and real-time sensor data from construction sites to predict and prevent issues such as equipment failures, delays, and cost overruns. Trends and patterns are observed, and this provides insight into proactive decision-making and risk management. Furthermore, based on the monitoring of real-time data from sensors, cameras, and wearables through IoT devices, AI can be used to analyse real-time data to detect potential safety risks and hazards on construction sites. Site workers and supervisors are therefore notified of potential danger, enabling proactive interventions to ensure a safer working environment. Additionally, maintenance needs, repairs, asset utilisation, etc., are being forecasted and predicted, leading to cost savings and improved operational efficiency.

5.3.1.1.5   VISUAL AND IMMERSIVE TECHNOLOGY (3D, 4D, VR, AR)

VR is increasingly improving the design, planning, and construction processes more efficiently. It facilitates design reviews, clash detection, on-site visualisation, and communication, and it also minimises rework. Construction companies use VR technology to create ultra-complex 4D models of buildings so that their clients can have a unique experience of the property before the construction work begins. VR models are assessed by construction operations teams to identify existing and potential areas of concern before work begins, and this leads to changes and modifications of designs as well as the incorporation of contingencies. Construction companies are also using AR to enable members of the construction team and their clients to experience the 3D version of augmented reality. This provides a much more realistic experience in comparison to VR as the client can experience and feel the space and its features such as the position of sockets on the walls, cabinets, and toilet furnishing.

An example of AR technology in construction is ATOM. It is a powerful custom-built engineering tool that combines a construction safety headset, augmented reality displays, and in-built computing power which enables construction teams to view 3D models with millimetre accuracy on site. The precision is high such that it accurately positions 3D models within construction site tolerances, and it enables the adherence to construction safety norms with protection of the headset and comfort. The models are positioned in absolute terms by physically tapping into the site coordinate system, and users can walk through construction sites viewing holograms of models positioned within construction tolerances. This can empower construction teams to take a proactive approach, validate in real-time, and build it right.

Drones and reality capture laser scanners can also improve the pace and cost of projects by providing easier, more comprehensive, and safer ways of performing property surveys, scans, and inspections, particularly for complex or high-rise structures and other remote assets like pipelines or rail corridors. The images from the drones and sensors can be inputted into reality capture software which stitches photographs together to create 3D models bringing the real world into the digital environment on a large scale. Drones are providing better safety for construction workers as there will be a reduced need to access problematic and relatively unsafe spaces on the construction site.

5.3.1.1.6   3D PRINTING, ROBOTICS, AND MODULAR CONSTRUCTION

There have been recent advances made in 3D printing in the construction industry and this is often integrated with robotics. Robotics and 3D technology in construction enable on-site construction to take place even in extreme conditions. They also provide a more effective response to black swan events (sudden events with devastating consequences) and disaster relief. Furthermore, they are used as solutions for the conversion of slums to better quality housing. This technology typically requires the creation of 3D design on a computer and the file is used to programme the machines to cut and prepare the materials to create an exact replica of the on-screen design. In March 2017, San Francisco start-up Apis Cor printed a 38 sq. m house in 24 hours using fibre concrete with a stated life of 175 years at an estimated cost of $10,000 (Molloy, 2017). It should, however, be noted that although the structure and weight-bearing walls were all 3D printed on-site, the fitting of the roof, windows, insulation, and appliances were fitted by workers. Another example is the collaboration of the US construction tech firm (ICON 3D) and NGO New

Story which resulted in the 3D printed house in 48 hours in Austin at the estimated cost of $10,000 (Brown, 2018). This attracts benefits such as low construction cost and time and also cuts down on-site injuries and deaths. In the UK, 3D-printed homes are also being promoted as potential approaches to solving the massive housing shortage and grassroots leaders are increasingly interested in this technology (Boland, 2020).

Robotics is also gaining traction in the construction industry and changing the way construction projects are planned, executed, and maintained. Construction robots are used to perform construction tasks such as bricklaying, masonry, concrete pouring, and steel beam assembly. These robots can work with better precision and speed, improving productivity, quality of construction, and workers' safety while minimising labour-intensive work. This technology can also be used for demolition, providing a safe and cheaper alternative to manually tearing down structures at the end of a property's lifecycle. These robots generate data that can be valuable for future work, and data can also be fed back into their control systems to adjust their operations and drive greater efficiency and higher accuracy.

Autonomous (self-driving) vehicles are also being used for material transportation on construction sites. These robotic vehicles can efficiently move heavy loads such as tools, equipment and, construction materials within the site, reducing the need for manual labour and improving logistics. Furthermore, robotic exoskeletons (wearable robotic devices like suits) are being used to provide physical support to construction workers, reducing the risk of injuries and enhancing their strength and endurance. They can also facilitate heavy weightlifting and repetitive motions, minimising the strain on construction workers. There are also other Unmanned Aerial Vehicles (UAVs or drones) for site scanning or inspections, and these can also be used for 3D and 4D printing, robotics, and AI. They capture high-resolution images and generate accurate 3D models of construction sites, helping project managers to monitor progress, identify potential issues, and improve overall project coordination.

Modular construction is also becoming widespread. It is known as prefabrication or off-site construction which involves individual components or modules of a structure being constructed in a controlled factory environment before being transported and assembled at the construction site. It begins with detailed design and engineering processes. Architects, engineers, and construction professionals collaborate to create modular designs that optimise efficiency, standardisation, and transportation logistics. The next stage is the fabrication which is conducted in a controlled factory. The individual modules are constructed and assembled, incorporating electrical, plumbing, and other necessary systems. This then goes to the quality control stage where the quality of the products is inspected to check for accuracy, adherence to design specifications, and compliance with building codes, and necessary adjustments or corrections are made. Finally, the modules are then transported from the factory to the construction site.

Meanwhile, the construction site is prepared and this involves clearing, excavation, levelling the ground, preparing the foundation for the installation, utility connections, etc. After this, the module is installed and secured to form the complete building. The finishing works are carried out and the site integration occurs which involves the integration of the building to the site infrastructure such as the connection of utilities, completing landscaping, and ensuring compliance with the local building codes and regulations. The interior fit-out and testing are completed, and the building is handed over for occupation. It should be noted that these stages may sometimes overlap, and some variation in processes may be observed depending on the project's complexity and specific requirements. The conditions in a factory can be managed and controlled a lot better than those on the site; prefabrication enables the streamlining of the entire process. Modular construction is becoming preferred to the traditional prefab method because the prefab method is associated with a range of health and safety issues and high costs of resolution, e.g., asbestos removal (Brown, 2018).

5.3.1.2   Challenges of Adopting ConTech and Digital Construction

Although ConTech and digital construction more broadly have positively transformed the construction industry, they also have their challenges. One of the core challenges is the adoption and integration of digital technologies and processes within the construction industry. Resistance to change, lack of awareness, and limited digital literacy among stakeholders can also hinder the successful implementation of digital construction and other sustainable practices within the built environment (Dauda and Ajayi, 2022). These digital technologies come with a high cost, usually requiring significant upfront investment in hardware, software, and training. Small and medium-sized construction firms may therefore have to deal with a significant financial challenge. There are also challenges in data management and integration. Furthermore, digital construction requires a skilled workforce with expertise in utilising the software and hardware, and training the workforce can be challenging, particularly in contexts where traditional practices and tools are prevalent.

There are several other challenges to digital construction including collaboration, coordination, cyber security, data privacy, scalability, standardisation, and legacy systems. Effective collaboration and coordination among different stakeholders including architects, engineers, contractors, and sub-contractors are essential for successful digital construction, and this may not always be easy to achieve particularly when there are complex teams and processes which require standardised protocols. Furthermore, implementing digital construction practices across multiple projects and organisations requires scalability and standardisation, and the development of common standards, protocols, and workflows that can be universally adopted within the construction industry is crucial to maximising the benefits of digital construction and enabling interoperability between different projects and stakeholders. Additionally, several construction firms still rely on legacy systems and traditional processes and methodologies. Integrating digital technologies into these existing systems and workflows can therefore be challenging, requiring a phased approach and careful change management to ensure smooth transitions and minimising disruptions.

It would be valuable to address these inefficiencies through industry-wide collaborations, investment in training, education, technological advancement, and willingness to embrace change. Overcoming these challenges can open the door to better integration of digital technology in construction, leading to increased productivity, improved project outcomes, and enhanced sustainability in the construction industry.

5.3.2   Building Information Modelling (BIM)

The drive for data-driven development has given rise to BIM with this system now evolving from being the industry outlier to the industry standard. BIM is a powerful digital technological methodology and process that is transforming the architecture, engineering, and construction (AEC) industry and altering the way buildings and infrastructure are designed, constructed, and operated by enabling the creation and management of comprehensive digital representations of assets. BIM is a collaborative process that allows project stakeholders to work together through the planning, design, and construction phases to deliver value across the entire building lifecycle

It is a digital representation of a building infrastructure project that includes 3D models as well as information about the project's physical and functional features. It enables architects to communicate and visualise their designs and ideas to engineers who have requirements on sustainability, structure, and mechanical, electrical, and plumbing (MEP). Contractors can see a simulation of the building and estimate the cost before construction, and facilities managers can use the digitised information in a central location to efficiently and efficiently maintain the building.

BIM has emerged from computer-aided design (CAD) and the development of digital technologies in the AEC industry. Key evolution stages of BIM are highlighted below:

- 1960–1970s: The development of CAD. Early CAD software focused on 2D drafting and geometric modelling providing a digital representation of building designs.
- 1980–1990s: The concept of object-oriented modelling emerged, laying the foundation for BIM. Research and development efforts focused on representing the building elements as intelligence objects within the properties and relationships. This advanced in the late 1990s when the term "Building Information Modelling" began to gain prominence as the digital representation of physical and functional characteristics of building facilities.
- 2000s: BIM continued to gain prominence with an increased adoption level with industry organisations such as the National Institute of Building Sciences (NIBS) promoting the benefits of BIM. Software developers also began to introduce BIM-specific software applications that allowed the creation and manipulation of intelligent 3D models with the properties and data attached to building elements.
- 2010s: BIM significantly advanced and became a lot more widespread with governments, organisations, industry bodies, etc. acknowledging its potential to improve project delivery, collaboration, and efficiency. Standardisation efforts also gained momentum in this period with guidelines being provided for BIM implementation and management.
- 2020s: BIM continues to advance in line with technological advancements and the increasing demand for digital transformation in the AEC industry. Cloud-based collaboration platforms, mobile applications and the integration of BIM with emerging technologies such as AI, VR, and IoT are reshaping the capabilities of BIM. The concept of digital twins is also gaining traction with BIM models being connected to real-time data from sensors, enabling continuous monitoring and analysis of building performance.

There are several widely used software applications for architectural design and documentation, including Autodesk Revit Architecture, Graphisoft Archicad, Nemetschek, Allplan Architecture, Digital Project, Nemetscheck, and Vectorworks. These programmes offer computer-aided solutions for various aspects of aesthetics and engineering during the entire planning and design phase of construction projects, covering building, interior, and urban areas.

Building performance analysis tools, such as Autodesk Green Building Studio, Graphisoft EcoDesign, IES Solution Virtual Environment, and DesignBuilder, enable designers to simulate the sustainability of the built environment. Architects and engineers can use these tools to calculate sustainability performance within the building model, optimising energy efficiency and reducing the carbon footprint to meet building energy requirements.

Interoperability is an important feature in this software realm, allowing the direct import of building layouts from different file formats and the smooth reuse of existing data without relying on third-party applications to interpret it.

Structural analysis software is utilised by structural engineers to efficiently create models of designs and buildings, enabling analysis of various aspects such as reinforced concrete, post-tensioned concrete, steel, wood, cold-formed steel, aluminium, and masonry. Popular software applications in this field include Autodesk Revit Structural Engineering, Bentley RAM Structural System, and Tekla Structures. These tools enhance the reliability of building structures by allowing the construction team to generate comprehensive project frameworks. Additionally, they enable the creation of schedules, material cut lists, framing elevations, and fully dimensioned 2D shop drawings according to various national standards. MEP engineers employ software solutions such as Autodesk Revit MEP, to visualise the installation details of components within the building. This aids in coordination with other contractors and stakeholders. Through the use of fabrication-ready components and objects like HVAC services and systems, engineers can draw 3D building models and calculate modules within the BIM model. This digital pre-planning reduces the effort required during on-site construction.

Simulation software allows construction planners to digitally review and analyse construction projects, estimating time and cost. Through 4D visualisation, which incorporates elements such as temporary works, logistics, costs, and resources, planners can edit and manage construction projects throughout their lifecycle. The goal is to enhance cost-effectiveness, quality, health, and safety before, during, and after construction. Utilising a unified visual interface, these software solutions, such as Autodesk Navisworks and Solibri Model Checker offer features like 4D time simulation, photorealistic rendering, and PDG-like publishing. Their aim is to improve coordination and communication, generate optimised construction schedules, maximise efficiency, and cater to the specific requirements of different trades and phases of the construction process. Once the construction work is completed, the digital data collected through BIM systems are transferred or handed over to the facility manager. These data facilitate maximising the asset's return and enable comprehensive facility management solutions throughout the building's lifecycle. They provide visibility into the construction process and ensure that design and construction information is retained for operational purposes. For example, when a contractor needs to replace filters or pipes, or conduct maintenance, they can access 3D models to accurately locate and identify the size, model, and construction year of the component, eliminating the need for guesswork.

5.3.2.1   Core Components of BIM

BIM models utilise digital components known as BIM objects, which can be in two categories: generic and manufacturer-specific. Generic BIM objects offer value through their geometry and basic component, and performance data, serving as placeholders for visual representation during the initial design phase. However, as there is a growing demand for accurate project and component information, the generic objects need to be substituted with manufacturer-specific BIM objects at a later stage. This implies that manufacturers who provide BIM objects are more likely to be specified and chosen due to their ability to represent the unique physical properties of the product, along with other essential data like energy consumption, operating costs, and product lifetime.

The core components of BIM are:

a   Data integration: BIM integrates data from multiple sources, including design, specifications, information, cost, and performance. This creates a comprehensive dataset that
provides a holistic view of the project and supports advanced data analytics, simulation,
and decision-making.
b   3D modelling: The core component of BIM is the creation of intelligent 3D models
that represent the physical and functional features of a building or infrastructure project. These models include geometry, properties, and relationships of building elements
such as floors, doors, windows, walls, and other building systems.
c   Analytics and simulation: BIM supports various dimensions of analytics and simulation
capabilities, including structural analysis, energy performance analysis, lighting analysis,
construction sequencing, etc. These tools enable the optimisation of designs, assessment of project feasibility, and evaluation of project performance.
d   Collaboration platform: BIM serves as a platform that enables the partnership of multiple disciplinary teams to work together, exchange information, and coordinate their
activities and tasks. Cloud-based BIM platforms further enhance accessibility, real-time
collaboration, and information sharing.
e   Documentation and drawings: BIM generates accurate and coordinated construction
documentation which includes drawings, specifications, and schedules. Due to the digitised nature of these documents, changes made in one element of the model can automatically be updated in associated documentation. This can enhance consistency and
minimise errors and mistakes.
f   Asset management and integration: BIM supports the integration of building systems,
equipment, and operational data for effective facilities management. It enables asset
tracking, maintenance planning and integration with CMMS (computerised maintenance management systems) or facility management software.

5.3.2.2   Benefits and Prospects of BIM

BIM has enhanced building construction and management in several ways. One of the
benefits is enhanced design and visualisation of buildings and infrastructure, allowing
stakeholders to better understand the project's design intent. Virtual walkthroughs and
renderings aid in making informed decisions, and optimising design solutions. This improves the construction planning and sequencing achieved through the improved capabilities for simulation and visualisation of construction sequences. This further enables project
teams to optimise schedules, identify potential logistics, issues and improve construction
efficiency with further cost and time-saving benefits. BIM's accurate quantity take-offs,
material tracking, and clash detection can help to streamline construction processes, reduce costly reworking, and ultimately reduce project delays. It also facilitated improved
cost estimation, project planning, and resource allocation. BIM also helps with clash detection and risk mitigation. Its clash detection capabilities enable the identification of conflicts
and clashes between various building systems such as structural, mechanical, electrical,
and plumbing. Early identification of potential clashes minimises rework, improves con-
structability, and reduces project risks. Additionally, BIM's data-rich models support the
management of buildings throughout their lifecycle and aid in asset tracking, maintenance
planning, energy analysis, and renovation or retrofitting projects, leading to improved
operational efficiency and sustainability.

BIM also improves the human resource element of the construction and maintenance of buildings and infrastructure. It facilitates collaboration and communication between project teams and stakeholders such as architects, engineers, contractors, and clients. It also provides a shared platform for the exchange of information, reduced errors, and enhanced coordination between disciplines.

Some of the further prospects of BIM are given as follows:

a   Further integration IoT: the further integration of BIM and IoT can enhance the capabilities of smart buildings and infrastructure, enabling real-time monitoring, predictive maintenance, and improved energy management. Digital technology can become a part of building and infrastructures' DNA making use of sensors and systems lined to IoT and allowing better use of information to make buildings more responsive to the needs of their owners, occupiers, and the broader environment (Autodesk and CIOB, 2020).

b   Further integration of AI: AI can be further leveraged to analyse and process BIM data leading to significant improvements in design optimisation, construction sequencing, and performance prediction.

c   Modular construction: BIM's ability to coordinate and optimise off-site construction processes will be a good alignment point for the growing modular construction model.

d   Sustainable design and construction: BIM can support sustainable design decisions by providing energy analysis, material lifecycle assessments, and simulation tools for evaluating the environmental impact of projects.

e   Digital twins: BIM can serve as the foundation for the further development of digital twins which are virtual replicas of physical assets. This will further enable real-time monitoring analysis and simulation of built environments for performance optimisation.

5.3.2.3   Challenges and Limitations of BIM Adoption

The implementation of BIM has several limitations and barriers. BIM adoption can be limited by high initial investment costs, lack of digital literacy, and resistance to change within organisations. The lack of standardised BIM protocols, file formats, and interoperability between software platforms can also hinder smooth data exchange and collaboration among stakeholders. There are also legal and contractual issues arising from the inability of traditional contracts and legal frameworks to sufficiently address the unique aspects of BIM projects such as intellectual property rights, liability, and risk allocation. Other challenges include skills development and data management. BIM requires a skilled workforce with expertise in BIM software, processes, and data management. Developing these skills and training the existing workforce can be challenging. Furthermore, handling and managing large volumes of project data, including intellectual property and sensitive information requires robust data management practices and cybersecurity measures.

As BIM continues to transform the AEC industry by enabling collaborative, data-rich, and efficient project delivery, it continues to improve collaboration, clash detection and mitigation, cost and time saving, and lifecycle support. Although challenges relating to implementation, standardisation, skills, and data management need to be addressed in order to fully maximise the potential of this innovation, BIM has the potential to further transform building and infrastructure delivery and management.

5.3.3 Digital Twins

There is an increasing desire to gain a 360-degree view of operations by combining Digital Twins technology with sensor data from IoT. Through digitisation and IoT, all the systems and components of buildings create a plethora of data that can be harvested, analysed, and used to provide actionable insights. This is leading to buildings being safer, more secure, more efficient, and more resilient. This enables buildings to continuously interact, learn, and adapt to create environments that are suitable for living.

Digital twins are virtual replicas or digital representations of physical assets, processes, systems, or environments. They are created by capturing and integrating real-time and historical data from sensors and other digital devices to provide a detailed and dynamic digital counterpart of the physical object or system. The "digital twins" concept takes its root in the field of aerospace engineering, where complex systems like aircraft are represented digitally to monitor their performance, simulate scenarios, and optimise maintenance. The concept has been further extended to other industries and sectors including manufacturing, healthcare, transportation, and building construction.

Integrating Building Twin across the entire building lifecycle (ideation, construction, and operation), data can be collected and analysed to improve performance and efficiency. Building Twin is the application of digital twins of buildings which provides the building data across the building lifecycle with a focus on the operation phase. It provides a real-time understanding of how a building is performing-enabling immediate adjustments to optimise efficiency and providing data to improve the design of future buildings.

5.3.3.1 Core Features, Components, and Applications of Digital Twins

The core features and components of digital twins are given as follows:

a Physical asset representation: digital twins aim to faithfully replicate the physical attributes, geometry, and behaviour of the real-world asset which includes capturing and modelling various aspects of the assets such as the shape, dimensions, materials, sensors, and other components.

b Real-time data integration: digital twins incorporate real-time data from sensors, IoT devices, and other sources to simulate the behaviour and condition of the physical asset. These data are continuously collected and used to update and enhance the digital twin's accuracy and integrity.

c Remote monitoring and control: digital twins provide remote monitoring and control of physical assets, allowing operators and engineers to assess and interact with the asset virtually. This capability is valuable for maintenance, remote operations, and troubleshooting.

d Lifecycle support: digital twins have the capability to cover the lifecycle of an asset, from design and construction to operation and maintenance. They provide a platform for collaboration, documentation, and decision-making throughout the asset's lifecycle.

e Analytics and simulation: digital twins leverage analytics and simulation techniques to analyse, monitor, and predict the behaviour, maintenance needs, and performance of the physical asset. By running simulations and scenarios, digital twins can provide insights into optimisation, operational efficiency, and failure predictions.

Digital twins can enhance the understanding, analytics, and optimisation of physical assets and systems. Some of the core applications to construction and infrastructure are:

a Manufacturing: they can be used to simulate manufacturing processes and optimisation of production of building materials. They can also be used to enable predictive maintenance and monitor machine performance, analyse sensor data, and improve overall production efficiency.
b Transportation: they can simulate and monitor vehicles, traffic systems, and transportation infrastructure. They also facilitate predictive maintenance, optimise routes, and enhance overall transportation efficiency.
c Building construction and operations: they can be applied to construction projects and existing buildings, enabling virtual design and planning as well as monitoring and optimisation of energy usage, occupancy patterns, and maintenance activities.
d Smart cities: they can be used to simulate and optimise urban infrastructure, including transportation systems, energy grids, and environmental monitoring. They also support urban planning, resource management, and predictive maintenance.

5.3.3.2   Benefits and Prospects of Digital Twins

Digital twins can improve asset performance. They enable real-time monitoring, analysis and simulation of assets which allows for predictive maintenance, optimisation of performance, and early identification of potential issues or failures. They can also enhance decision-making through the valuable insights they provide by way of data analytics and simulation. They support scenario testing, risk assessment, and optimisation of operations, leading to better outcomes.

Digital twins enable remote monitoring and access to control systems of assets, allowing asset operators to remotely access and control a property's systems, reducing the need for physical presence and enabling a more efficient operation in remote or hazardous environments. It also promotes more effective and efficient asset lifecycle management. As digital twins typically cover the entire lifecycle of an asset, supporting activities ranging from design and construction to operation and maintenance can facilitate collaboration, documentation, and knowledge transfer across the various lifecycle stages. Ultimately, digital twins have the potential to minimise maintenance and management cost and time through the optimisation of maintenance activities, predicting failures, and reducing downtime. Furthermore, cost savings can be achieved through improved asset performance and more efficient resource allocation.

5.3.3.3   Challenges and Limitations of Digital Twins Adoption

While digital twins hold immense promise for improving asset performance, enabling predictive maintenance, and enhancing decision-making in building and infrastructure delivery and maintenance, they are not without challenges and limitations. One of these is data availability and integration. Building a digital twin requires access to accurate, comprehensive. and reliable data from various sources. Ensuring data quality, compatibility, and integration can be challenging, particularly when legacy systems or multiple stakeholders are involved. Data privacy and security concerns are also inhibiting the adoption of digital twins. Digital twins rely on collecting and analysing vast amounts of data and this often raises concerns about security, privacy, and compliance with data protection regulations. Thus, protecting sensitive information and maintaining data integrity are important considerations. As digital twins become more sophisticated, ethical considerations are also arising such as transparency, fairness, and accountability in their use. This raises questions about the responsible and ethical use of data, privacy, and the potential for unintended consequences that need to be addressed.

The implementation and integration of digital twins can also pose complex challenges often requiring the integration of diverse technologies and systems. Organisations may also need to invest in infrastructure, expertise, and make changes to their operations in order to successfully deploy, utilise, and integrate digital twins. There is also limited domain expertise in this area and developing and managing digital twins requires specialised domain expertise, including data analytics, modelling, and simulation. Organisations may therefore face challenges in finding and retaining skilled professionals who can effectively utilise and interpret digital twin capabilities.

Scalability and interoperability are also limitations to digital twins' adoption. Scaling digital twins across multiple assets or systems can be challenging and interoperability issues may arise when integrating digital twins from different domains or platforms, hindering data exchange and collaboration. There are also challenges relating to model accuracy and representation. The accuracy and integrity of a digital twin depend on the quality of the model and the data used to create it. Deviations between the digital twin and the physical asset can occur, potentially affecting the reliability of predictions and decisions. Additionally, limitations relating to cost and returns on investment. Developing and maintaining digital twins can involve significant costs including data acquisition, software, hardware, and ongoing maintenance. Organisations would usually evaluate the potential return on investment to justify the expenses made to acquire and maintain these tools.

Organisations, therefore, need to thoroughly evaluate the costs implications, data privacy and security, and challenges relating to interoperability that come with digital twins to decide on the nature and timing of its adoption. The challenges highlighted must be critically evaluated and mitigated to fully harness the potential of digital twins and ensure their successful implementation and effective utilisation.

5.3.4   Smart Buildings

The concept of "smart buildings" has become more prominent in recent years as a promising solution to enhance energy efficiency, occupant comfort, and operational efficiency in buildings. Smart buildings have revolutionised architecture and construction, leveraging advanced technologies to optimise building performance and efficiency. The smart building concept refers to buildings that combine space with technology and captures digital technological advancements that facilitate the control and efficient management of space. Smart buildings incorporate a network of interconnected devices, sensors, and systems that enable intelligent control, automation, and data-driven decision-making, and the key objective is to create an environment that adapts and responds to the needs of occupants while minimising energy consumption and maximising resource efficiency. By integrating various technologies such as the IoT, AI, data analytics, and cloud computing, smart buildings enable real-time monitoring, analysis, and control of various building systems. IoT technology particularly has the ability to collect massive amounts of data for localisation and tracking (Tan and Miller, 2023). Big data produced from smart buildings can also create a set of new financial assets. The big data include data generated from building use, tenancy types, histories, tenancy applications and rejections, transactions, and preferences (Porter et al., 2019).

One of the core benefits of smart buildings is energy efficiency. Through the integration of sensors and automation systems, smart buildings can optimise energy usage by adjusting lighting, heating, ventilation, and air conditioning (HVAC) based on occupancy patterns, weather conditions, and energy demand. This not only reduces energy consumption but also lowers operational costs and carbon footprint. These buildings also attract social benefits by prioritising occupant comfort and well-being. By monitoring indoor air quality, temperature, and lighting levels, smart buildings can automatically adjust environmental conditions to create a comfortable and productive space. Additionally, personalised control interfaces and mobile applications enable occupants to customise settings according to their preferences, enhancing their overall satisfaction and experience in using buildings and related facilities. Smart buildings also enhance safety and security through advanced monitoring and surveillance systems. Video cameras, access control systems, and motion sensors can detect and respond to potential threats, ensuring the safety of occupants and assets. Integration with emergency response systems allows for swift and coordinated action in critical situations.

Operational efficiency is another significant aspect of smart buildings. They enable the collection and analysis of data from building systems, such as HVAC, security, and maintenance, and this further enhances predictive maintenance, early fault detection, and efficient asset management. They further help to optimise resource allocation, reduce downtime, and extend the lifespan of building equipment. Furthermore, smart buildings offer benefits at the urban scale. Through data sharing and integration with smart city infrastructure, such as transportation and energy grids, smart buildings contribute to the overall efficiency and sustainability of the city. They can participate in demand response programmes, optimise energy usage based on grid conditions, and support the integration of renewable energy sources.

Businesses like TAP (Tenant Assistance Programme) provide tenants with information about the buildings that they are renting with details ranging from energy consumption to insurance, health, and safety compliance. As the commitment to ESG continues to increase, particularly for built-environment-related companies, sustainability and ESG are becoming core industry considerations. This requires the building service input such as air, water, transportation, and power and outputs such as emission, refuse and sewage to be strategically managed for optimum environmental efficiency (environmental), physical and mental well-being of the occupants (social), and spatial efficiency (physical); suffice to note that all of these are expected to be implemented while ensuring the efficient utilisation of financial resources and time (economic). Key players in all market segments (users, investors, and developers) now generally expect that buildings will operate from both cost and functional perspectives. It is expected that lowering operational costs can potentially enhance the property occupancy rate and income, and by extension value with the potential to further enhance the investors' returns.

Traditional market pricing models do not account for sustainability and ESG. One of the reasons for this is that in the past, the drive for sustainability and environmental efficiency was usually a public concern and hardly transmitted into market pricing (University of Oxford Research, 2017). Leases were traditionally typically based on space utility with little or no regard for the actual energy consumption and other associated utilities such as energy, water, etc. In most cases, the utility bill was the responsibility of the tenant, and running cost was within the remit of tenants and not priced in the initial lease negotiation. This meant that the landlord did not make any special consideration for efficiency, and the tenants had to accept the burden of putting in measures to minimise the cost of their energy consumption. It has taken a while for the awareness of the importance of ensuring energy efficiency in buildings and the subsequent use cost to be incorporated into the corporate real estate decision process. There is a continuing drive to see that buildings' performances are with minimal waste in a bid to balance what is used with what is expelled. There is the general expectation that smart buildings will be clean, environmentally friendly, and supportive of the long-term use of the asset for its occupants or purpose.

5.3.4.3   Dimensions of Smart Buildings

According to the University of Oxford Research (2017), smart buildings have two dimensions: the first is the use of technological tools to enhance building efficiency and the second is the need for buildings to support technological platforms such as big tech data centres in the US and UK. These two dimensions are further explored.

5.3.4.3.1   BUILDINGS THAT UTILISE TECHNOLOGICAL TOOLS

Smart buildings enable the efficient operation and management of real estate assets through various means. They provide information on building and urban centre performance and also enhance and control building services. These services find applications in property maintenance, asset management, and facilities management. Mobile technology, including smartphones, tablets, and other portable devices, is utilised and integrated into dashboards for controlling electronic functions through apps and internet-based services. The IoT is used to monitor and measure objects and remotely control existing infrastructure using machine learning algorithms. Remote control of heating, lighting, and room temperature through apps, as well as interaction with cameras via mobile apps, is now a common feature in residential, commercial, and operational real estate. It should be noted that sensors do not automatically transform a building into being "smart". A building may only be classified as "smart" when there is a "brain", typically entailing the individual components, facilities, and corresponding sensors having the ability to communicate with each other and when the systems are interoperable (Brown, 2018).

In recent years, Apple and Google have introduced their smart home ecosystems, Apple Home and Google Home, respectively. Amazon dominates the smart home speaker market with products like Eco and Alexa. Companies like Distech Controls are delivering innovative building automation and energy management systems to maximise comfort, environmental quality, and sustainability. The evolution of smart buildings is leaning towards unmanned, robotic buildings involved in activities like 3D printing and warehousing. Companies such as Honeywell, Siemens, and CAME have transitioned from supplying mechanical products to home automation and intelligent systems. One key challenge for developers and real estate managers, particularly in office and workspace settings, is the provision and management of parking spaces. The shared economy concept may offer potential solutions in this regard.

There are various tools available to address inefficiencies in home management. The global smart home market size was valued at $79.16 billion in 2022, and it is expected to grow at a compound annual growth rate (CAGR) of 27% from 2023 to 2030 (Grand View Research, 2023a). The security element of smart homes has been a significant area of focus over the years. Technology companies are competing to develop the most advanced and competitive smart home security devices. For example, in 2012, Vivint Smart Home was launched after being acquired by Blackstone. Google also launched a smart home security product, the Nest Doorbell a few years ago. This offers round-the-clock video surveillance and integrates other smart features related to entryway security. The Nest Doorbell, along with similar embedded internet devices exemplifies the application of IoT in real estate and opens possibilities for similar smart solutions throughout the property value chain. Market leaders like Google's Nest Doorbell, Amazon's Blink, and Arlo's video doorbell are vying for market share, providing homeowners with enhanced security even when they are away and also managing package delivery services in the absence of the resident(s). The smart home security camera market is projected to reach $30.10 billion by 2030 with an expected CAGR of 19.2 from 2023 to 2030 (Grand View Research, 2023c). More broadly, the smart home market size is projected to reach $537 billion by 2030 (Grand View Research, 2023b) a reflection of the positive sentiments and prospects in this market.

5.3.4.3.1   BUILDINGS THAT SUPPORT TECH PLATFORMS (DATA CENTRES)

A data centre is a centralised facility that houses a vast amount of computer servers, networking equipment, storage systems, and other related components with the main purpose of securely storing, managing, processing, and distributing large amounts of data. Data centres are designed to provide a controlled environment with optimal conditions for the reliable operation of computer systems, including regulated temperature, humidity, and power supply and they play a crucial role in supporting various digital services and applications, such as websites, cloud computing, streaming platforms, online storage, and more. They serve as the bedrock of contemporary digital infrastructure, enabling organisations to store and access their data, run critical business applications, and deliver online services to users around the world.

The data centre market is becoming a core segment of the operational real estate asset class. Initially, high-tech buildings were mixed-use office/industrial buildings primarily found in science parks. The term "high-tech" initially referred to the operations taking place within the building, with minimal focus on the building itself. However, it now refers to buildings specifically designed to house technology-based platforms, including data centres (the University of Oxford Research, 2017). High-tech buildings accommodate data centres and provide specialised logistics such as online distribution hubs, click-and-collect stores, Amazon locker sites, and co-working spaces tailored for tech companies that require fast broadband access. These buildings may also feature amenities like autonomous car charging stations, drone hubs, and more. Additionally, there are hyperscale data centres operated by large service providers. With the increasing internet traffic and the growing need for new servers, hyperscale data centres will remain in high demand, especially as the use of cloud computing and related facilities continues to expand.

Connectivity stands as a foundational element for large-scale data centres. The demand for such facilities is primarily driven by tech giants and their extensive operations. Facebook, for example, witnesses millions of media uploads (songs, photos, and videos) across its various platforms, resulting in a substantial need for data storage and cloud infrastructure. With the acquisitions of platforms like WhatsApp, Instagram, and Messenger, and the recent launch of Meta (including Meta Workplace, Meta Quest, and Meta Portal), the demand for data centres in the US market has further escalated. This becomes particularly significant as Meta expands beyond traditional 2D screens and ventures into immersive technologies like augmented and VR. Meta's diverse product offerings connect over three billion individuals globally (Meta, 2021) and similar platforms share a comparable narrative. Amazon, for instance, has transformed its Amazon Web Services (AWS) data centres into a vital component of its operations, serving both internal needs and providing a competitive advantage in the external market.

5.3.4.2   Challenges and Limitations of Smart Building Adoption

While the concept of smart buildings is promising, several critical aspects need to be carefully considered and mitigated. One of the core challenges with smart buildings is the complexity of implementation. Integrating various technologies and systems, such as sensors, data analytics, automation, and connectivity, requires substantial planning, coordination, and expertise. This complexity can lead to higher costs, longer deployment timelines, and potential interoperability issues between different components and systems. Another critical aspect is data privacy and security. Smart buildings collect and process vast amounts of data from various sources, including occupancy patterns, energy consumption, and personal preferences. Safeguarding these data against unauthorised access, hacking, or misuse is paramount. Additionally, ensuring transparency and obtaining informed consent from occupants regarding data collection and usage is crucial to maintaining trust and respecting privacy rights.

Furthermore, the long-term sustainability and adaptability of smart buildings are subjects of concern. Technology evolves rapidly, and the lifespan of building infrastructure is often much longer than the lifespan of individual smart devices or systems. This raises questions about the scalability and compatibility of future technologies with existing smart building infrastructure. Upgrades and retrofits may be required to keep up with evolving standards and emerging technologies, posing financial and logistical challenges. Interoperability is another critical issue. Smart buildings rely on various technologies and systems from different manufacturers, each with its proprietary protocols and interfaces. Achieving seamless integration and communication between these disparate systems can be challenging, resulting in fragmented solutions that limit the full potential of a smart building ecosystem.

The complexity of smart buildings can lead to increased reliance on external technical expertise. Building owners and occupants may need specialised skills to operate and maintain the systems effectively. This can create a dependency on service providers and potentially increase operational costs, hindering the widespread adoption of smart building solutions. In addition, while smart buildings aim to optimise energy consumption and enhance sustainability, there is a risk of over-reliance on technology without addressing fundamental design and architectural principles. Neglecting the importance of passive design strategies, energy-efficient materials, and occupant behaviour can challenge the overall sustainability goals and limit the effectiveness of smart building solutions.

In conclusion, smart buildings have the potential to deliver significant benefits in terms of energy efficiency, occupant comfort, and operational optimisation. However, careful attention must be given to addressing the challenges of complexity, data privacy and security, long-term sustainability, interoperability, dependence on external expertise, and the integration of smart technologies with fundamental design principles. By addressing these critical aspects, smart buildings can reach their potential and transform the built environment.

5.3.5   Digital Property Management and Facilities Management

5.3.5.1   Digital Property Management

Digital property management refers to the utilisation of technology and digital resources for the management, supervision, and administration of real estate assets. It entails the use of software, platforms, and other digital tools to optimise and automate various aspects of property management, including tenant communication, rent collection, maintenance inquiries, lease agreements, financial reporting, and record-keeping. Digital property management empowers property owners, managers, and tenants to conveniently access and exchange information, carry out tasks, and monitor property-related data with greater efficiency. By leveraging digital solutions, it aims to enhance operational effectiveness, foster improved communication, and deliver a smooth experience to all individuals involved in the management of real estate properties.

The use of digital systems and tools can enable property managers to obtain data from sensors and IoT devices in the building and to analyse the data to gain insight into property use, maintenance issues, occupancy, etc. Asset managers can now take advantage of this contemporary technology to be virtually closer to the building and gain more insight using data from different sources. This can be used to forecast rental income, capital value appreciation, operational expenditure, and outgoings. Purchase of equipment, fittings, renovations, etc. can be projected in advance with better precision, making the process of asset management more transparent, scalable, and efficient. This digital approach can ultimately enhance their reporting system through advanced data analytics, data visualisation as well as digital reporting, generation, storage, and transmission. This can further enhance efficiency, transparency, and communication and ultimately minimise costs, save time, and significantly reduce workload and work-related risks and hazards. These advances are particularly important considering the growing demand for contemporary leases and flexible working such as mixed uses, co-working co-living, and short-term let.

Brown (2018) provides a useful classification of property management systems based on the various real estate investment strategies and investors:

Opportunistic investment: Assets owned by opportunistic investors typically have significantly higher levels of risk while also promising higher returns following improvements and renovations. This investment strategy often involves extensive renovation and upgrading efforts, as well as other uncertainties that the asset manager must address. BIM makes it possible to visualise and simulate the refurbishment process before it takes place, gaining insights into the specific components and facilities that will be installed and enhancing the planning phase. Furthermore, BIM can also assist in optimising energy consumption and reducing the likelihood of errors being discovered after completion, which may cost time and money to be rectified.

Value-add investments: The assets within the value-add portfolio require enhancements in terms of management and operations. There is also a significant risk of losses if the improvement plan cannot be executed; however, the properties offer moderate to high returns to offset this risk. In addition to comprehending macroeconomic and microeconomic market conditions, asset managers must possess knowledge about the building itself, including its deficiencies and potential strategies for increasing the property's value. Data plays a crucial role in this process, and the utilisation of sensors through the IoT can provide the necessary information and data for effective space utilisation planning and cost optimisation. By installing sensors in facilities such as elevators, water supply systems, and windows, predictive maintenance requirements can be identified, and the building automation system within the property can minimise utility expenses, maximise returns, and ultimately enhance profitability. Additionally, the integration of mobile technology, cloud computing, and internet connectivity allows for the provision of additional services beyond the building's premises and enables users to access them remotely through their smartphones.

Core and core plus investments: Core property portfolios are characterised by low risk and typically offer correspondingly lower returns, and the primary objective for asset managers is to maintain stability and maximise the occupancy of the property. Managing these assets involves dealing with a substantial volume of data that needs to be analysed for crucial decision-making related to property acquisition, occupancy rates, rental rates, and customer satisfaction. The use of IoT and advanced data analytics can enhance insights into the property's performance, and forecast future performance and pricing trends.

Generally, technology can be used to track tenants, their behaviour utility consumption, etc. and invoice them appropriately. Aspects of property management that align with brokerage will be discussed further in Chapter 7.

5.3.5.2   Digital Facilities Management

Digital facilities management refers to the utilisation of digital technologies and tools for the effective management and upkeep of a facility's physical assets and infrastructure. It encompasses the implementation of software, sensors, and data analytics to monitor and regulate various facets of facility operations, which include energy management, equipment maintenance, space utilisation, security systems, and environmental conditions. IoT devices and sensors collect information and provide data from the facilities, and these can be tracked and used to improve the services provided. The automated systems integrated into these systems allow for the facilities to communicate with each other and a dashboard. Sensors are used through mobile devices, headsets, carbon monoxide detectors, thermostats, smart watches, cars, etc. The transfer of data and communication makes the utilisation of information gathered by the sensors actionable. For instance, a thermostat senses an increase in temperature and transfers this information to a central system which activates the air conditioning system or extractor. These sensors can either be integrated into building components or separately installed on the wall in a component, at a desk, or even a chair.

Ineffective and inefficient preventive maintenance of buildings and facilities often leads to increased costs and reduced returns. However, due to the utilisation of sensors and digital platforms, valuable information about the condition of assets and their components can now be collected and analysed. Digital platforms that leverage big data can now predict the operational and capital expenditure requirements of individual buildings, which was previously a complex task. BIM can also play a role in this process, as the data collected can be shared with facility managers for day-to-day operations. For instance, using 3D models of the building, it becomes easier to detect leaks in concealed valves and guide contractors to the exact location of the faulty component for repairs. The utilisation of these digital tools empowers facility managers to access real-time data and assess performance indicators for more effective decision-making and prompt action which can significantly reduce the likelihood of complete breakdown, optimise resource utilisation, improve operational efficiency, and enhance the well-being and safety of occupants. By harnessing digital solutions, facility management processes can be streamlined, facilitating proactive maintenance and asset management.

5.4   Case Study Analysis

Following the review of core areas of digital transformation in construction, building maintenance, and facilities management, this section analyses specific PropTech platforms and companies within this area of the built environment value chain. Each case study begins with an overview and summary of key details of the platform/company; the solutions it provides are then analysed using the ESEP efficiency framework introduced in Chapter 1. The analysis concludes with a brief overview of the limitations and further prospects.

5.4.1   Case Study 1: ALICE Technologies

5.4.1.1   Background, Key Details and Technologies Utilised

ALICE Technologies is a construction operating platform that enhances the creation, exploration, and update of construction schedules with the aim of significantly reducing construction risk, cost, and time. The company was built to address four core issues: how technology can be used to develop faster construction schedules; how idle time can be reduced and how construction crews can more effectively be utilised and profit maximisation; how waste can be reduced while achieving better building quality; how to develop construction plans to increase the probability that contractors will deliver on time and on budget.

ALICE Technologies was founded in 2015 and it has raised a total of $48.5 million in funding over ten rounds. Their latest funding was raised on June 13, 2022, from the series B round. It is funded by 18 investors with Gaingels and Future Ventures being the most recent investors. It is applied to construction, smart cities, and building engineering using Artificial Intelligence (AI). It works with construction leaders in the infrastructure, industrial, and commercial construction segments such as Parsons, HDCC, and Kajima Corporation. It uses 18 tech products and services including HTML5, Google Analytics and jQuery, and 36 technologies for its website including Viewport Meta, iPhone/Mobile Compatible, and SPF.

ALICE offers tools designed to assist with planning stages from "pre-construction" through project delivery. For example, it can explore scenarios that make the most efficient use of project resources like labour, equipment, and materials and test the impact of changes in these variables on project outcomes. The software thus helps construction companies plan projects including bridges, tunnels, high-speed rail systems, and mixed-use towers. The company serves clients in the hospitality industry. Their services especially post-COVID-19 enable hotel staff to make their roles more fluid as staff work with smaller teams, adhere to new procedures, and cover more ground. Hotel management software from ALICE Tech aids to streamline staff communication, eliminating the risk of losing paper notes and ensuring that departments and the entire hotel are aligned. All team members can find all the notes and tickets in the same place for each request to be attended to.

5.4.1.2   Key Efficiencies and Solutions the Platform Provides

The solutions that ALICE Technologies provides are itemised using the ESEP efficiency framework (Table 5.1).

Table 5.1 ESEP Efficiency Analysis of ALICE Technologies

ESEP	Inefficiency with the Traditional Approach	How the Tool/Platform/Product Addresses the Inefficiency
Environmental	Due to the analogue nature of traditional construction schedules, construction may take a longer time and this can lead to more intensive use of equipment and machinery, increasing carbon emissions.	The ability of the platform to automate project scheduling can significantly reduce construction time and, by extension, reduce use of equipment and machinery. This can further lead to a reduction in carbon emissions.
	Traditional construction scheduling requires extensive use of stationery.	Automated construction scheduling will significantly minimise the use of stationary. This is important, particularly when changes are to be made to the schedule as these changes can be implemented and updated in the system without the usual paper usage.
Social	Traditional project scheduling entails the use of products such as Microsoft Excel, Microsoft Project, and Oracle's P6. This process is cumbersome.	The process of construction scheduling is now easier with a better ability to create multiple scheduling variations. Furthermore, user-friendliness can also enhance the social aspect of the construction process.
	In traditional scheduling, the tools used can only create a few variations of a project schedule, and attempting to create multiple variations can take a long time. Making changes to schedules can also be tedious.	The technology adopted provides multiple variations which provides the contractor and project managers with several options to select from using the AI technology embedded. For instance, one thousand project schedules can be generated, and clients can run "what if" analysis to test potential variations in how the construction is implemented. Furthermore, changes can be made to the schedule more easily and these changes can be automatically updated across the system. This increases the ease of simulations and corrections.
	The traditional approach is prone to errors which can affect the project implementation.	The automated approach to project scheduling minimises errors that can be made in scheduling, and when errors are detected, corrections can be made more effectively and efficiently. This can also improve building quality.
Economic	Traditional project scheduling is time-consuming, particularly when there are corrections to be made or various scenarios are being simulated.	This tool allows for the use of technology to develop faster construction schedules. This can enhance project delivery within time and budget.

(Continued)

Table 5.1 (Continued)

ESEP	Inefficiency with the Traditional Approach	How the Tool/Platform/Product Addresses the Inefficiency
	Due to the inefficiency in the traditional construction schedules, there is a reasonable amount of idle time in the construction process which can create financial wastage.	The platform reduces idle time, and it ensures that the construction crew are more effectively utilised. This minimises wastage and enhances profitability.
	It is difficult to develop construction schedules that can deliver on time and budget using traditional approaches.	The machine learning of the tool means that it can improve its capabilities to better deliver on time and within budget.
Physical/spatial		

Source: Authors analysis.

5.4.1.3   Limitations

The use of this tool requires IT and digital skills. The adoption may therefore be currently limited by the generally low level of digital technology across the construction industry. Project and construction managers (and workers) will need to undergo training and upskill to be able to utilise this tool and to get the best out of it.

Further details on ALICE Technologies can be accessed from their website using this link: https://www.alicetechnologies.com/home.

5.4.2   Case Study 2: Converge

5.4.2.1   Background, Key Details and Technologies Utilised

Converge is a platform that digitises and optimises construction through the use of AI and cloud-based technologies that are powered by a suite of smart wireless sensors. The company was founded in 2014 and it is based in Central London, UK. One of the core products of the company is ConstructionDNA- an AI-based platform for understanding and optimising construction holistically on a project, portfolio, and enterprise level. By digitising the physical world using sensors, contextualising it with construction data, and applying AI, ConstructionDNA generates ground-breaking end-to-end intel and actionable insights. These insights are generated through an intelligent algorithm that combines Converge's proprietary concrete performance dataset with local weather feeds, allowing clients to understand and predict their strength of material at a >95% accuracy rate.

ConstructionDNA is the culmination of SaaS products, ConcreteDNA, StructureDNA, and LogisticsDNA. It understands construction holistically which makes it much more than a data aggregator or digital twin; it thus has the capabilities to improve sustainability, unforeseen efficiencies and improve workplace safety. ConcreteDNA for instance, improves the quality, cycle times, and efficiency with real-time wireless concrete strength data and more accurate AI predictions, while StructureDNA measures structural behaviour to optimise construction quality, speed, and whole-life performance. Additionally, LogisticsDNA optimises the logistics flow of prefabricated construction by tracking the modular assets and identifying bottlenecks, and responding to them accordingly.

Converge has other tools such as the Signal and Mesh systems. The signal system is underpinned by the Signal Sensor v2 which is the latest iteration of the fully wireless embeddable sensor that measures in-situ concrete temperature and compressive strength. This is then uploaded to the cloud in real time via the Signal Live Hub. The v2 provides an added ability to connect Tails and embeddable sensor extensions that connect to a new port v2. This further unlocks the ability to measure novel multiple data types. The Mesh system is a concrete strength sensor that is designed to collect and transmit concrete strength data to the cloud even in the toughest networking conditions. This is driving the design of a data relay system to bypass network obstruction through a cabled thermal probe, node transmitter, repeaters, and a hub.

5.4.2.2   Key Efficiencies and Solutions the Platform Provides

The solutions that Converge provides are itemised using the ESEP efficiency framework (Table 5.2).

Table 5.2 ESEP Efficiency Analysis of Converge

ESEP	Inefficiency with the Traditional Approach	How the Tool/Platform/Product Addresses the Inefficiency
Environmental	Traditional construction processes require extensive use of stationery and paperwork.	The use of AI and cloud-based storage significantly reduces the paperwork in the construction process, particularly when there is the need to make alterations and updates to designs and construction schedules.
	Due to the analogue nature of traditional construction, the process can take a long time, and the use of equipment and machinery increases carbon emissions.	The use of automated construction process significantly reduces the construction time and by extension, reduces the use of equipment and machinery. This can reduce carbon emissions.
	It is difficult to adequately predict and simulate the carbon footprint of a construction process in order to identify the areas with high emission prospects and to mitigate them.	The tool provides a holistic insight into a construction project which makes it easier to identify environmentally inefficient aspects of the project and to mitigate them.
Social	There is a high level of uncertainty associated with construction projects and changes to one aspect of the construction can affect other aspects. This can cause anxiety for project and construction managers.	By digitising the construction planning and management process and using AI predictions, the project timeline and processes are more accurately predicted and potential issues can be quickly identified and mitigated effectively and efficiently. This can minimise the concerns and anxiety that the project and construction managers may have.

(Continued)

Table 5.2 (Continued)

ESEP	Inefficiency with the Traditional Approach	How the Tool/Platform/Product Addresses the Inefficiency
	Ensuring that the quality of work and standards are adhered to is difficult using traditional construction planning and management approaches. Furthermore, monitoring the level of work and adherence to standards and predefined targets can also be challenging.	The tool can enable the project management team to monitor the quality of work and other standards in real-time which will ensure that predefined targets and set standards are met and adhered to.
Economic	Construction planning, management, execution, and monitoring take considerable time.	With sensors and AI, the construction planning, management monitoring, and the entire process can be completed in a shorter time due to more streamlined activities and processes. Has the capabilities to improve sustainability, unforeseen efficiencies and improve workplace safety. ConcreteDNA for instance, improves the quality, cycle times, and efficiency with real-time wireless concrete strength data and more accurate AI predictions, while StructureDNA measures structural behaviour to optimise construction quality, speed, and whole-life performance.
	Due to the inefficiency in the traditional construction processes, there is a reasonable amount of idle time in the construction process which can create financial wastage.	The platform reduces idle time, and it ensures that the construction process is more effectively executed. This minimises wastage and enhances profitability.
	Construction plans hardly deliver on budget using traditional approaches.	The machine learning of the tool means that it can improve the construction process to better deliver on budget.
Physical/spatial		

Source: Authors analysis.

5.4.2.3   Future Prospects and Limitations

With milestone alerts, materials performance analytics, BIM integration, and AI predictions, Converge has the potential to deliver real-time digital twin of concreting to digitise, optimise, and further decarbonise construction materials and processes. Project and construction managers (and workers) will however need to undergo training and upskill to be able to utilise this tool and to get the best out of it. Furthermore, the use of complex digital systems, if not holistically reviewed, may increase carbon emissions generated through heightened digitisation such as cloud-based systems.

Further details on Converge can be accessed from their website using this link: https://www.converge.io/.

5.4.3 Case Study 3: Built Robotics Exosystem

5.4.3.1 Background, Key Details, and Technologies Utilised

Exosystem is a heavy equipment robotic system developed by Built Robotics. It is a building solution system that employs the use of bulldozers that have computer systems attached to them to perform highly sensitive and potentially dangerous tasks that the human workman would be unable to safely perform on-site. The robot is an excavator with a software and hardware combination mounted, making it a fully autonomous robot.

Built Robotics started studies into autonomous work back in 2016 and the robotic integration at that time was done on a continuous track loader. From 2017 to 2018, the technology was tested in bulldozers, and excavators were automated between 2018 and 2019. Due to the tedious and divergent tasks that Exosystem performs, it is sturdy, adaptable to different uses, and easy to service and maintain. This equipment is also valuable in light of the growing use of excavators in the built environment and the popularity of trenching as a useful earth-moving process. For optimal performance, there are special sensors attached to the body of the excavator. These special sensors include gadgets like GPUs, IMUs, and cameras. They enable it to have a better 360-degree pack of sensibilities. The software for the machine is installed and then interfaces with the robot via the Electro-Hydraulics system which directs the robot that is controlled by the user. The robot is controlled through a remote software called Everest. The robot becomes fully autonomous once the Exosystem is installed in the excavator.

5.4.3.2 Key Efficiencies and Solutions the Platform Provides

The solutions that Exosystem provides are itemised using the ESEP efficiency framework (Table 5.3).

Table 5.3 ESEP Efficiency Analysis of Exosystem

ESEP	Inefficiency with the Traditional Approach	How the Tool/Platform/Product Addresses the Inefficiency
Environmental	Difficulty in ensuring and measuring the adherence to environmental standards on construction sites.	Due to the automated nature of the tool, it can be programmed to operate in accordance with environmental standards and this can be monitored in real-time and also retrospectively.
Social	Construction workers taking part in excavation and demolition are exposed to accidents resulting from cave-ins, falling objects, equipment accidents and loss of life.	This tool addresses the danger in construction types where human lives are put at risk by minimising accidents and loss of lives.
	Fatigue of site workers can lead to poor handling of traditional mechanical bulldozers which can lead to poor judgement and decisions with potentially fatal consequences.	The tool is able to work for longer hours and still perform optimally with minimal chances of error and a higher level of precision and accuracy further minimising dangerous and potentially fatal mistakes.

(Continued)

Table 5.3 (Continued)

ESEP	Inefficiency with the Traditional Approach	How the Tool/Platform/Product Addresses the Inefficiency
	Construction workers particularly those operating the traditional excavators and other related tools are exposed to dust and carbon inhalation.	The tool can be operated remotely which minimises the exposure of the operator to dust and carbon inhalation.
	Repetitive actions on construction sites such as excavation and demolition can become boring, and it can lead to the operator losing interest in the task; this can lead to job dissatisfaction.	With the programming of the tool, the operator does not need to engage in the repetitive task and he/she can spend time engaging in a more diverse range of higher-level tasks that require innovation and intelligence-driven tasks.
Economic	Operators of mechanical bulldozers will need to stay on the equipment throughout the whole length of the task.	With the programming feature, the operator does not need to stay on the equipment throughout the length of the task. This can enable the operator to engage in other tasks and activities saving time and human resources.
	Generating and transforming data from the traditional approach to excavation and demolition is difficult and expensive.	Data is automatically generated and stored during the operation of the equipment at no extra cost and more efficiently. This can save time and money that would have been expended on manual data collection and the insight from the data can be used to improve the economic efficiency of operations.
Physical/ spatial	In traditional site preparation and excavation, it is difficult to achieve the levelling of the land with precision. Furthermore, this process may sometimes affect soil stabilisation and can also lead to the destruction of underground utilities.	The tool removes debris and levels the land with a higher level of precision, ensuring the stability of the site, neighbouring buildings and slope stability.

Source: Authors analysis.

5.4.3.3   Future Prospects and Limitations

The programming of the Exosystem requires a high level of skill, and specialised hardware and software functions are required. This also creates roles such as Robotic Equipment Operator (REO) saddled with the responsibility of getting the robot to function correctly and appropriately. The operator also is the maintenance officer for the robot and makes sure that it is always in the best condition. Identifying individuals and professionals with the appropriate skill set and experience is therefore a challenge. Furthermore, building robotics is expensive, and requiring high financial commitment.

This suggests that equipment such as Exosystem is still being used on relatively few construction sites.

Further details on Exosystem can be accessed from their website using this link: https://www.builtrobotics.com/technology/exosystem.

5.4.4   Case Study 4: CarbiCrete

5.4.4.1   Background, Key Details, and Technologies Utilised

CarbiCrete is a carbon removal technology company that is developing innovative and cost-effective construction solutions to reduce greenhouse gas emissions associated with construction. The company has developed a patented technology, initially created at McGill University, which allows for the production of cement-free concrete with a carbon-negative impact. Instead of using cement, CarbiCrete's process incorporates slag, an industrial by-product from steel factories, as a binding ingredient in precast concrete products. During the process, carbon dioxide (CO_2) is injected into the fresh concrete, providing strength while permanently sequestering CO_2 within the resulting products. CarbiCrete licences its technology to concrete manufacturers and monitors the necessary retrofitting to implement the process. The company's main offices and research and development facility are situated in Lachine, Quebec. Additionally, a pilot project is currently underway at Patio Drummond, a hardscape manufacturer in Drummondville, Quebec, where production is being scaled up to 25,000 CMUs (concrete masonry units) per day.

CarbiCrete was born out of a need to reduce greenhouse gases produced during the construction development process. Cement which is the binding ingredient in concrete is responsible for close to a tenth of all greenhouse gas emissions stemming from building construction materials. Mehrdad Mahoutian was a PhD student at McGill University when he founded CardiCrete in 2016. While embarking on his PhD research, he developed a process to eliminate the use of carbon in the concrete mix. The company entered a competition run by Carbon XPRIZE in a bid to get innovators to enhance the quest for carbon emissions. Harsco (a steel materials processing company) and Innovobot (a venture capital firm) got interested and contributed to the globalisation of CarbiCrete. The company received further support and funding from Transition Énergétique Quebec and Sustainable Development Technology Canada. About four years after it was founded, CarbiCrete opened its corporate headquarters in Montreal, Canada in late 2020 and also opened its research and development facility in the same year. The company has grown from nine employees in 2016 to 24 employees in 2022. A pilot facility was opened in Drummondville in January 2021, and the company has begun its second stage of its pilot. This is the commercialisation stage, and it will ensure that carbon-negative CMUs (Concrete Masonry Units) go out globally.

5.4.4.2   Key Efficiencies and Solutions the Platform Provides

The solutions that CarbiCrete provides are itemised using the ESEP efficiency framework (Table 5.4).

Table 5.4 ESEP Efficiency Analysis of CarbiCrete

ESEP	Inefficiency with the Traditional Approach	How the Tool/Platform/Product Addresses the Inefficiency
Environmental	Cement is a core building material and it makes a significant contribution to carbon emissions.	The carbon removal technology in CarbiCrete enables more environmentally friendly concrete use. It offers concrete quality with a carbon-negative footprint.
	Traditional use of concrete for building construction does not have a methodology to accurately capture the carbon emission associated with the concrete in use.	Validating CarbiCrete's net carbon-negative impact involves performing an industry-standard Lifecycle Analysis (LCA) which accounts for transportation and conversion impacts.
Social	The use of cement in concrete has problems such as durability, water absorption, and lower compressive strength.	It offers a reliable building material with better mechanical and durable properties, the same water absorption properties, higher compressive strength, and better freeze/thaw resistance.
	Building construction using cement-based concrete has ethical challenges.	The company works with local concrete manufacturers to produce cement-free CMUs. The use of steel slag instead of cement makes for ethical and affordable mitigation in construction projects.
Economic	Building construction using cement-based concrete is expensive.	Working with local concrete manufacturers and the use of steel slag instead of cement provides an affordable solution to the high costs associated with cement-based concrete. Slag is inexpensive and in high supply.
Physical/spatial		

Source: Authors analysis.

5.4.4.3   Future Prospects and Limitations

Alongside tree planting, the reduction in the use of gas-powered cars, the shift from using fuels in power plants to using renewable energy and other sundry climate change actions backed globally, the adoption of CarbiCrete as a construction development solution is gaining traction and it can play an important role in driving down the carbon footprint associated with building construction. There are however socio-cultural challenges as contractors and builders may not be ready to transit away from cement, considering that this has been a core concrete ingredient for so long. The adoption of this technology may also involve higher costs which contractors and their clients may not be ready to incur.

Further details on CarbiCrete can be accessed from their website using this link: https://carbicrete.com/.

5.4.4.5   Other Cases (Exercises for Students)

This section contains a short summary of case studies. Readers are encouraged to develop these case studies and analyse them using the ESEP efficiency framework (taking a similar approach to the previous section).[1]

5.4.4.5.1   ARCHDESK

ARCHDESK is an adaptable project management platform that tracks the performance of projects on real-time dashboards, builds processes, and creates reports. It focuses on integrating processes and data on construction and aids in communication and resolution of issues on a construction site. It also enhances finance management, operations, estimation, and scheduling on a site. Furthermore, its cloud-based feature enables various professionals that are working on a project to work seamlessly and collaborate on a project in a decentralised way from various locations.

Further details on Archdesk can be accessed from their website using this link: https://archdesk.com/.

5.4.4.5.2   MEASURABL

Measurabl is a digital tool that enhances building efficiency by identifying various ways through which building operating costs can be reduced. It offers a wide range of features and functionality to power sustainability measurement, management, and reporting across a real estate portfolio. The services include data automation, performance tracking, target setting, customised reports on demand physical climate risk insights, and benchmarking and certification. The insights provided by the platform can improve the decision-making process on a building based on the data generated and analysed. These data can further support ESG target setting, monitoring, and assessment.

Further details on Measurabl can be accessed from their website using this link: https://www.measurabl.com/.

5.4.4.5.3   OPENSPACE

OpenSpace is a platform that provides construction and project management support and enhances efficiency and collaboration with intuitive tools that elevate teamwork to new heights. Using cutting-edge technology akin to AI navigation systems in self-driving cars, it empowers construction project teams to work faster and collaboratively. The image-based workflows replace text-based systems, eliminating confusion and unlocking unprecedented efficiencies. The Vision Engine tool enables project teams to easily document the progress of a project by attaching a camera and recording in the app and mapping photos to the plans automatically while the project manager or his/her designate walks through the site. This provides a complete as-built record of the building's journey from pre-construction to handover and operation. OpenSpace's capabilities encompass tools like BIM Compare, Split View, and AI-powered OpenSpace Track, ensuring progress tracking, work verification, improved coordination, and reduced risk. The integration of features like Field Notes facilitates communication, QA/QC documentation, and streamlined workflows, whether on-site or in the office.

Further details on OpenSpace can be accessed from their website using this link: https://www.openspace.ai/.

5.4.4.5.4   PROCORE

Procore is a cloud-based building construction information data management platform that provides all members of the building project team with all information and data necessary for the successful execution of the project. This digital system minimises errors in communication and enables access to several members of the construction project team. The cloud-based feature particularly enhances access to all members of the construction team from remote locations. The platform uses an intelligent AI system to integrate data from various sources including site plans and drawings, and the output can enable construction teams to take control of their construction outcomes and minimise risk, cost, and wastage while saving time.

Further details on Procore can be accessed from their website using this link: https://www.procore.com.

5.4.4.5.5   RABBET

Rabbet enables developers to manage their entire portfolio in one complete system. Reliable project data in real-time ensures the ability, clarity, and foresight necessary for growth. This technology applies to real estate development, project financing, budget management, construction lending, and contract tracking. Rabbet reduces the time spent managing project finances, digitises information and automates paper processes to improve accuracy, and centralises project finances for better reporting across projects, thereby improving the efficiency, visibility, and collaboration for the project team and reducing the administrative burden on developers.

Further details on Rabbet can be accessed from their website using this link: https://rabbet.com/.

5.4.4.5.6   SOIL CONNECT (EREGULATORY)

Soil Connect is a software and construction service platform that helps contractors, builders, and haulers with their bulk material supply, demand, and transportation needs. It was launched in 2019. the platform is a SaaS-enabled marketplace for construction professionals to locate, transport, and acquire soil aggregates. The platform provides services and products such as digital dirt brokerage, eTickets (capturing, tracking, and sharing the details of hauling materials from one destination to the next), Marketplace (finding and messaging bulk material suppliers and buyers), and eRegulatory (capturing and saving essential truck, load, and route data). The eRegulatory is particularly useful for enabling users to solve issues with paper manifests and regulatory needs. Soil Connect leverages technology to make construction-related dirt transactions easier and it has been applied to landscaping, waste management, the environmental management.

Further details on Soil Connect can be accessed from their website using this link: https://www.soilconnect.com/.

5.4.4.5.7   TRIMBLE VIEWPOINT

Trimble Viewpoint is a cloud-based construction management app that is designed to help contractors better manage construction projects. The scope of relevance of the app spans materials management, building methods management, project timeline management, real-time accounting, human resource management, data analysis, and business intelligence solutions, all in real-time. It employs a communication solution and a data analysis edge to enable stakeholders to study the analytics and data while tracking the progress of other departments on the project. Because it is also mobile-based, it is easy for all members of the building construction team to be up to date with the necessary information concerning the project to be well-informed and to make better-informed decisions. This tool is used for commercial and residential properties.

Further details on Trimble Viewpoint can be accessed from their website using this link: https://www.viewpoint.com/.

5.5   Chapter Summary

There is an increasing need for built environment professionals particularly those in building construction and maintenance to improve their digital competency, awareness, and practical skills that support them to cope with the ever-changing technological advancements in the construction industry. This requires deep knowledge of the built environment and specifically construction, but also some computer science knowledge, skills, and awareness. This chapter provides an understanding of issues of digital construction and property maintenance and their economic, social, environmental, and physical implications throughout the lifecycle of a building (ideation, design, construction/retrofit, operation, and decommissioning). It has also provided some foundational knowledge on digital design and optimal solutions to enhance productivity and achieve a sustainable, smart, and responsive built environment.

Note

1 Readers are further encouraged to analyse other products, tools and platforms in this area using the ESEP efficiency framework.

References

Autodesk, & CIOB. (2020). *Reimagining construction.* https://www.autodesk.co.uk/campaigns/ciob-reimagining-future-of-construction/paper/reimagining-construction-paper.

Baiche, B., Walliman, N., & Ogden, R. (2006). Compliance with building regulations in England and Wales. *Structural Survey, 24*(4), 279–299. https://doi.org/10.1108/02630800610704427.

Barrelas, J., Ren, Q., & Pereira, C. (2021). Implications of climate change in the implementation of maintenance planning and use of building inspection systems. *Journal of Building Engineering, 40*(May), 102777. https://doi.org/10.1016/j.jobe.2021.102777.

Bilal, M., Oyedele, L. O., Qadir, J., Munir, K., Ajayi, S. O., Akinade, O. O., Owolabi, H. A., Alaka, H. A., & Pasha, M. (2016). Big Data in the construction industry: A review of present status, opportunities, and future trends. *Advanced Engineering Informatics, 30*(3), 500–521. https://doi.org/10.1016/j.aei.2016.07.001.

Bimobject (2022). Trade Shows are Cancelled. How to Navigate the New Digital Landscape (Part 2). Available online at https://business.bimobject.com/resources/?f=1544&s=resources (accessed 4 May 2023).

Boland, H. (2020, October 20). How 3D printed homes could solve Britain's housing crisis. *The Telegraph.* https://www.telegraph.co.uk/technology/2020/10/20/3d-printed-homes-could-solve-britains-housing-crisis/.

Brown, R. (2018). *PropTech: A guide to how property technology is changing how we live, work and invest.* Casametro.

Cao, Y., Wang, T., & Song, X. (2015). An energy-aware, agent-based maintenance-scheduling framework to improve occupant satisfaction. *Automation in Construction, 60,* 49–57. https://doi.org/10.1016/j.autcon.2015.09.002.

Ching, F. D. K. (2023). *ARCHITECTURE: Form, space and order* (5th ed.). John Wiley & Sons, Ltd. https://books.google.co.uk/books?hl=en&lr=&id=aNy4EAAAQBAJ&oi=fnd&pg=PP11&dq=importance+of+architectural++design&ots=G0dlFR6XPi&sig=Ve_h316GLVF7Gy5at-KxU2xE-4z0&redir_esc=y#v=onepage&q=importance of architectural design&f=false.

Chiponde, D. B., Ziko, J. M., Mutale, L., & Chipalabwe, V. (2017). An investigation of health and safety practices for the public during construction in the Zambian construction industry. In F. Emuze, & M. Behm (Eds.), *Towards better safety, health, wellbeing and life in construction.* 134–142 CIB, Department of Built Environment, Central University of Technology.

Dauda, J. A., & Ajayi, S. O. (2022). Understanding the impediments to sustainable structural retrofit of existing buildings in the UK. *Journal of Building Engineering, 60*(June), 105168. https://doi.org/10.1016/j.jobe.2022.105168.

Dauda, J. A., Rahmon, S. A., Tijani, I. A., Mohammad, F., & Okegbenro, W. O. (2022). Design optimisation of reinforced concrete pile foundation using generalised reduced gradient algorithm. *Frontiers in Engineering and Built Environment, 2*(3), 133–153. https://doi.org/10.1108/febe-12-2021-0059.

Du, Z., Fan, B., Jin, X., & Chi, J. (2014). Fault detection and diagnosis for buildings and HVAC systems using combined neural networks and subtractive clustering analysis. *Building and Environment, 73*, 1–11. https://doi.org/10.1016/j.buildenv.2013.11.021.

Grand View Research. (2023a). *Smart home market size, share & trends analysis report by products (lighting control, security & accesscontrols), by application (new construction, retrofit), by protocols (wireless, wired), by region, and segmentforecasts, 2023-2030.* https://doi.org/10.18356/9ba8eaeb-en.

Grand View Research. (2023b). *Smart home market size to reach $537.01 billion by 2030.* https://www.grandviewresearch.com/press-release/global-smart-homes-market#.

Grand View Research. (2023c). *Smart home security camera market to reach $30.10 Billion by 2030.* https://www.grandviewresearch.com/press-release/global-smart-home-security-camera-market#.

Han, S., Na, S., & Lim, N. G. (2020). Evaluation of road transport pollutant emissions from transporting building materials to the construction site by replacing old vehicles. *International Journal of Environmental Research and Public Health, 17*(24), 1–15. https://doi.org/10.3390/ijerph17249316.

Hauashdh, A., Jailani, J., Rahman, I. A., & AL-fadhali, N. (2021). Structural equation model for assessing factors affecting building maintenance success. *Journal of Building Engineering, 44*(May), 102680. https://doi.org/10.1016/j.jobe.2021.102680.

Hauashdh, A., Jailani, J., Rahman, I. A., & AL-fadhali, N. (2022). Strategic approaches towards achieving sustainable and effective building maintenance practices in maintenance-managed buildings: A combination of expert interviews and a literature review. *Journal of Building Engineering, 45*(October 2021), 103490. https://doi.org/10.1016/j.jobe.2021.103490.

Health and Safety Executive. (2023). *Statistical count of in-year work-related deaths where HSE is the relevant enforcing authority.* Statistical Count of In-Year Work-Related Deaths Where HSE Is the Relevant Enforcing Authority. https://www.hse.gov.uk/statistics/fatalquarterly.htm.

Jeyamohan, D. C., & Clarke, S. (2022). *Green quadrant IoT platforms for smart buildings 2022* (Issue January). https://www.siemens.com/global/en/products/buildings/contact/green-quadrant-iot-platforms-smart-buildings.html.

Kim, W., & Katipamula, S. (2018). A review of fault detection and diagnostics methods for building systems. *Science and Technology for the Built Environment, 24*(1), 3–21. https://doi.org/10.1080/23744731.2017.1318008.

KPMG. (2015). Climbing the curve. In *2015 global construction project survey.* https://kpmg.com/building.

Kreiger, E. L., Kreiger, M. A., & Case, M. P. (2019). Development of the construction processes for reinforced additively constructed concrete. *Additive Manufacturing, 28*(October 2018), 39–49. https://doi.org/10.1016/j.addma.2019.02.015.

Lee, H. H. Y., & Scott, D. (2009). Overview of maintenance strategy, acceptable maintenance standard and resources from a building maintenance operation perspective. *Journal of Building Appraisal, 4*(4), 269–278. https://doi.org/10.1057/jba.2008.46.

Lin, M., Afshari, A., & Azar, E. (2018). A data-driven analysis of building energy use with emphasis on operation and maintenance: A case study from the UAE. *Journal of Cleaner Production, 192,* 169–178. https://doi.org/10.1016/j.jclepro.2018.04.270.

Madureira, S., Flores-Colen, I., de Brito, J., & Pereira, C. (2017). Maintenance planning of fa-cades in current buildings. *Construction and Building Materials, 147,* 790–802. https://doi.org/10.1016/j.conbuildmat.2017.04.195.

Meta. (2021). *2021 sustainability report.*

Molloy, M. (2017, March 3). This incredibly cheap house was 3D printed in just 24 hours. *The Telegraph.* https://www.telegraph.co.uk/technology/2017/03/03/incredibly-cheap-house-3d-printed-just-24-hours/.

Muthumanickam, N. K., Duarte, J. P., & Simpson, T. W. (2023). Multidisciplinary concurrent optimization framework for multi-phase building design process. In *Artificial intelligence for engineering design, analysis and manufacturing: AIEDAM, 37.* https://doi.org/10.1017/S0890060422000191.

Pan, W., & Garmston, H. (2012). Building regulations in energy efficiency: Compliance in England and Wales. *Energy Policy, 45*(2012), 594–605. https://doi.org/10.1016/j.enpol.2012.03.010.

Park, S., & Kim, I. (2015). BIM-based quality control for safety issues in the design and construc-tion phases. *Archnet-IJAR, 9*(3), 111–129. https://doi.org/10.26687/archnet-ijar.v9i3.881.

Porter, L., Fields, D., Landau-Ward, A., Rogers, D., Sadowski, J., Maalsen, S., Kitchin, R., Dawkins, O., Young, G., & Bates, L. K. (2019). Planning, land and housing in the digital data revolution/ the politics of digital transformations of housing/digital innovations, PropTech and Housing–the view from Melbourne/digital housing and renters: Disrupting the Australian rental bond system and Te. *Planning Theory and Practice, 20*(4), 575–603. https://doi.org/10.1080/14649357.2019.1651997.

Roth, K. W., Westphalen, D., Llana, P., & Feng, M. (2004). The energy impact of faults in US com-mercial buildings. *International Refrigeration and Air Conditioning Conference* (pp. 600–609). http://docs.lib.purdue.edu/cgi/viewcontent.cgi?article=1664&context=iracc&sei-redir=1&ref erer=http://www.google.ru/url?sa=t&rct=j&q=Energy+Consumption+Characteristics+of+Com mercial+Building+HVAC+Systems+Volume+1&source=web&cd=6&ved=0CFAQFjAF&url=http %253A.

Sha'ar, K. Z., Assaf, S. A., Bambang, T., Babsail, M., & Fattah, A. M. A. El. (2017). Design–construction interface problems in large building construction projects. *International Journal of Construction Management, 17*(3), 238–250. https://doi.org/10.1080/15623599.2016.1187248.

Shin, H., Lee, H.-S., Park, M., & Lee, J. G. (2018). Facility management process of an office building. *Journal of Infrastructure Systems, 24*(3). https://doi.org/10.1061/(asce)is.1943-555x.0000436.

Sierra, F. (2022). COVID-19: Main challenges during construction stage. *Engineering, Con-struction and Architectural Management, 29*(4), 1817–1834. https://doi.org/10.1108/ECAM-09-2020-0719.

Tan, Z., & Miller, N. G. (2023). Connecting digitalization and sustainability: Proptech in the real estate operations and management. *Journal of Sustainable Real Estate, 15*(1), 1–15. https://doi.org/10.1080/19498276.2023.2203292.

Turner, M., Mills, T., Kleiner, B., & Lingard, H. (2017). Suicide in the construction industry: It's time to talk. In F. Emuze, & M. Behm (Eds.), *Towards better safety, health, wellbeing and life in construction.* CIB Department of Built Environment, Central University of Technology, 45–55.

University of Oxford Research. (2017). *PrepTech 3.0: The future of real estate.*

Venkatarama Reddy, B. V., & Jagadish, K. S. (2003). Embodied energy of common and alternative building materials and technologies. *Energy and Buildings*, *35*(2), 129–137. https://doi.org/10.1016/S0378-7788(01)00141-4.

Witra (2022). Construction. Witra IoT Out of the Box Case Study.

Yan, H., Shen, Q., Fan, L. C. H., Wang, Y., & Zhang, L. (2010). Greenhouse gas emissions in building construction: A case study of One Peking in Hong Kong. *Building and Environment*, *45*(4), 949–955. https://doi.org/10.1016/j.buildenv.2009.09.014.

Zhou, W., Whyte, J., & Sacks, R. (2012). Construction safety and digital design: A review. *Automation in Construction*, *22*, 102–111. https://doi.org/10.1016/j.autcon.2011.07.005.

Chapter 6

Digital Technology and Innovation in Real Estate Funding, Finance, and Investment

6.1 Overview of Real Estate Funding, Finance, and Investment

Real estate is capital intensive, and investment in the sector, either through the investment or through the development market, requires a significant level of funding. This section provides an overview of three core areas of this segment of the real estate value chain, namely, funding, finance, and investment.

6.1.1 Real Estate Funding and Finance

Real estate funding and finance refer to the various methods that are used to acquire, develop, and invest in real estate assets and the financial analysis and management associated with these activities. Funding involves the process of securing capital or funds, and it plays an important role in the real estate industry, enabling individuals, developers, and investors to participate in the market and achieve investment objectives. The two-core real estate funding approaches are equity and debt financing. Equity finance involves the use of personal equity such as savings or assets to fund real estate projects. It may also entail private equity that involves obtaining capital from institutional or high-net-worth investors in exchange for equity, profit sharing, or equity ownership. Equity financing may also be through joint ventures that involve partnering with other investors to pool financial resources and share risk and rewards in a real estate project (Glickman, 2014a). In order to attain an optimal capital structure, a suitable debt-to-equity ratio is important.

Funding real estate through debt is often considered one of the most efficient approaches to funding. For a real estate market to be efficient and functional, it must be built upon a robust debt market (Phillips, 2009). Real estate debt finance entails the use of borrowed funds or loans to finance a real estate project. In addition to the tax advantage it provides, it has been found to have about twice the impact on firms' leverage as other types of tangible assets when it is used as collateral (Giambona et al., 2014). Real estate finance may be accessed through loans from banks and financial institutions, usually requiring collateral and repayment with interest. There are also construction loans tailored for financing construction or renovating properties usually with funds disbursed at the various stages of the project. Additionally, mortgage loans can be used specifically for property acquisition or refinancing, with the property serving as collateral. In cases where the traditional debt financing options are unavailable or for quick acquisition, hard-money loans (short-term high-interest loans based on the property's value) and mezzanine financing (combines debt and equity financing) can also be sought. In the case of mezzanine financing, lenders

DOI: 10.1201/9781003262725-8

provide loans that are subordinate to primary mortgages by senior to equity investments, and the interest rates will usually be higher than traditional loans.

In many cases, real estate projects are also financed through the government and public institutions. This includes loans from housing administrations and authorities, which usually require a lower down payment and other more favourable credit criteria. An example of this is the Help-to-Buy equity loan in the UK, which allows prospective homeowners to purchase their homes with a five-year interest-free loan. It also includes government initiatives, grants, subsidies, tax credits, low-interest loans for affordable housing or urban development projects, etc. Some countries such as the United States also have Veteran Affairs (VA) loans that enable eligible military veterans to access mortgage facilities at more favourable terms and lower down payment options. Additionally, the government is also increasingly getting involved in public–private partnerships (PPPs), which are collaborative arrangements between public entities and private developers to finance and develop real estate projects.

There are several other alternative sources of real estate finance. An example is Real Estate Investment Trusts (REITs). REITs are companies that own, operate, or finance income-generating real estate, allowing investors to purchase shares and receive dividends in return. There are also private placements (where direct offerings of securities are made to a select group of accredited investors to raise capital for real estate ventures) and seller financing where property owners act as lenders, providing financing options to buyers, typically through mortgages or instalment contracts. Many REITs are public companies whose securities are traded on the national stock exchange (Glickman, 2014a).

Real estate finance deals with the funding and management of real estate properties and projects. It includes financial analysis, strategies, and decision-making that are involved in acquiring, developing, operating, and investing in real estate. This enables individuals, developers, investors, and institutions to maximise returns, manage risks, and optimise the use of capital. Real estate funding includes investment analysis comprising cash flow analysis, return on investment (ROI), and risk assessment.[1] It also entails real estate valuation comprising the property appraisal and cap rates determination[2] (real estate valuation and appraisal are extensively discussed in Chapter 7). Furthermore, it involves financial instruments and structures, such as mortgages, securitisation, REITs, and syndication,[3] and risk management and financial analysis, such as financial modelling, risk mitigation, and due diligence.[4]

Real estate funding and finance are complex and dynamic, and they generally depend on the overall economic climate, institutional factors, political factors, legal factors, the nature of the project, the property location, real estate market conditions, borrower's credit worthiness, etc. Glickman (2014b) provides further details on property finance.

6.1.2   *Real Estate Investment, Portfolio, and Asset Management*

Real estate investment refers to real estate assets or a congregate of assets that are owned or operated to generate income. Although the global real estate value as of 2020 was estimated at $228 trillion, the largest proportion of the stock is a privately owned residential estate ($90 trillion); land, public assets, and owner-occupied commercial real estate make up $103 trillion (Pi Labs, 2020a). Pi Labs (2020a) further reveals that the global investable stock of commercial property owned by institutional investors is approximately $35 trillion – about half the size of the global stock market capitalisation. While 13% of these funds are kept by publicly listed REITs and property companies, 8% of these funds are owned by private funds (core and private equity formats), and 79% of these funds are held directly (either in separate accounts or self-managed). These data highlight the importance of the real estate investment market and underscore how important it is that real estate investment funds are efficiently allocated and managed.

The largest global real estate investors are pension funds, insurance companies, and sovereign wealth funds (government funds), and they are attracted to real estate investment due to the reliable returns, healthy yield premium over bonds, and relatively low correlation with broader capital markets. When funds are raised, investors require the fund managers to keep providing regular and periodic updates on their investment activities and performance. The larger institutional investors particularly demand strategic relationships with their managers, leveraging their access to research, analytics, and advice. Institutional real estate investment managers traditionally control the structuring of investment vehicles and limit access to those vehicles to qualified investors. They are statutorily mandated to undertake know your customer (KYC) and anti-money laundering (AML) checks on investors who wish to participate in a fund to ensure the legitimacy of the funds that they attract.

The real estate investment market is broad and complex. "It brings a supply of existing and new investments coming into the market into equilibrium with the demand for investments backed by money" (Blackledge, 2016: 15). One of the core functions of investment and fund management is capital deployment. This refers to the strategic allocation and utilisation of funds or capital in real estate investment activities. This process entails identifying and acquiring real estate assets as well as managing and optimising the deployment of capital to generate returns for investors. For capital to be deployed, the investment manager needs to define the investment strategy that outlines the objectives, risk tolerance, and target returns of the real estate investment management firm or investor. This strategy guides the allocation of capital towards specific types of real estate assets, such as residential, commercial, industrial, or mixed-use properties. Stock selection significantly affects portfolio performance, and portfolio managers ensure stock selection and tactical asset allocation to successfully manage their funds (Jackson and Orr, 2011). The investment or fund manager is also responsible for deal sourcing and acquisition, which involves identifying and sourcing potential real estate investment opportunities. The fund manager will need to conduct market research, analyse property performance, assess the financial viability of potential acquisitions, etc. to ensure that the properties that are acquired align with the investment strategy and offer favourable risk-return profiles.

Investment managers also need to ensure due diligence on potential real estate investments before deploying capital. This includes property inspections, financial analysis, legal reviews, and assessment of market conditions. They will also need to undertake underwriting that involves evaluating the investment's financial projections, cash flows, potential risks, and sensitivity analysis to determine its suitability for capital deployment. The fund manager supervises the negotiation in the acquisition stage, which, according to Saull and Baum (2019), is from the time of the first advertisement of a proposed sale to the agreement of a price. After negotiation has taken place and a price has been agreed, the transaction itself takes place. Property transaction requires a lot of legal work in sales and purchase. Solicitors are required for the conveyance, which is the administrative process through which rights over landed property are created and transferred. This includes the period of due diligence and post-exchange. Some aspects of commercial conveyance such as workflow are digitised, and this offers a more efficient transfer of information than a paper-based process.

When an investment has been selected, the fund manager supervises the capital deployment and the funds allocation. This decision may be based on factors such as the investment's size, investment strategy, market conditions, risk profile, and investment concentration. The investment manager is also responsible for determining the finance and funding options to be adopted such as equity finance, debt finance, or a hybrid, depending on the investment strategy and risk profile. After an asset has been purchased, the investment manager ensures that the asset's returns are maximised while risks are minimised and mitigated. This involves actively managing the property improvements, renovations, tenant enhancements, and other initiatives aimed at enhancing the property's value and generating higher returns.

Ultimately, the investment managers ensure a general portfolio optimisation, which includes the allocation of capital across the entire portfolio of investments. This involves continuously evaluating the performance of individual properties and projects and adjusting capital allocation based on market conditions, investment objectives, and risk considerations. Portfolio management requires an in-depth understanding of risks and their impact on the cash flow potential of a property that can meet the fund's goals. Asset returns within a portfolio have varying correlations with each other, and it is often argued that the total portfolio risk should be less than the sum of each asset's individual risk. Furthermore, although diversification is often adopted as a risk mitigation or risk management strategy, particularly in international real estate investment, some risks are systemic and therefore cannot be "diversified away". Investors and fund managers in a recent survey indicated that the biggest systemic risks currently facing real estate are climate change and government regulation (SEI, 2020). This, therefore, calls for a high level of expertise and experience in carrying out investment management functions.

6.2   Inefficiencies in Traditional Real Estate Funding, Finance, and Investment

Due to risk and uncertainty and high capital associated with real estate investment, finance, and development, these financial segments of the real estate value chain require effective and efficient approaches, techniques, and methods. There are several challenges and inefficiencies associated with the traditional approach to the various financial aspects of real estate; these will be discussed in line with the three core service areas, namely, funding, finance, and investment.

6.2.1   Real Estate Funding and Finance

Real estate funding and finance have several challenges and limitations, which can be classified as capital, systemic, and technical.

6.2.1.1   Capital Constraints

Limited access to capital and high capital costs are some of the primary challenges in real estate funding. Banks and lending institutions generally have strict and often bureaucratic lending criteria and processes, making it difficult for individuals and organisations to secure funding for their projects or investments. Some of these criteria may be more stringent for individuals who are usually required to have high credit scores and make significant down payments. Individuals with a limited credit history or insufficient funds for a down payment often find it challenging to secure finance for development or investment. A significant proportion of real estate finance applications are made by small-scale investors or individuals who will typically engage with the mortgage market (Chandra, 2019). The traditional approach to accessing mortgage finance is that the individual will save up the down payment (equity) and apply for a loan (debt) for the outstanding property value. This is with the exception of countries like the UK where the government provides an equity loan (known as Help to Buy). Saving up for the down payment is a major challenge, particularly in big cities where there are rising house prices and lower real income.

There are also issues associated with the traditional mortgage search and application process. This process is often lengthy and complicated. It usually begins with a comparison of interest rates across financial institutions, as people move from one bank to the other. As Chandra (2019) points out, this is not ideal as several people may not have the time and resources to visit several financial institutions and will therefore just settle for the mortgage product that they have access to or one that they are able to first identify.

The traditional system includes the following:

1　Visit the bank to make a first enquiry and book an appointment for a full consultation.
2　Attend full consultation and take documentation for the meeting.
3　Fill out necessary forms and questionnaires which are usually very lengthy.
4　Wait for a response from the bank, usually via post, check documents, and sign the remaining documents.
5　Return the documents to the bank.
6　Await a response from the bank.

Regardless of whether the funding being requested is for individuals or corporate entities, traditional property funding and finance typically require detailed documentation and paperwork in most cases, and there is a very lengthy approval process. Valuation and appraisal are important elements of real estate funding, and this also adds to the time taken for the funding/finance application to be completed. These factors can negatively affect real estate transactions, expansion, and opportunities for growth, particularly in markets where demand is significantly stronger than supply; they can also serve as entry barriers to real estate investment (Saull and Baum, 2019).

Some of the challenges and limitations associated with access to capital often lead to high finance costs, which may manifest through higher interest rates. Real estate loans typically involve higher interest rates relative to other loan types. In addition to these, loan origination fees and other financing costs may also be incurred, and these can impact the ROI particularly for smaller investors or developers with limited resources. Furthermore, the profitability of a project can also be affected, particularly if the property's rental income or value does not increase at a similar rate. Additionally, real estate finance is often obtained through long-term loans that make them susceptible to interest rate fluctuations which can impact the cost of borrowing, particularly for those with variable interest rates. In the UK, for instance, the increase in interest rates between 2021 and 2023 has significantly increased individuals', households', and organisations' borrowing costs and loan payments, affecting the cost of living and the profitability of real estate investments. These challenges are exacerbated by the indivisibility and illiquidity of real estate assets, making it difficult for properties to be sold off quickly to mitigate increases in borrowing costs.

6.2.1.2   Systemic and Institutional Limitations

Systemic limitations are constraints in real estate funding and finance that are underpinned by factors and issues that are macro in nature, i.e., they are driven and dictated by countries' systems and processes such that individual projects and investments can hardly influence them. The macroeconomic climate, geopolitical structures, institutional factors, and legal framework in a country can significantly create bottlenecks in real estate funding and finance. Countries with volatile economies, unstable interest rates, and high inflation may have higher base borrowing costs due to the higher risk that would be associated with investing in them. Real estate markets can be subject to significant volatility, which introduces risks for lenders and investors. Fluctuations in property values, interest rates, and market demand can affect the ability to secure funding or repay loans. Furthermore, economic recessions can lead to a decline in property prices and an increase in loan default rates. High volatility and risk in a country can negatively affect the value and performance of the property market within an economic system. Real estate funders in such markets would therefore seek to integrate risk management protocols, and this may sometimes be associated with higher interest rates. Countries that lack transparency may also have higher borrowing costs, and access to capital may also be restrictive as lenders and investors may not possess a full understanding of the country's systems and structures. This may also lead to hidden fees, complex loan structures, and ambiguous contractual clauses, which can further cause disputes.

Financial regulation also plays a crucial role in creating checks and controls in systems and organisations. Real estate funding is subject to several regulatory and legal requirements that often create additional challenges for borrowers and lenders. The financial and banking sector is the nervous system of an open-market-based economy as it plays the roles of safekeeping of deposits, provision of credit facilities, and investment (Stiglitz, 1998). The interconnectivity between national and international financial systems (Cetorelli and Goldberg, 2012) underscores the need to have strong national financial systems. Financial institutions are expected to be shock absorbers and not shock amplifiers, hence the need to control and monitor them to ensure that they operate within the dictates of policy and regulation.

Although regulations are important aspects of financial stability (Blundell-Wignall and Atkinson, 2010), they can have adverse effects on financial systems (Dermine, 2013). The liquidity constraints that may result from financial regulations may discourage the funding of long-term real estate investment and impede portfolio diversification. This may further lead real estate funds to move to less risky sectors with moderate returns such as bonds. Additionally, compliance with zoning laws, environmental regulations, and building codes can increase project costs and timelines. Litigation and other forms of legal disputes can further exacerbate delays in real estate funding and projects. Heightened credit conditions may also reduce financial institutions' lending capacity, which can further lead to a reduction in the supply of funds that are available for projects and, by extension, heightened borrowing costs.

6.2.1.3   Technical Challenges

The technical challenges relate to issues based on competencies, knowledge, and experience. Real estate is heterogenous, and financing can involve complex financial structures such as joint ventures, mezzanine financing, or syndicates. These structures can further create challenges such as coordination among multiple practices, complex legal agreements, and differing investment objectives. Furthermore, the fact that they are long-term financial commitments makes the search for a suitable strategy difficult. Relatively new investors or those with limited experience in real estate may face challenges in securing funding. Lenders often prefer to work with experienced investors with a proven track record, making new entrants into the market difficult.

Accurately determining the worth of an interest in an asset is essential for real estate finance. However, determining the accurate value of a property is a complex and subjective task. Appraisal discrepancies can lead to challenges when seeking financing or refinancing, as lenders rely on appraisals to access loan amounts and loan-to-value ratios. Real estate finance also requires efficient cash flow management, and this can significantly affect the viability of an investment, particularly assets that are for rental purposes. It is often challenging to ensure a consistent and sufficient cash flow to cover mortgage payments, property maintenance costs, and other expenses. The cash flow, profitability, and viability of a project can be affected by unexpected repairs, vacancies, and delinquent rent payments.

The challenges highlighted can vary based on the specific market, location, and economic conditions.

6.2.2   Real Estate Investment, Portfolio, and Asset Management

Real estate investment managers are saddled with the responsibility of managing huge amounts of funds, and this is accompanied by varying issues. Although many of these investment challenges apply to small-scale and large-scale investors, some are peculiar based on the size of the investment portfolio being managed. For instance, the world's largest real estate investors are a group of 12 sovereign wealth funds and pension funds estimated to pool between $30 billion and $60 billion invested in real estate (Pi Labs, 2020a). Furthermore, the two biggest investment managers each have approximately $200 billion of assets under their management (Pi Labs, 2020a). Although this may seem like an abundance of financial resources, classic economic theory suggests that efficient resource allocation, more specifically, the allocation of funds to certain properties, may become an issue. The largest global real estate investors are pension funds, insurance companies, and sovereign wealth funds, and they are attracted due to the reliable returns, healthy yield premium over bonds, and relatively low correlation with broader capital markets. However, the market complexities, information asymmetry, high entry cost, illiquidity, difficulty in diversifying, multiple parties involved in transactions, and associated fees and cost of transactions make investment management complex. Investment in highly valued properties further exacerbates the problem of illiquidity as properties of such magnitude cannot be easily and quickly traded.

Furthermore, the huge capital in real estate investment is accompanied by peculiar systemic risk. For instance, the 2007 global financial crisis exposed some of the fundamental issues and challenges in investment and funds management. Institutional investors are frustrated by gating, closing, or valuation cuts in open-ended retail funds in 2007, 2016, and 2020, and there has been a decline in the natural supply of investment with defined benefit schemes being replaced by less real estate-friendly defined contributions. Increased regulation has also increased the stress on the middle and back office of real estate investment managers.

Regardless of the fund size, there are several challenges that all real estate investment portfolio types are susceptible to which affect their optimal performance and efficiency. These will be discussed under two broad themes, namely, economics and ESG.

6.2.2.1   Economic Issues

Economic challenges in real estate investment relate to issues that are either driven by factors relating to finance or those that affect the overall returns, costs, and time efficiency in the investment processes. Some of the economic inefficiencies are discussed below:

1 **Market volatility**: Real estate markets are dependent on the performance and stability of the economic and political environment in the country of operation. Thus, countries with volatile economic, political, and institutional systems will pose a higher level of risk to the investor. Real estate markets can go through various fluctuations, and these can be exhibited through changes in demand, supply, vacancy rates, rental values, yield and prices, and capital values. Economic downturns can lead to declines in property values, and rapid appreciation in values can lead to bubbles that may eventually culminate in bursts. It is difficult for an investor to accurately predict future market volatility using traditional modelling approaches, and although conducting market research and due diligence may help to minimise the risks associated with market volatility, it is still difficult to predict black swan events such as pandemics, natural disasters, and changes in local or national policies (Higgins, 2015, 2013, 2014; Oladiran et al., 2023). This further increases the risks of investing in a market and the higher cost of finance that may result.

2 **Finance constraints**: Sourcing and securing real estate investment capital is often challenging, particularly for new and smaller investment funds with limited resources. Lenders will usually have more stringent requirements and conditions, and these can hinder investment opportunities and growth potential.

3 **Cash flow management**: Cash flow fluctuation is one of the key problems for investment managers, and being able to accurately predict cash flows is problematic, particularly for rental properties. Rental income can fluctuate because of vacancies, rent payment defaults, and unexpected repairs. Insufficient cash flow can negatively impact the profitability and viability of an investment, and it can affect the investor's ability to meet financial obligations; thus, effective and efficient cash flow management is crucial in real estate investment management. This requires systems to be put in place to ensure consistent rental income, monitor expenses, and maintain adequate reserves for property maintenance and unexpected costs.

4 **Liquidity**: Real estate investments are typically illiquid compared to other types of investments. Disposing (selling) of a property can take a long time, and it will often cost a significant amount of money. The process is generally complex, particularly in unfavourable market conditions where the demand indicators are unfavourable. These factors generally combine to make real estate investment less liquid, and this can limit the investment manager's ability to access funds when they are needed for the adjustment of investment strategies. Furthermore, a "wrong" investment decision may become too costly as the investor will need to continue to maintain assets even when they are not.

5 **Investment strategy**: Developing and maintaining an investment strategy can prove to be difficult using traditional investment approaches. For instance, it is difficult to modify investment strategies when sudden-outlier events occur and, in some cases, when some market fundamentals change rapidly. Most investors can work ahead of the curve due to the inefficient and ineffective forecasting models, and they have a significant impact on both property value and investment performance. Changes to investment strategies can affect cash flow and risk management with implications on financing and debt management. The investment manager has to monitor and seek to adjust loan terms, interest rates, payment schedules, and refinancing, all of which need to be synchronised for optimal performance. Conducting these tasks using manual approaches or spreadsheets takes a long time, and if mistakes are made, there is a significant financial resource cost, an increase in credit cost, a reduction in financial flexibility, and a potential default risk. Furthermore, developing and implementing exit strategies and portfolio optimisation using traditional approaches are challenging because they require appropriate timing, method of sales, reinvesting strategy, and diversifying the portfolio – all of which need to be well synchronised and executed.

6 **Risk management**: This includes market, tenant-related, property-specific, and legal and regulatory risks. In cases where these risks are not well identified and mitigated, they can lead to huge financial losses, legal issues, and reputational damage. There are also issues when attempting to diversify a portfolio for smaller funds due to the high capital involved. This can lead to the concentration of a portfolio in a location or asset class, thereby increasing systemic risk and exposure. On the contrary, a well-diversified portfolio can lead to higher costs of management resulting from complexities associated with multiple real estate assets. Traditional methods of investment make holistic and comprehensive risk management inefficient and, in some cases, costly.

7 **Asset management and maintenance**: In cases where investment management is also combined with asset management, additional complexities are introduced. These include tenant screening, rent collection, property maintenance, and attending to tenants' complaints. Tenant relations and retention are also important. Building positive relationships with tenants, addressing concerns promptly, and providing quality customer service can contribute to tenant satisfaction and retention. There is also the need to monitor the property's performance with a focus on indicators such as rental income, occupancy rates, property expenses, and property value appreciation. These responsibilities are time-consuming, and they require effective communication and organisational skills, and basic legal knowledge and exposure (particularly those relating to landlords and tenants). This is essential in ensuring that the returns on the asset are maximised and costs are minimised. There will also be the need to carry out repairs and maintenance which will incur expenses and potentially affect the profitability and viability of the investment. Ineffective and inefficient maintenance can lead to unexpected repairs or major renovations which can disrupt cash flow. Inefficient and ineffective property management and maintenance can lead to unexpected vacancies and poor tenant selection, which can affect the property's performance. Maintenance, repairs, upgrades, maintenance schedules, budgeting repairs, coordinating with contractors, etc. can be time-consuming and challenging, and neglecting these functions can lead to tenant dissatisfaction, decreased property value, and increased expenses in the long run. All of these systems are difficult to manage, coordinate, and monitor using analysed and non-digital systems.

The fourth revolution[5] has further challenged the traditional economics of occupier markets, and these changes are affecting the returns on real estate investment as well as risk. Remote work has affected office investment, and this has been exacerbated by the COVID-19 pandemic. This is leading to changes in the quality and quantity of space demanded demonstrated through higher vacancy rates, lower rents, less space per desk, home working, and shorter and more flexible leases (Oladiran et al., 2023). This has increased uncertainty and risk in this market. These are likely to further lead to occupancy top-ups delivering better quality office spaces. This further suggests that a premium or discount may be the result of new assets that enable better customer experience. Furthermore, online shopping has also affected high street and shopping centres, and by extension, the demand for retail spaces. This is placing pressures on retail investment with landlords having lower rents and higher occupancy rates. According to Pi Labs (2020a), this is likely to lead to turnover-indexed retail rents, which suggests that more data will be created and shared, providing the opportunity to understand retail location economics better.

6.2.2.2   Environmental, Social, and Corporate Governance (ESG) Issues

Conversations around real estate investment now transcend financial and economic issues, and they are now evolving towards social, environmental, and corporate issues as the consciousness around ESG issues (Pi Labs, 2020a). These issues have gained more prominence in real estate investment as a result of a dramatic cultural shift driven by the global financial crisis, the pressures of climate change, rapid urbanisation, and the recent pandemic. Real estate investment now transcends high, stable, and reliable returns to investors using various ESG criteria. Real estate investors and fund managers are now increasingly under pressure to meet high ESG criteria. It should be noted that although these issues may have core social, governance, or environmental fundamentals, they may also have economic impacts and footprints.

One of the corporate-governance-related criticisms of traditional real estate investment is the socio-economically induced taste-based,[6] statistical[7] or disparate impact[8] discrimination and exclusion of individuals and groups from investing. Institutional real estate investment managers can control the structuring of investment vehicles in a way that limits access to qualified investors based on certain pre-defined criteria. This can be achieved through the setting up of a minimum investment sum, effectively trimming down the investment base. This has led to the clamour for 'democratisation'. Companies that are targeting this democratisation of asset ownership will ultimately be reliant on innovations in the legal technology (LegalTech) and regulatory technology (RegTech) markets in order to carry out their activities with economic efficiency. As compliance increasingly gets digitised through platforms such as Vauban, the whole process of setting up a fully regulated, easily divisible corporate packaging around any single asset or future fund will potentially lay the foundations necessary for a breakthrough in commercial real estate democratisation while also achieving higher time and cost efficiency.

Real estate investment also has regulatory and legal challenges resulting from the various regulations and legal requirements. Compliance with zoning laws, building codes, landlord-tenant laws, property tax regulations, securities laws (if dealing with investment funds), and environmental regulations is essential. Violation of these laws and regulations can lead to fines, legal battles, reputational damage, and penalties. It is difficult to have a comprehensive knowledge of these regulations, particularly for new entrants in the market and international investors. For instance, institutional investors are statutorily mandated to undertake KYC and AML checks on investors who intend to participate in a fund to ensure the legitimacy of the funds that they attract. Although a valid security measure, this process introduces a barrier to more efficient investment. According to Pi Labs (2020a), 64% of financial institutions claim that siloed systems and lack of consistent cross-referenced IDs were common challenges in onboarding, with numerous tools to access and verify client information. This further generates economic inefficiency with the process taking approximately 20%–30% of a relationship manager's time being consumed by onboarding and KYC-related tasks (Innovate Finance and Young, 2016). Additionally, current internal processes result in duplication of work documents, delaying progress and damaging the ROI.

Another aspect of corporate governance in real estate investment is reporting and performance evaluation. Real estate investment management requires accurate and timely reporting of financial performance and investor communications. This requires financial statements, asset performance tracking, etc. to provide transparency to investors. Failure to provide comprehensive and timely reporting can strain investor relationships, and errors in reporting can also affect trust and integrity. Using a manual collation and analysis of data and using analogue reporting style, it is difficult to avoid errors, and there may sometimes be delays in preparing reports.

Several aspects of traditional real estate investment have a high environmental impact. For instance, high travel is required in sourcing funds, asset inspection, purchases, and other related transactions. Furthermore, the process traditionally required reliance on local experts, in-person site visits, hand-signed documents, human valuations, manual surveys, site visits, etc. These high volumes of travelling increase aviation and automobile-based emissions, and the use of paper-based marketing brochures has a negative impact on wildlife, biodiversity, and environmental sustainability.

Finally, the traditional approach to real estate investment and portfolio management will make the assessment and evaluation of the ESG performance of an investment fund difficult, ineffective, and inefficient because several indicators and signals in this process may be missed out; thus, valuable insights may easily be missed.

Figure 6.1 provides a summary of key inefficiencies associated with real estate funding, finance, investment.

6.3 Innovative/Emergent Systems and Services in Real Estate Funding, Finance, and Investment

As real estate funding, finance, and investment become more complex and complicated, traditional approaches to these services are becoming more sophisticated with a heightened level of innovation and automation. Several digital tools and systems have been integrated with these elements of the real estate value chain, and this has led to the emergence of contemporary real estate digital tools, systems, and platforms. This sub-section provides a review of innovative and emergent digital technological-driven innovations and buzzwords that have emerged in the realm of real estate funding, finance, and investment. Some of the phrases and buzzwords that are the focus of this section were more or less unheard of two decades ago; some others have been used by the authors for the lack of an existing phrase or buzzword.

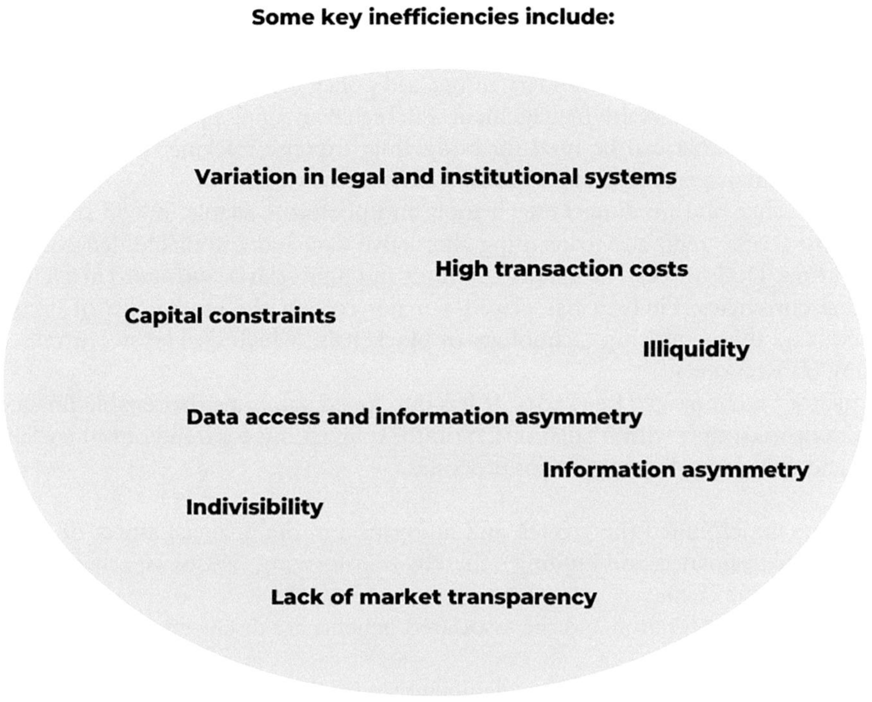

Some key inefficiencies include:

Variation in legal and institutional systems

High transaction costs

Capital constraints

Illiquidity

Data access and information asymmetry

Information asymmetry

Indivisibility

Lack of market transparency

Figure 6.1 A summary of some of the key inefficiencies associated with real estate funding, finance, investment.

6.3.1 FinTech in Real Estate

"FinTech" is a contemporary buzzword that combines "financial" and "technology". It is the intersection of financial services and technology where technology-focused start-ups and new market entrants innovate the products and services that are hitherto provided by the financial services industry. It has a complex ecosystem that comprises various industry segments and players. This is an emerging area that has gained significant momentum in the last decade, and it is causing disruption to traditional financial services. FinTech companies have leveraged innovations in IT and computing including data, analytics, software, internet, mobile apps, and other digital technologies to provide user-friendly financial services. According to PriceWaterhouseCooper (2016), FinTech start-ups attracted $12.2 billion in 2015 globally (up from $5.6 billion in 2014), and according to Innovate Finance (2022), the global capital flow into FinTech was $92.2 billion in 2022, albeit it is a decrease from $130 billion in the preceding year (2021).[9] The significant growth in FinTech investment capital is an indication of the growing relevance and growth in this contemporary financial sector.

FinTech covers a wide range of financial services. They include the following:

- Open banking and financial aggregation: It has promoted open banking where customer financial data are shared securely between financial institutions and third-party entities in order to enhance integrated and personalised financial services.
- Transactions: These are promoted via the automation of payment and money transfers. They come through the provision of digital payment systems, peer-to-peer money transfers, mobile wallets, and electronic remittance services.

- Insurance: FinTech firms have transformed the insurance industry creating a new system that is encapsulated in the buzzword "InsureTech". This has improved insurance processes, claims, management, underwriting, and policy management.
- Personal finance and wealth management: FinTech companies provide online platforms, tools, and apps that can be used for budgeting, expense tracking, financial planning, investment management, and AI-enabled advisory services.
- Online finance and funding: Fintech tools and platforms enable individuals and businesses to access credit and loans using alternative data and automated decision-making algorithms. Decisions can be made in minutes and funds can be disbursed in a few hours.
- Digital currencies: FinTech has played a major role in the emergence of cryptocurrencies and the underlying technology of blockchain, which enables decentralised but secure transactions.
- Regulatory technology (RegTech): It has developed solutions that enable financial institutions to comply with regulations, monitor transactions for fraud or money laundering, and enhance risk management processes.

FinTech has underpinned the growth and innovation in the financial aspect of PropTech (i.e., investment, finance, and funding), thereby transforming various aspects of these segments of the value chain.

Major areas of integration and the associated benefits are discussed further:

1 **Investment**: FinTech solutions are democratising real estate investment through access to a wider range of investment opportunities that would hitherto have been difficult or impossible without this financial innovation. For instance, peer to peer and crowdfunding (to be discussed further in Section 6.3.2) enable individuals and companies to invest in real estate projects with lower investment thresholds, making real estate investment more accessible and proving options for diversification. The crowdfunding and P2P models were initiated following the global financial crises of 2008 and the resultant difficulty in raising and accessing mainstream real estate funding (Brown, 2018). FinTech has also provided online marketplaces for the trading of real estate investment shares through real estate investment platforms. These platforms provide a streamlined process, financial information, transparency, access to property information, data, and analytics. Investors can also conduct due diligence, compare properties, analyse investment opportunities, and use those to make informed real estate investment decisions. These can further enable investors to explore and select opportunities based on their investment goals, strategy, preferences, risk appetite, etc. and for them to monitor and manage their real estate portfolios, track performance, receive regular updates, and access financial reports. This opens up the market to smaller investors who can participate with lower investment thresholds, promoting greater financial inclusion. These can further streamline administrative tasks and enable investors to have better control and visibility over their investments, ultimately enhancing the efficiency of financial projections, risk assessments, and analytics for more effective investment decisions.

2 **Funding**: FinTech solutions have provided alternative real estate financing options. For instance, online lending platforms, peer-to-peer lending, and crowdfunding platforms enable developers and investors to access capital that is outside traditional financial institutions. These platforms, tools, and products offer streamlined application processes, quicker funding decisions, automated underwriting, and faster credit disbursement,

which effectively provide borrowers with quicker access to financing. The emergence of real estate crowdfunding platforms has connected real estate developers and sponsors with potential investors, enabling them to raise capital for projects. Furthermore, the concepts of fractional ownership in real estate have been enhanced through tokenisation. Through the use of blockchain technology, properties can be divided into smaller shares, and these shares can be tokenised, enabling investors to buy and trade fractional ownership shares in real estate. This enhances liquidity and flexibility in funding and finance. In addition, FinTech has significantly transformed traditional mortgage lending by introducing online lending platforms and alternative finance options. Peer-to-peer mortgage platforms also connect borrowers directly with the lenders, and online mortgage platforms and digital lenders provide borrowers with convenient tools for submitting their applications, uploading their documents, and tracking their applications. These platforms provide more accessible loans, speed up the application process, automate and streamline underwriting, credit assessment, and streamline the approval process making real estate funding and finance more efficient.

3 **Advisory and services**: FinTech tools and platforms have facilitated the collection and analysis of real estate data such as property prices, market trends, rental values, demographics, and property characteristics. FinTech tools and platforms utilise big data and AI to analyse property data, rental rates, market dynamics, and economic indicators, providing valuable insights for investment strategies and risk management. These can be used to provide investors with valuable insights into an investment opportunity or a market and support them to make data-driven investment decisions to assess potential risks and returns more accurately. Furthermore, it determines creditworthiness through access to non-traditional data points such as rental history, utility payment, and social media activity. FinTech also enables escrow services and smart contracts. FinTech-supported escrow services ensure that funds are securely transferred protecting the interests of buyers and sellers, and FinTech-enabled smart contracts automate and enforce the terms of the agreement, enhancing transaction efficiency. The compliance checks and verification processes that these solutions provide ultimately ensure regulatory compliance and reduce fraud.

The integration and application of FinTech to real estate operations and service delivery have several limitations and challenges. They include the following:

1 Market and financial risks: Real estate services that are underpinned by FinTech (such as crowd-funding or fractional ownership) may expose investors to higher levels of market volatility or financial risks. This may manifest through investors facing challenges in assessing the quality of investment opportunities and accurately evaluating risk factors. It is therefore advisable that real estate investors, particularly new entrants, should gain some foundational knowledge in real estate in order to understand the risk and fundamentals of the real estate markets before engaging with real estate investment and finance platforms that are facilitated by PropTech. Similarly, innovators and inventors of real estate FinTech tools should have a thorough understanding of real estate markets and their mechanisms in order to develop effective and efficient tools.

2 Legal and institutional framework: Due to the significant pace of FinTech integration in real estate, some parts of the current real estate investment and finance regulatory frameworks may be getting obsolete. This can create challenges relating to investor

protection, data privacy, money laundering, compliance, KYC regulations, etc. There is a need for the current regulation to adapt by developing appropriate regulations to address the gap in the regulatory framework to support FinTech-driven real estate activities.

3 Data Security, Cybersecurity, and Privacy: FinTech platforms (including those that power real estate) rely heavily on data capture and analytics, including personal and sensitive financial (and other information). This raises huge concerns regarding data security and privacy. FinTech companies must implement robust security measures and comply with data protection regulations to mitigate the risk of data breaches and unauthorised use of personal information. Furthermore, because FinTech platforms rely heavily on technology infrastructure, technical glitches or disruptions can have significant implications for real estate investment. Additionally, there are risks such as cyber-crime and threats, which may lead to system hacking, data breaches, and unauthorised access to sensitive financial information.

4 Impact on professional services: FinTech has disrupted traditional real estate services in many ways. Traditional lenders have seen higher levels of competition and loss of market share to more contemporary funding approaches. Although these traditional lenders still hold significant market shares, the power dynamic within the industry could be shifting, and traditional players are beginning to adapt and innovate in order to re-capture their market share. FinTech platforms often prioritise automation and efficiency, and in some cases, individual professional service and personalised guidance may be absent. This limited professional guidance and reliance on technology have the potential to reduce personalised customer service.

5 Inequality: Increased use of FinTech-backed platforms can exacerbate inequality in access to real estate investment and finance. Firstly, AI-based systems may have been trained (sometimes unintentionally) to either ignore personas or to attribute certain discriminatory weightings to certain personas, which can lead to unfavourable decisions or advisories that can adversely impact some disadvantaged groups. Furthermore, investors with higher levels of technological literacy and financial resources may have a competitive advantage in navigating FinTech platforms and, by extension, taking advantage of the most suitable investment opportunities. This further has the potential to widen the wealth gap and create a digital divide in real estate investment and finance.

FinTech has significantly changed financial services and the finance industry promoting competition, innovation, and financial inclusion. In real estate investment and finance, it has improved access to investment opportunities and fractional ownership models, and streamlined investment management and alternative options. These advancements have made real estate investment more accessible, transparent, and efficient.

6.3.2   Crowdfunding, P2P, and Digital Marketplaces

Although Section 6.3.1 highlighted some of the contemporary advanced funding systems in real estate, this was done solely in the context of FinTech. Section 6.3.2 provides a more holistic review of automated real estate funding and finance tools, specifically crowdfunding and P2P (peer-to-peer) and other digital lending platforms.

Automated real estate funding and financing systems refer to a wide range of digital approaches to funding and financing real estate. These systems leverage data, analytics, algorithms, and IT systems to enhance the efficiency, effectiveness, accuracy, and transparency in real estate finance transactions. Bridging the information gap between what the customer needs and what the lender is providing is important in addressing some of the inefficiencies in traditional real estate funding and finance, and digital technology has played a significant role in improving the speed, efficiency, and effectiveness of documentation and information access for the lender and the borrower.

Crowdfunding refers to a financing model that involves raising small amounts of money from different individuals or organisations, usually through a web-based platform. This approach to funding requires individuals to pool their financial resources to support a specific project or investment in exchange for some form of returns. This approach to funding has become popular as a means of democratising access to funding, enabling individuals and organisations to pursue their ideas and ventures with the support of the broader community. It has been used to support innovative projects, start-ups, social causes, and the creative industry. Crowdfunding is sometimes referred to as P2P lending because it connects individuals (and businesses) with relatively similar investment goals through online platforms and enables them to use their pooled resources to achieve their goals with several benefits for the borrower and lender.

Crowdfunding and P2P have become commonplace in real estate finance. The real estate crowdfunding model is typically requiring a platform that receives financial resources from several small-scale investors, and these investors can access high-yielding investment options that would hitherto have been accessible only to accredited investors. The mechanism of the P2P model is that investors' funds are matched to a loan for an individual or business, while crowdfunding is typically either a reward-based or equity-based system. For private and non-accredited investors looking to diversify their real estate portfolio, digital and online real estate crowdfunding platforms offer a fast and relatively straightforward option that connects them with previously inaccessible investment opportunities. Furthermore, real estate investors looking to raise capital using flexible options can access equity and debt that they would likely have taken a long time to secure. Traditionally, investment in real estate shares was through REITs, open and closed-ended funds. These usually have intermediaries who all receive a commission from the investors, and this minimises the returns on investment. Real estate crowdfunding platforms are now replacing these intermediaries. Thus, an individual who is inexperienced in real estate investment, an institutional investor, a borrower in search of a flexible loan, or a real estate investment sponsor looking to raise joint venture equity can invest. In most cases, the only intermediary is the crowdfunding platform.

According to a recent survey, 61% of commercial real estate (CRE) executives believe that online marketplaces will continue to disrupt the real estate financial service, and the majority of CRE firms revealed that they have an online marketplace for a recent transaction, with most intending to increase their use of technology in the future (Altus Group, 2020). This system of funding is however yet to mature in several countries. In the UK, for instance, crowdfunding makes up between 0.1% and 0.25% of all real estate capital raised (Oxford Future of Real Estate Initiative, 2020). There are however predictions that this share will increase in tandem with the development of blockchain technology and its potential to tokenise real estate investment.

Online mortgage lenders and other digital loan marketplaces are also providing real estate finance through online application processes. Although these funds may not be pooled using crowdfunding, they also use similar internet-based systems to fast track the loan application processes, making the process quicker and generally more efficient. This automated system, therefore, expedites loan origination in real estate finance. It provides online platforms and tools that borrowers can use to submit loan applications, upload the required supporting documents and track the progress of the applications. There are companies that collect and provide aggregated data to facilitate mortgage applications. These data are fed into the lending algorithms and automated underwriting processes to assess borrower creditworthiness and streamline the approval process. Automated systems are also used for data analytics and algorithms to assess the risk profit of borrowers in real estate financing. Furthermore, core functions such as income verification, credit history check, property value, and market trends are used to determine the applicant's creditworthiness. These systems will then adopt advanced credit scoring models to make more accurate and efficient risk assessments, ultimately leading to a fast track of the real estate loan application and a final decision.

Generally, these platforms have been facilitated through mainstream FinTech such as mobile wallets, prepaid cards, and other online digital banks such as Monzo and Revolute. They are attractive to individuals and businesses with relatively small capital and who would like to invest in the property market through non-conventional channels. Although these methods of financing promise moderate to high returns, they present several risks. A major problem is the loss of capital or the risk of losing the entire investment capital or a significant portion of it. This may be the result of fluctuations in real estate markets, misappropriation of funds, platform failure, or cybersecurity risks. The platform risk may result from the crowdfunding platform encountering financial difficulties, regulatory issues, or the lack of competence to adequately administer the project. Furthermore, there are technological risks such as online platform downtime, data breaches, and cyber-attacks which can compromise investors' personal and financial information, disrupt communication, and lead to financial losses.

Real estate crowdfunding and P2P have other project-specific risks. These risks to the location, market conditions, project type, development plans, management expertise, etc. of a project. There may also be a lack of liquidity, as some of the projects may be illiquid, depending on the structure and operation of the platforms. Also, due to the capital-intensive nature of real estate projects and the relatively lower capital that may be raised through crowdfunding, diversification potential may be limited, exposing investors to the risk associated with a poorly diversified investment portfolio. These problems are exacerbated by platforms with limited information and communication which can cause an imbalance in decision-making and risk assessment.

It is therefore important for investors to assess and thoroughly evaluate a platform's reputation, track record, stability, and general systems, as well as the real estate projects that are being financed to understand the capital risk and the suitability of the investment. This due diligence is quite important particularly because changes in regulations or legal frameworks could impact the crowdfunding industry, potentially leading to additional compliance burdens or limitations on investment opportunities.

6.3.3   Tokenisation and Blockchain in Real Estate

The general concept and ideas behind blockchain technology were introduced in Section 2.5. This sub-section provides specific insight into this technology as it relates mainly to real estate finance and investment.

"Virtual asset" (VA) is a broad term that describes systems of capturing value in digital form including tokens, digital currencies, and cryptocurrencies. VA is increasingly gaining attention and popularity around the world. In 2021, the International Monetary Fund (IMF) provided some legal and practical considerations for VA and AML (Fernando et al., 2021). The World Bank also has a "virtual assets and virtual asset service providers risk assessment" toolkit, and in a recent World Bank Policy Research Working Paper (Feyen et al., 2022), it was noted that although VA possesses high risks, there are also huge opportunities and an emerging asset class by financial market participants and policymakers, reaching about $2.8 trillion in market capitalisation in November 2021. It is therefore no surprise that various world governments in the last three years have issued statements and made publications stating their stances and consideration for this emerging phenomenon. For instance, in 2022, the US Secretary of the Treasury, in consultation with the Secretary of State, the Secretary of Commerce, and the Administrator of the US Agency for International Development (USAID), and the head of other relevant agencies presented a framework for interagency engagement as part of efforts towards ensuring the responsible development of digital assets (US Department of the Treasury, 2022). Similarly, the UK issued publications/reports on virtual assets (HM Treasury, 2023; Law Commission, 2023). These reflect the increasing attention that countries are placing on DA. The Hong Kong SAR government also issued a policy statement on VA development in Hong Kong in 2022. This statement contains the government's policy stance and approach towards VA development. This has drawn the attention of investors and service providers to Securities Token Offering (STO). The fact that these are government-backed will support various business models to be created in real estate businesses more efficiently, from commercial real estate retail sector to shopping malls and hospitality services and operations. This further has the potential to create new types of services and products.

Traditional real estate investment is illiquid and not inclusive. It is becoming increasingly important for real estate to have a broad base that enables investors of various backgrounds to participate in the ownership of institutional-grade assets, allowing them to build diversified portfolios with low capital. One of the core advantages of the VA is the intangibility with endless possibilities. For instance, real estate businesses can adopt blockchain technology to tokenise assets, products, and services through STO to enhance customer experiences and raise funds from a broader base of investors. Tokenisation is therefore being increasingly utilised to address this problem. Real estate tokenisation is a new capitalised method that indicates an asset's economic value or fractional ownership by releasing blockchain tokens that can be traded on a secondary market (Laurent et al., 2018).

Tokenisation refers to the process of converting an asset such as real estate, artwork, or financial instruments into digital tokens that can be securely stored and transferred on a blockchain or distributed ledger technology platform. Laurent et al. (2018) define it as the process of issuing a blockchain token that digitally represents a real tradable asset in a similar way to the traditional process of securitisation. It is however important to highlight the difference between a token and a security token. A token can be viewed as a digital representation of a real estate asset utilising blockchain technology. Although a basic token digitally presents a property, a security token digitally represents security (Starr et al., 2021). Smith et al. (2019) provide further insight by discussing the differences between two types of blockchain-based representations of securities: security tokens (blockchain-native tokens that are non-operational outside the blockchain) and tokenised securities (blockchain-embedded representations of real-world securities). They further argue that although security tokens and tokenised securities confer benefits that supersede traditional means of representing securities (e.g., paper, or digitally siloed databases), they differ largely in the type of legal and regulatory frameworks that they may require to achieve effective adoption.

With the real estate tokenised model, each token effectively represents a fraction or unit of the underlying asset, providing ownership or access rights to the individual or entity that owns the token.[10] Real estate tokens are digital finance instruments, represented through blockchain tokens, that grant exposure to an underlying real estate asset or real estate development project (University of Oxford Research, 2020). By fractionalising ownership interest in an asset through tokenisation, interests can be traded via their digital representation on online exchanges. It should be noted that the tokens entitle investors to receive digital receipts that represent their (fractional) interest in an entity that owns the asset, rather than receiving part ownership of the physical asset.

Tokenisation has several benefits and prospects for real estate investment and finance which are summarised in Table 6.1.

Indeed, VA has infinite possibilities and prospects. Real estate businesses can adopt blockchain technology to tokenise assets, products, and services through security token offerings (STO) to enhance customer experience and raise funds from a broader base of investors. This will typically cost a lot less, and the associated transactions are stored in a

Table 6.1 Benefits of Real Estate Tokenisation

Benefits	Further Details
Fractionalisation	• Assets such as real estate have a high barrier to entry due to the high upfront capital required
	• Fractionalising such assets democratises their access for smaller investors
Liquidity	• Fractionalisation increases the pool of potential investors and can unlock the global investor base
	• Secondary markets also facilitate additional liquidity
	• Liquid assets command a premium and can increase asset value
Automation	• Smart contracts can automate steps such as compliance, document verification, trading, and escrow
	• Dividends and other cash flows can be programmatically paid when due
Settlement time	• Tokens can settle in minutes or hours (depending on the underlying blockchain)
	• This unlocks the capital that is tied in the market which currently settles T+3/T+2
Customisability	• Tokenisation enables exposure to individual real estate assets. Thus, instead of investing in the whole sector, portfolios can be customised to single buildings
Structured products	• Additional value can be realised once assets are tokenised and that enables the creation of additional layered financial products such as baskets of assets and derivatives
	• since the underlying is tokenised, creating complex products becomes simpler through coded smart contracts
Data transparency	• Secure and visible recordkeeping on the blockchain can increase transparency to the underlying data
	• Especially in complex derivative products, it can link security to its underlying value drivers
Cost efficiency	• By removing certain intermediaries and increasing the efficiency of processes, costs can be lowered

Source: Authors' illustration based on Smith et al. (2019).

decentralised blockchain network with a higher level of efficiency in comparison to traditional fund-raising methods. The tokens that represent values in assets and services may also reach out to a wider investor base that includes digital native investors who may prefer to own and store their wealth in digital assets. Furthermore, tokens that carry utilities such as parking spaces and hotels can support operators to manage their loyalty programmes and increase consumer stickiness.

The blockchain is the backend infrastructure that enables digital assets to mimic the features of physical assets and enables digital transactions to take place securely. In simpler terms, it is a comprehensive databank that records transactions. It functions as a computer program that can be coded using computer language, and because it is coded, it can be programmed to execute actions based on specified rules and conditions using the "if-this-then-that" type of logic.

The emergence of cryptocurrencies has also changed the financial and investment landscape including real estate finance (Brown, 2018). Cryptocurrency has decentralised transactions and minimised the impact of middlemen. Thus, digital currency can be transferred directly between two individuals with the transactions being recorded on a ledger. This system, developed by Satoshi Nakamoto, is gaining traction mainly due to the high level of perceived decentralisation, transparency, and security it offers through blockchain technology (as described in Section 2.5). One of the real estate innovative funding tools is the initial coin offering (ICO) generated by the combination of blockchain, crowdfunding, and FinTech. A vast majority of ICO platforms play the role of third parties, which, in a sense, sustains some of the inherent inefficiencies in traditional real estate funding, while the blockchain value is weakened. The FIBREE 2022 report showed that the number of blockchain-related real estate products increased by 20% in 2022 (FIBREE, 2022), with the previous report showing that the vast majority of block-chain backed products have targeted real estate investment and financing applications.

Cryptocurrency-enabled payments and transactions are also not uncommon anymore. Cai Capital was one of the first companies to facilitate the sale of property priced in Bitcoin in 2014 (Palmer, 2014) and in 2017, Knox Group of Companies – a Dubai Developer also announced that they would accept payment for an off-plan apartment in Dubai in Bitcoin (2017). In the same year, Go Homes in the UK also were reported to be the first to sell a home using a cryptocurrency in 2017 (Renshaw, 2017). In terms of cryptocurrency-enabled property exchange, a platform called "ClickToPurchase" was reported to execute over £200 million worth of property transactions across their blockchain-enabled online property exchange platform. This enhanced transaction speed and also minimised uncertainty because the sales transactions between verified and trusted parties were made using electronic signatures with the information recorded on a transparent record or blockchain. There are also property trading and fund-raising platforms such as Brickblock, Dominium, Harbor, and TrustMe. These are blockchain trading platforms that allow an asset-backed token, including property to be traded. They do not have standard middlemen, which significantly reduces the transaction time to seconds or minutes rather than weeks or months. It should be noted that fractional ownership and secondary market liquidity do not necessarily require blockchain technology. REITs are examples of alternatives to this.

IPSX, a London-based exchange and marketplace for CRE securities, has attained investment exchange status in the UK. It has become the first regulated securities exchange to be specifically focused on facilitating initial public offerings and secondary market trading for companies that own single or multiple high-quality institutional properties. This has made it possible for various investors to access income-producing commercial real estate assets with secondary market liquidity through the purchase of shares traded on a regulated stock exchange. The tax approval by HMRC has enabled the trading of tax-transparent single-asset REITs, which function similarly to tokenised assets with the potential to replicate tokenised assets that provide an avenue for investors to trade shares in single-asset-owning real estate firms with higher cost efficiency and transparency. An attempt was made to replicate this in the United States with the introduction of LEX, a marketplace for commercial real estate securities. However, the LEX brokerage platform shut down due to financial constraints in February 2023. Pi Labs (2020a) however cautions that this may not necessarily translate to high traction as unqualified investors may not necessarily be attracted to individual asset offerings when they could buy shares in reputable REITs.

For VAs in real estate finance and investment to be further deepened, several issues need to be resolved. The prospect of this system is dependent on market participants being more comfortable and knowledgeable about blockchain technology and its use. Security and fraud prevention also needs to be improved, and standards and regulatory framework should be improved to enhance trust in the system. Blockchain could also be best integrated into administrative tasks in investment management such as replacing the settlement systems at Central Banks and clearing houses, suggesting that this could lead to further integration in the debt market. For this to be achieved, RegTech and LegalTech will also have to be developed simultaneously. Tools for instantaneous consumer-facing AML and KYC checks, for instance, will help facilitate this.

There is a need to increase research interest in this area. The current research into real estate tokenisation focuses mainly on investors, and most of the existing research is conceptual. It is therefore valuable to increase the scope of scholarly contributions and debates in this area, and this can be enhanced through experimentation, the development, and rigorous testing of conceptual innovation, although, as evidenced in Souza et al. (2021), the inability of real estate companies to utilise the hybrid flipped form of organisational structures may limit advancement in this respect.

6.3.4 Automated Investment Management Systems in Real Estate

Automated Investment Management Systems (AIMS) ultimately attempt to automate human real estate investment processes through mechanical tasks. For instance, digital platforms such as Capital Brain Intelligence (CBI) technologies work with underwriting and financial professionals to receive macroeconomic data directly into their underwriting and budgeting models. The COVID-19 pandemic led to the biggest change in the capital raising process because of restriction to movement during the lockdown. Virtual meetings became more entrenched in various forms of collaborations and deals that hitherto had been conducted face to face, and the real estate industry also adapted in the same way. As of 14 June 2020, the MS Team subscription base had grown by 894% (relative to 17 February 2020 at the start of the pandemic); similarly, Zoom grew by 677% (O'Halloran, 2020).

The mass adoption of virtual meeting platforms laid the foundation for a dramatic cultural shift in real estate capital sourcing. Real estate companies are now seeking ways of replacing their analogue systems with specific, digital, cloud-based solutions to remain productive in the face of increased remote working (Pi Labs, 2020a). The pandemic has also driven changes in other areas of real estate investment, and the general development of web-based platforms and mobile applications has further led to the development of secondary markets. Real estate investment managers are also increasingly seeking to identify sectors that they can invest in. According to (Pi Labs, 2020a), the typical portfolio mix in the 1990s was 50% retail, 40% office, and 10% industrial. However, the retail sector has experienced a rapid decline from the mid-2000s with the balanced funds reducing their retail weighting by all means. This has led to a significant increase in logistics investment, while rental residential, although on a growth trajectory, has limited supply and is difficult to find on a sufficient scale. As seen in recent times, the office sector has also been challenging. It is reported that 82% of real estate companies are looking to invest in home working and digital tools in the immediate future, while the second strongest preference (50%) is for technologies that foster stronger digital links with the marketplace (Pi Labs, 2020b).

The ultimate aim is to provide investors with a suite of tools that can help investment managers more precisely and accurately gain market knowledge for more informed and evidence-based acquisition and portfolio management. There are four core areas of investment management that AIMS are enhancing, particularly with institutional grade investors: advanced market insight, revenue optimisation, cost minimisation, and efficiency systems and processes, all with promising environmental, social, economic, and physical benefits.

6.3.4.1   Market Insight and Intelligence

Market insight is a fundamental aspect of real estate investment and portfolio management. It provides the investment manager with the requisite details and information for making vital investment decisions. As real estate assets and asset classes become more complex and as investment options and strategies become increasingly complicated, digital systems are being leveraged for market insight and intelligence. Current real estate investment requires the use of multiple data points from various sources, and these are often combined into data files and worksheets as a basis for investment decisions to be made. The often fragmented structure and nature of the datasets can lead firms to purchase data from multiple vendors and then used to create proprietary data, combining these with internal data collected from their management systems and models and then analysed for investment insights. This lack of standardisation often leads to time lags and structural errors, and the functions are often carried out by professionals with varying insights, experience, and knowledge. AI and advanced analytics are enabling the streamlining of workflows and processes for better efficiency, comfort for the investment manager, and peace of mind for the investors.

Various digital platforms are providing advanced data and analytics support to enable investment managers to identify and explore asset acquisition opportunities, and portfolio risk management strategies. They can receive real-time competitive intelligence for benchmarking. Many property fund managers pride themselves on their access to stock, and sourcing for stock is an important aspect of this area. Purchasing properties through off-market channels has become a codeword for securing exceptional deals, and the effectiveness of fund managers is now increasingly evaluated based on their ability to access market information and identify buying opportunities (Baum, 2015). Specialised platforms such as Datscha (the Landsite) and Reonomy provide access to investment opportunities for a wide range of commercial, multi-family, and land assets within a specific geographical area. These platforms enable investment managers to discover off-market deals by utilising various filtering functions, such as asset type, size, sales history, debt history, tenants, ownership, and tax history.

AI-enabled market insight intelligence can be obtained through automated data collection techniques. One of the ways this can happen is through the accessing of APIs, i.e., the link between software packages. These can then be linked to various databases, after which the data are analysed. The ability of these processes to extract patterns and use the resulting intelligence can support the design of new market-entry strategies. AI and machine learning systems are increasingly being used to reveal complex patterns and make sense of vast data. For instance, this can be valuable in modelling the revenue potential of a city centre hotel which traditional models may not sufficiently capture. Furthermore, the underwriting process is significantly enhanced using data-backed advanced market intelligence that enables the underwriter and investment manager to identify future growth locations, and more specifically, asset selection based on risk/return insight. These insights are particularly essential for investment in second-tier cities, out-of-favour sub-markets, emerging or developing national markets (such as economies in the global south), and emerging speciality market sectors (such as operational real estate). Two identical cities with similar credit quality have been reported to show stark variations when proprietary analyses are conducted using FBI crime data, EPA climate change data, housing affordability data, etc. (Pi Labs, 2020a). This underscores the value that digital technology-enabled market insight can provide.

There are various examples of real estate market intelligence tools that leverage the capabilities of artificial intelligence (AI). For instance, Skyline AI employs an extensive commercial model that integrates 130 data sources spanning 50 years. These comprehensive data include information on stock market performance, interest rates, financial indicators, commercial real estate analytics, demographic data, and more. By utilising this vast data set, Skyline AI constructs an algorithmic representation of a property within its urban environment, enabling firms to make long-term predictions regarding the property's performance for up to 15 years.

In the residential real estate sector, AI is also being utilised to identify suitable acquisition targets for iBuyers (instant buyers) like Opendoor. These iBuyers aim to facilitate completed transactions within a remarkably short timeframe of seven days. To achieve this, they employ advanced algorithms to analyse various factors and determine suitable residential properties for acquisition. An example of such technology is REalyse, which specifically targets residential investors. REalyse brings together an extensive range of available data sets about residential properties and presents them through powerful and user-friendly analytics tools. This enables investors to create definitive and quantifiable assessments of both risk and opportunities within the residential real estate market.

6.3.4.2   Enhanced Investment Systems and Processes

AIMS are increasingly being used for various aspects of real estate investment systems and processes. Real estate has the potential to become a mainstream asset class with a greater level of transparency and information, as well as more measurable returns and benchmarks, stronger legal and tax frameworks, and more efficient methods for investors to acquire, lease, manage, and sell their property assets. This is leading to a high level of portfolio diversification and decentralisation of systems and operations. With investment management teams getting increasingly decentralised, there is an increasing need for online platforms to centralise critical portfolio data to become even more apparent.

The structure of traditional real estate investment is non-operational, meaning that there is a separation between the building owner and the customer or space user. This decentralisation is facilitated by intermediaries such as asset managers, facilities managers, and property managers. The procurement approach in this model is speculative, as developers have no knowledge or control over who will ultimately occupy the space or who will purchase the property. Consequently, once the building is sold, the developer's obligations and responsibilities typically come to an end. However, the real estate investment landscape is changing. For instance, operational real estate (e.g., hotels, purpose-built student accommodation) is growing as an active asset class (Oladiran et al., 2022), costs are now less transferrable to tenants through FRI lease and volatile net cash flows, lease terms are getting shorter, etc.; thus, the investment landscape is increasingly becoming uncertain and probabilistic. Traditional approaches to investment management are therefore becoming obsolete given the complexities of contemporary real estate investment and the need for instance access to real-time data to push deals forward, manage tenant relationships, and make informed decisions. Some areas of innovation and digitisation in real estate investment are summarised below:

6.3.4.2.1 SOURCING OF FUNDS

Investment management now sources funds using various digital tools. Modern software solutions have now evolved to provide valuable insights into how investment managers are perceived by potential investors, introducing an additional layer of understanding for those seeking to secure capital. An example of such innovation is the Appfolio Investment Manager, which empowers real estate investment management teams to monitor individual investor interactions through their integrated investment management and investor portals. By tracking these interactions, personalised and targeted offers can be made, saving time in the capital raising process and increasing the likelihood of securing investment. Once a connection has been established, the use of secure cloud-based data rooms and investment portals like Drooms eliminates the necessity of sending information via traditional mail or conducting in-person presentations. This enables the entire onboarding process to take place online.

Contact with potential investors is now being done virtually. Virtual meeting platforms (e.g., Google Meet, Microsoft Teams, and Zoom) are now used to discuss and conclude deals; documents are signed using digital platforms (e.g., DocuSign and E-sign), and funds are transferred using various FinTech tools and platforms. The combination of data rooms, e-signature software such as DocuSign, RegTech for KYC compliance, and onboarding solutions facilitates a future where capital can be raised from major institutional investors without the need for face-to-face meetings or physical document transfers. This has significantly reduced commute-related time and costs and further reduced commute-related carbon emissions. Furthermore, the risk associated with commuting (accidents, injuries) is significantly reduced, and the process of raising funds using these sourcing and transaction approaches has become more convenient and secure.[11]

6.3.4.2.2 CAPITAL DEPLOYMENT AND PORTFOLIO MANAGEMENT

The use of paper-based marketing brochures, reliance on local experts, in-person site visits, hand-signed documents, human evaluations, and manual surveys/inspections are gradually making way for innovative and emerging technologies. This enables investment managers that adopt these technologies and other innovative approaches to source for better assets, negotiate more favourable terms, forecast risk, and conclude transactions more efficiently. Access to advanced market intelligence using digital tools is improving fund managers' ability to identify investment opportunities and appropriately deploy capital, and better manage assets within the portfolio to achieve investment objectives.

Digital technology is also offering risk mitigation benefits in portfolio management. For instance, Kamma offers a hedge against regulation risk through the identification of property licencing and regulatory requirements at the asset or regional level. This can be used to resolve challenges relating to securitisation and achieving net zero through property licensing and compliance. Investment can use this platform to minimise the need for manual checks which often have high costs. Furthermore, it can also ensure that property licencing obligations are met. Another example is Cherre – a real estate data and analytics platform that connects real estate data within an organisation, improving the organisation's ability for better investment, management, and underwriting decisions. Furthermore, they provide services that automate operator and partner data collection, advanced data mapping, and reporting. New asset management reporting packages automate the validation process and provide more advanced responses. Additional platforms like Proda and SpaceQuant utilise AI to improve the effectiveness of rent roll processing by extracting information from unorganised rent rolls. The data are first gathered, analysed, consolidated, and standardised before being prepared for further processing or transmission of the refined and verified rent roll data. This technological advancement allows for the automation of the labour-intensive manual processing of rent rolls in Excel format, which has historically been susceptible to errors.

6.3.4.2.3   NEGOTIATION AND TRANSACTION

A challenging aspect of real estate transactions is the negotiation process. It is a core cause of the time inefficiency of real estate transactions and the illiquidity of real estate. It requires excellent communication skills, relevant experience, and a high level of diplomacy. In most cases, the seller of the property has more information than the buyer, and this often places the buyer at a disadvantage. Furthermore, property auctions are riddled with a high level of information asymmetry. By moving property auctions online, they become more accessible and convenient. For example, BidX1 offers registered online users the opportunity to purchase land, residential properties, and small-scale commercial properties almost immediately, increasing the attractiveness and efficiency of the purchase process.

Conveyancing is also often tedious and complex using traditional approaches. However, online portals such as eConveyancer provide investors with a range of firms that provide conveyancing services. This is even more important for international investment where the investors have limited information on legal and conveyance systems in a country. e-Conveyance is becoming a widely used process globally. In 2018, the UK government approved changes to the Land Registration Rules which laid the foundation for digital land registration, and by extension, e-conveyancing. e-Conveying can be described as an accurate, secure, and efficient way of conducting the settlement and lodgement stages of a conveyancing transaction, thereby replacing manual and paper-based processes in traditional property transactions (Knight, 2022). Using e-conveyancing, all documents needed for a transaction are made accessible to the parties involved through an online platform. PEXA, for instance, integrates e-conveyancing with other property purchase services in Australia.

Blockchain also offers the potential for further digital transformation in real estate investment transactions. The decentralisation feature of the blockchain makes it secure, efficient, and accurate. The majority of CRE investors expected that blockchain would be more applicable to real estate transactions in comparison to other real estate use cases; however, these same respondents also ranked blockchain low when asked which emerging technologies will be influential in the future of their firm and their intention to adopt it (Altus Group, 2019). This suggests that there may be a mismatch between optimism and the actual preparedness of real estate investors to adopt these emerging technologies.

6.3.4.2.4   ADMINISTRATION AND OPERATIONS

Risk and returns are two key indicators requiring greater understanding by investors, and the end goal is typically to minimise risk exposure and maximise returns; thus, performance analysis can improve investment administration and operations. Investment managers place significant emphasis on strategies that can generate and monitor sustainable income, profits, and market share. Consequently, they are profoundly concerned about competition and business risks. It is not uncommon for these managers to evaluate their returns and performance by comparing them to industry benchmarks (Baum, 2015). However, one of the major challenges with existing benchmarking approaches lies in designing a system that can adapt to changing market conditions while effectively capturing both risk and return through appropriate measurement and attribution of performance (IPF, 2018).

Benchmarking serves as a valuable method for limited partners and general partners to assess the performance of individual funds and portfolios about their peers, as well as to evaluate broader performance metrics. However, historical limitations in benchmark transparency, accuracy, and timeliness have created the potential for measurement discrepancies. This issue is particularly a challenge for investors regarding the payment of performance-related fees. To address these, contemporary benchmarking models such as VTS Marketview are reshaping the landscape by enabling real estate professionals to compare their performance in real time against market benchmarks across critical operational, financial, supply, and demand indicators (Businesswire, 2018).

Approximately 41% of real estate investment firms in the survey conducted by Altus Group were reported to be implementing automation in their benchmarking and performance analysis processes, which surpasses the adoption rate in any other area of the real estate industry (Altus Group, 2019). This automation offers time-saving advantages, leading to economic benefits. However, there is still a need for these firms to uncover new insights informing their investment decision-making. AI systems hold the most promise in this regard, although their utilisation remains relatively low due to the limited availability of standardised digital data upon which these systems rely. Companies such as Architrave have recognised this need, and they offer a comprehensive solution that includes document scanning, storage, and analysis, all within a single package. This addresses the requirement for converting analogue data into a structured and specific database, paving the way for the adoption of more advanced technologies that can deliver added value. Increased market transparency enhances the accuracy and reliability of performance analysis and benchmarking. There is a correlation between market transparency and the adoption of PropTech in areas such as smart buildings, the Internet of Things (IoT), and big data (JLL, 2020). Being able to access data on the various elements of a building can help with benchmarking the performance of an asset and a portfolio beyond financial indices. For example, the real estate industry faces mounting pressure to incorporate environmental, social, and governance (ESG) criteria into investment benchmarks, as more investors and corporations evaluate sustainability credentials at both the asset and city levels when making their investment decisions.

6.3.4.2.5   REPORTING AND RELATIONSHIP MANAGEMENT

When funds are raised, investors require the fund managers to provide regular and periodic updates on their investment activities and performance. The larger institutional investors particularly demand strategic relationships with their managers, leveraging their access to research, analytics, and advice. Although several products and tools offer AI-based analytics sourced through the aggregation of disparate data and platforms, 60% of firms are reported to still use spreadsheets as their main primary tools to fulfil investor reporting functions (Pi Labs, 2020a). There is therefore an increasing need for a one-stop shop for the data needed to compile reports. To tackle the issue of data being confined within separate and isolated software solutions, the data standard agency OSCRE is supporting the development and implementation of real estate standards and effective data governance practices to promote improved transparency and connectivity within technology stacks throughout the investment management value chain. By establishing this integration, data portability challenges can be addressed, allowing for more seamless and interconnected workflows across various software platforms. Furthermore, platforms such as Cherre also provide automated data mapping and reportage. Additionally, reporting is generally more enhanced, and investors may have access to real-life portfolio performance.

With more streamlined property investment management processes, operations costs can be significantly minimised, and returns can be maximised more efficiently and effectively. Professionals can now focus more on strategic investment decisions, making use of an extensive array of insights and streamlined processes.

6.3.4.3   Revenue Optimisation

With increasing insight and improved correlations among data points, revenue can be enhanced for brokers and investment managers. AIMS are used to determine future price projections for specific projects and neighbourhoods and to help define the factors that determine the likelihood of a higher ROI. Factors that could be accounted for include local employment growth trends, new planning permissions for competing development, and new construction permits for infrastructure e.g., rail or metro station or other transport infrastructure which can affect yields, rents, and values in general. Furthermore, the power of market data and advanced analytics can enable the investment manager to make accurate predictions to improve asset selection for purchase, tenant selection, etc. AIMS can also integrate these into firms' internal asset management systems, enabling automatic disparate data processing and advanced analytics and generating automated reports to dashboards and ultimately the investment decision-makers. For instance, by tracking the hyperlocal submarket activity around a property and determining relationships between local market data and recent events, some AIMS (such as CBI) can automatically optimise the asking-effective rent to align with current market trends and conditions. This can improve rental revenue and a higher achieved property sales price.

According to Souza et al. (2021), AI application to real estate investment management has the potential to push up net-of-fee returns directly to the firms' institutional, accredited, and qualified investors. Thus, sovereign fund advisors, pension funds, and endowments can pass these savings on to their clients, resulting in higher nominal risk adjustment net returns. This further enhances the realignment of principal-agent relationships including the fiduciary, moral, and ethical responsibilities of the manager/adviser to the client.

6.3.4.4   Cost Minimisation

The enhancement of various aspects of the investment manager's functions through digital technology can directly or indirectly lead to cost minimisation. For instance, advanced analytics and improved access to information can provide insights into the areas of the asset management process that can be streamlined, and the asset costs can be minimised and areas that cost can be cut down. Furthermore, automated systems will reduce the cost of labour translating into higher returns for individual real estate properties, portfolios, and companies. Inefficient portfolio and asset management can lead to higher vacancy rates which can lead to higher physical and economic costs because of space underutilisation and utility bills for the vacant spaces.

Massive costs are also directly and indirectly associated with inefficient investment risk management and underwriting. AIMS can process complex analytics such as identifying when a market is heating up or cooling off in real-time and enable investment managers to see these trends, investigate them, and use them to make decisions.

6.3.4.4   Challenges and Limitations of Automated Investment Management Systems

The adoption of AIMS has several challenges and limitations. The ReimTech 2020 roundtable comprising 30 global participants from leading property firms provided valuable insights on the barriers and limitations of technological implementation in real estate investment. Some of the key issues are highlighted in Table 6.2:

Pi Labs (2020a) further raises an important point. They argue that users cluster on the platforms that have the biggest user base, and the effect of this can be a virtuous cycle of growth for the bigger platforms while the smaller platforms are edged out. These bigger

Table 6.2 Challenges and Barriers to Implementing AIMS

Challenge	Detail
Fragmented data sources for due diligence	The due diligence process in assessing acquisitions is fragmented and there are multiple information sources and diverse stakeholders; these require a high level of data and analytics expertise.
Variation in KYC processes	Separate KYC processes involving various investors and joint venture partners are time-consuming; having a more centralised system with the process only occurring once to the highest standard required in the market will be economically efficient.
Integrating various systems for investment decisions is challenging	Misalignment between professional and client's data format and requirement often leads to the costly omission of vital data. Although expert platforms can get data from the right sources, it is challenging to integrate all of those systems into a single decision-making process where all the data is combined. Furthermore, there should be standardised data formats for better integration.
Insufficient digital literacy	Investment managers generally do not possess skills in advanced analytics and digital systems; this is limiting the adoption of AIMS across a wide spectrum.
High cost of digital tools and systems	AIMS are quite expensive to acquire or subscribe to; this will particularly be challenging for smaller funds.

Source: Authors (2023).

platforms can better leverage their access to data and advanced analytics to provide products and services than their competitors. The combination of huge funding requirements to gain market share and the potential monopoly gains associated with market leadership may lead to unhealthy sectoral control. This is a valid concern; however, it can also be argued that real estate services firms already have huge sectoral control, and there is no evidence that this has affected other real estate services start-ups.

6.3.5   Virtual Capital (VC)

Venture capital is a type of financing provided to start-up and early-stage firms that have high growth potential and perceived long-term growth potential. Venture capitalists (VCs) are investors who provide funding to these companies in exchange for an equity stake or ownership in the business. They take on higher risks compared to traditional investors in hopes of earning substantial returns on their investments. Venture capital investment is usually classified based on the volume of funds raised, and these are usually referred to as funding rounds. The different categories are illustrated in Table 6.3.

Table 6.3 Venture Capital Funding Rounds[12]

Funding Amount	Funding Round	Details
$0–1 million	Angel round/ family and friends Pre-seed	This is the earliest stage of funding when a company is still in its ideation phase or has just started. Funding is often provided by the founders themselves, friends, family, or angel investors to help the business get off the ground. It is used to conduct market research, develop prototypes, and validate the business concept.
$1–4 million	Seed	The initial investment received by a start-up after the pre-seed stage. It is usually provided by angel investors, venture capital firms, or early-stage investment funds. Seed funding helps start-ups develop their products or services, conduct further market research, build a team, and establish a strong foundation for growth. This is sometimes referred to as the time 'proof of concept'.
$4–15 million	Series A	When a start-up has progressed beyond the early stages and it has achieved certain milestones, demonstrated market traction, acquired customers, and generated initial revenue, VCs typically invest in the company to help it expand its operations, hire more employees, develop its product further, and scale its business. Product market fit and traction are also assessed.
$15–40 million	Series B	This usually takes place when a start-up has successfully achieved significant growth and is seeking additional capital to further expand its market reach, invest in marketing and sales efforts, and strengthen its position within the industry. Series B funding often involves larger investment amounts compared to earlier stages.
$40–100 million $100–250 million $250 million +	Series C Series D Series E	These stages occur as the company continues to grow and expand. These rounds are aimed at supporting further scaling, entering new markets, acquiring other companies, or preparing for an initial public offering (IPO) or acquisition.

Source: Authors analysis.

Venture capital plays a crucial role in driving innovation and the development of technology solutions within the real estate industry. PropTech venture capital focuses specifically on investing in start-ups and companies that are leveraging technology to improve various aspects of the property market. These investments support the growth and advancement of technologies such as property management software, online marketplaces, smart home technologies, data analytics platforms, virtual reality tools, and other innovations that aim to transform the way real estate is bought, sold, managed, and experienced. PropTech venture capital firms actively seek out promising start-ups in the real estate technology sector and support them with the funding required to accelerate their growth. These investments not only help these start-ups scale their operations but also enable them to develop and refine their digital solutions. Additionally, VCs often provide strategic guidance, industry expertise, and valuable networks to support the success of the companies they invest in.

PropTech VC has been instrumental in driving innovation and disruption in the real estate industry, pushing the boundaries of what is possible and shaping the future of property transactions, management, and experiences. They effectively contribute to the advancement of technology solutions that enhance efficiency, transparency, sustainability, and overall customer experiences in the real estate sector. Investment in PropTech has accelerated in the last ten years, and 53% of real estate firms are reported to be directly investing in at least one type of PropTech firm (Altus Group, 2019). These are indicators of the growth trajectory of PropTech VCs and an upward trajectory.

6.4   Case Study Analysis

Following the review of core areas of digital transformation in real estate funding, finance, and investment, this section analyses specific PropTech platforms and firms within this area of the built environment value chain. Each case study begins with an overview and summary of key details of the platform/company; the solutions it provides are then analysed using the ESEP efficiency framework introduced in Chapter 1.[13] The analysis concludes with a brief overview of the limitations and further prospects.

6.4.1   Case Study 1: Cherre

6.4.1.1   Background, Key Details, and Technologies Utilised

Cherre is a data analysis platform that takes all the internal and external data of real estate companies, "warehouses" them, and creates actionable insights into how best the data can be useful for companies. The platform can be used for assets in the various real estate sectors. Many real estate investment firms currently have a vast array of data without effective and efficient integration systems; thus, in planning, analysis, and forecasting, they may fail to consider important information and details that can enhance their portfolio and asset performance. Cherre thus stands out as a platform that provides powerful data integration and analytics, with actionable insights that enhance business and business decisions for companies that subscribe to it. Furthermore, the platform connects decision-makers to accurate property and market information, enabling them to make faster, smarter decisions. By providing what they refer to as a "single source of truth", they empower their clients to evaluate investment opportunities and trends with higher speed and accuracy while also saving financial resources that would have been expended on manual data collection, integration and analysis.

Cherre was founded in 2018 with a seed funding of $9 million. They have gone on to develop tools such making such as API integration and geospatial and mapping services using AI and various other digital technological tools. Their products include CoreConnect (real estate industry data connection and management platform), Cherre API (access to connected data analytics warehouse), SFR Data Kit (data integration, analytical models, and single-family residential API), Connection Services (data connection services and proprietary data model), CoreExplore (analytics and insight platform), Reporting and Analytics (providing insights with extensive library of reporting and analytics dashboards), and Property Search (analysis and evaluation of market properties for faster decisions). The company announced a $50 million growth funding round in 2021.

6.4.1.2   Key Efficiencies and Solutions That Cherre Provides

The solutions that Cherre provides are itemised using the ESEP efficiency framework (Table 6.4):

Table 6.4 ESEP Efficiency Analysis of Cherr

ESEP	Inefficiency with the Traditional Approach	How the Tool/Platform/Product Addresses the Inefficiency
Environmental	The analogue nature of data sourcing and collection using traditional approaches can lead to more extensive use of computing power and energy (e.g., electricity).	The integrated data sourcing and analytics dashboard reduces the energy and computing power that would normally be used to process information. By reducing the time and manpower required to source and analyse data, the computing power and energy that would have been used to complete these tasks are significantly minimised.
	Collaboration with investors and other stakeholders requires a high level of travel and commute.	Collaboration is done with reduced commuting; meetings can be held remotely with reference to the shared dashboard information, leveraging remote meeting tools. This platform can also support remote working which will further minimise carbon emissions.
	Traditional investment analysis and reporting require a high volume of stationery.	Cherre provides a platform that can be shared with various stakeholders without the need to print out the report or output. This can significantly reduce the use of stationery.
Social	There is a high level of uncertainty in real estate investment.	Cherre provides important insights that investors can leverage to make more informed decisions. Although it does not totally eliminate uncertainty, the advanced analytic capabilities provide insight that would hitherto have been difficult to observe using traditional approaches.
	Lack of transparency in and trust real estate in investment.	The platform provides a higher level of transparency through the shared dashboard. This enables the investors and stakeholders to understand the market they are investing in, and the relevant trends and forecasts. This reduces investors' and stakeholders' anxiety.

(Continued)

Table 6.4 (Continued)

ESEP	Inefficiency with the Traditional Approach	How the Tool/Platform/Product Addresses the Inefficiency
	Data sourcing and analysis is cumbersome and complex using traditional approaches.	The data sourcing process is a lot more streamlined and less stressful.
	Due to the complexities of data sourcing and analysis using analogue methods, as well as the potential of hidden insights, investment decision-making is often difficult.	With advanced analytics and improved insights using this tool, investment decisions can be taken with a higher level of accuracy and confidence, and this can make investment strategies more effective and efficient.
	Collaboration, updates and discussions with investors and stakeholders are usually more difficult, often requiring a high level of commute.	This platform provides an avenue for collaboration, discussions, and conversations to be conducted remotely, allowing for a higher level of comfort.
Economic	Investment decisions such as funds allocation, asset selection etc. take time. Considering that timing is essential in real estate investment, investors may miss out on deals or make lagged investment.	The real-time data and analytics capabilities of this platform enable the investment manager and investor to make timely decisions and take advantage of opportunities. According to the platform, their clients have recorded a 50% time-saving in arriving at actionable points.
	In addition to the cost of the time taken for delayed investment decisions, data sourcing and analytics cost money (personnel, software, equipment etc.).	With this platform, the data collection and analytics are streamlined, leading to huge savings on time, equipment, labour and other associated resources.
	Investment management and reporting are time-consuming.	Time can be saved using these integrated systems and the time saved can be allocated to other economically beneficial tasks and activities.
	Ineffective investment strategies resulting from poor or insufficient insight can lead to loss of revenue and unforeseen costs.	The platform improves investment insights which increase revenue and minimises unforeseen costs. These can also lead to further client retention which becomes an added benefit for the investment manager.
Physical/ spatial	Optimising spatial utility is difficult to achieve using traditional investment approaches.	Cherre provides important insights that can enable the investment and asset manager to minimise the vacancy rate on assets and thus utilise the space more optimally. It can also provide insights on emerging trends which the portfolio or asset manager can use to modify the space uses to accommodate changing trends e.g., changing work patterns being accommodated with flexible leases or spaces.

Source: Authors analysis.

6.4.1.3   Limitations, Challenges, and Prospects

The adoption of this digital platform has been limited. For real estate firms to be able to use it, they need to pay the requisite fee, and their staff will need to undergo training and upskill to be able to utilise this tool and to get the best out of the systems. This will require huge financial resources, and it can be a challenge particularly, for smaller fund managers. Although the platform has the potential to attract environmental benefits by minimising high computing power and energy in analysis, it can also be argued that due to the high performance of the platform, high computing and energy usage will still be required to integrate all the systems, and the subscription to the cloud and other data systems will also lead to an increase in the demand for data centres which, as discussed in Section Two, is also significantly contributing to the carbon emission and global energy consumption.

With a wider level of adoption and access to a wider pool of data, the machine learning capabilities of Cherre can enable it to make a remarkable improvement over time and to more improved data analytics services.

Further details on Cherre can be accessed from their website using this link: https://cherre.com/.

6.4.2   Case Study 2: CoStar

6.4.2.1   Background, Key Details, and Technologies Utilised

CoStar is a real estate data and analytics platform that provides industry information and new analytics services that real estate investors and service providers need to make smart decisions. CoStar provides extensive research operations and comprehensive data, providing their clients with an enhanced understanding of market trends, transactions, assets, and industry players. They currently service approximately 188,000 industry professionals, providing insights on commercial, residential, and operational real estate asset classes.

CoStar Group and its CoStar product were founded in 1987 by Andrew C. Florance in Washington, D.C. At the time, as a computing consultant and aspiring real estate developer, Andrew was frustrated with the quality of real estate information that was available and realised that he could use his knowledge of computers to develop a commercial multiple-listing database service. The concept was to centralise the process, putting economies of scale to work so that brokers could reduce their own research efforts and instead obtain more accurate and comprehensive information at less expense from CoStar. This information was originally presented to CoStar's customers on paper: Andy printed ten copies of the 1,200 pages of information he had compiled, using a rented laser printer, and sold them for $800 each. Later, the information would be distributed on compact discs. And in 2000, CoStar phased out physical formats and became a web-based application. In 2004, CoStar Group acquired London-based FOCUS, shortly followed by Scottish Property Network, expanding its market coverage into the United Kingdom, and by 2012, it acquired LoopNet, followed by Apartments.com in 2014 – these acquisitions diversified CoStar's activities from information services into marketplaces.

Since its creation, the data in CoStar's information product have evolved in its breadth and depth, assembling and presenting data from a wide array of sources including primary research, public record, agent submissions, aerial and drone research, data partners, and more. The platform also provides commercial, multifamily, residential, land, and business marketplaces offering for sale and/or for-lease listings for public consumption. The product's features have also been further developed, with continuous product development resulting in enhanced analytics, forecasting, custom search and reporting, machine learning, and more, providing real estate market intelligence, news, and analytics.

CoStar currently has over 85 million unique monthly visitors to their online marketplaces and 170k+ subscribers to their information service. The platform currently tracks 6.9 million commercial properties and the total revenue generated in 2020 was estimated at $1.66 billion. Its services include the provision of property records and search, lease and sale transactions, market, submarket and custom analytics, market KPI forecasts, market and submarket reports, and tenant, owner, and broker information. Other services include for-sale and for-lease listings, loan and financial information, public record, demographic and environmental data, land parcels, registered true owners, and industry news.

6.4.2.2 Key Efficiencies and Solutions the Platform Provides

The solutions that CoStar provides are itemised using the ESEP efficiency framework (Table 6.5).

Table 6.5 ESEP Efficiency Analysis of CoStar

ESEP	Inefficiency with the Traditional Approach	How the Tool/Platform/Product Addresses the Inefficiency
Environmental	The traditional approach to data sourcing and collection can lead to the extensive use of computing power and energy (e.g., electricity).	The integrated data sourcing and analytics dashboard in CoStar reduces the energy and computing power that would normally be used to access, integrate and process information. By reducing the time and manpower required to source and analyse data, the computing power and energy that would have been used to complete these tasks would be significantly minimised.
	Collaboration of real estate professionals requires a high level of travel and commute.	This platform enables collaboration with a significantly reduced commuting level; meetings can be held remotely with reference to the shared dashboard information, leveraging remote meeting tools such as Zoom and Microsoft Teams. The CoStar platform also supports remote working which will further minimise carbon emissions.
	Traditional investment analysis and reporting require a high volume of stationery.	CoStar provides a platform for information and data to be shared with various stakeholders in remote locations; teams can therefore access the digital versions of data without the need to print out the report or output. This can significantly reduce the use of stationery.

(Continued)

Table 6.5 (Continued)

ESEP	Inefficiency with the Traditional Approach	How the Tool/Platform/Product Addresses the Inefficiency
Social	There is a high level of uncertainty in real estate investment.	CoStar provides valuable insights and information that investors can utilise to make more informed decisions. Although it does not eliminate uncertainty, it provides information that can be utilised to minimise the uncertainty in investment and advisory services.
	Lack of transparency in real estate markets.	The platform provides a higher level of transparency through access to various forms of data from different markets. Platform users are provided with access to robust, detailed and reliable intelligence on a diverse range of data points. This enables the investors and real estate service providers to understand the market they are investing in, and the relevant trends and forecasts in those markets. This further reduces stakeholders' anxiety. The News services also contribute to improving the transparency in real estate markets.
	Data sourcing and analysis is cumbersome and complex using traditional approaches.	The data sourcing process is a lot more streamlined and less stressful. The data from the platform can be accessed and downloaded with relative ease.
Economic	Investment decisions such as funds allocation, asset selection etc. take time. Considering that timing is essential in real estate decisions (investment, development, valuation, letting, sales, etc.), market opportunities may be missed and wrong decisions may be taken.	The extensive data, update, analytics and news services provided by CoStar enables the market participants and stakeholders to make well-timed decisions.
	In addition to the cost of the time taken for delayed decisions, data sourcing and analytics using traditional approaches cost money (personnel, software, equipment etc.).	With CoStar, the data collection process is streamlined, leading to huge savings on time, equipment, labour and other associated resources. Data sets can be downloaded from the platform, making data use easier.
	Sourcing data manually requires a high level of consultation and searching which is time-consuming.	The time taken to collect data is significantly minimised, and the time that is saved can be allocated to other tasks and activities.

(Continued)

Table 6.5 (Continued)

ESEP	Inefficiency with the Traditional Approach	How the Tool/Platform/Product Addresses the Inefficiency
	Ineffective investment strategies resulting from poor or insufficient insight can lead to loss of revenue and unforeseen costs.	The platform improves valuable insights that can be used for investment and for other advisory services. The higher level of predictability and forecasting capabilities can lead to an increase in revenue and a reduction in unforeseen costs. These can also lead to further client retention which becomes an added benefit for the investment manager. Furthermore, through the data and news provided, portfolio and asset managers can take advantage of investment opportunities and also minimise risks in their portfolios and assets with access to unique market insights.
Physical/ spatial	Optimising spatial utility is difficult to accomplish using traditional investment approaches.	CoStar provides important insight that developers, investors, and asset managers can use in their development, marketing and property management strategies to minimise the vacancy rate on properties and thus more optimally utilise the space. It can also provide insights on emerging trends which the portfolio or asset manager can use to modify the space uses to accommodate changing trends e.g., changing work patterns being accommodated with flexible leases or spaces.

Source: Authors analysis.

6.4.2.3   Limitations, Challenges, and Prospects

Although CoStar provides free access to some of the services, there are other services and tools that require a subscription. This will require a financial commitment from firms and service providers which some firms, particularly smaller ones, may find challenging. From an environmental point of view, the potential benefits that are accrued from less use of computing power and energy may be eroded by the high computing power and energy required for the platform to be maintained and utilised. The subscription to cloud-based systems and other data systems will also lead to an increase in the demand for data centres which, as discussed in Section Two, is also significantly contributing to carbon emission and global energy consumption.

CoStar has the potential to remarkably increase its market share. With a team of approximately 1,400 marketing and research advisors, they can continue to improve through research. Their research capabilities also include aerial surveys by drones and planes, and they also have architectural photographers who tour markets and capture high-quality images and video footage of properties which can further advance their ability to provide local market insights. The platform also invests in software development, which enables them to constantly update and introduce new features and tools on their platform. With a continued investment in data and technology, CoStar has the potential to make significant improvements to its services.

Further details on CoStar can be accessed from their website using this link: https://www.costar.com/.

6.4.3   Case Study 3: EVORA

6.4.3.1   Background, Key Details, and Technologies Utilised

EVORA is an ESG data management and advisory platform used for real estate investment. It utilises software that automates the acquisition, cleansing, and reporting of sustainability data. EVORA was founded in 2011 by Chris Bennett, Paul Sutcliffe, and Ed Gabbitas because of their passion for environmental sustainability. The increasing scale of global awareness and prioritisation of ESG has resulted in a growing demand for EVORA's expertise. The company has a 200+ person workforce, with 72 of them hired in 2021. The company has offices in the UK, Germany, Italy, Romania, India and New York and it provides consultancy for large-scale real estate asset owners, asset managers, pension funds and investment banks, including LGIM, Hines and M&G, providing a full-service sustainability offering encompassing strategy, data management and reporting and action planning, while utilising their proprietary ESG data management software, SIERA.

SIERA is EVORA's proprietary sustainability data management software that was launched in 2013 to address the data needs of the real estate investment market. It enables real estate investors to make crucial decisions around ESG capital allocation. ESG-related targets continue to emerge in the real estate industry, and its acceleration is at an exceptional pace creating the need for more ESG data and more complex, regulated ESG reporting. Initially used as a consultant tool to drive internal productivity and ensure the quality of data, SIERA has now been developed to be used externally by a range of different users, servicing over 200 real estate investment and debt funds. In 2020, it expanded its ESG services to infrastructure and in 2021 developed its capability to support debt providers and offer a green finance service line. In the same year, SIERA+ was launched, built specifically to support engagement between property, asset and investment teams, to improve ESG data quality, and to provide insights at all stages of the investment lifecycle. SIERA+ has a design that is practical, easy to use, and fast which allows users to have better results quicker and to create tailored solutions aligned with businesses' best practices.

6.4.3.2   Key Efficiencies and Solutions the Platform Provides

The solutions that Evora provides are itemised using the ESEP efficiency framework (Table 6.6).

6.4.3.3   Limitations, Challenges, and Prospects

There are several factors limiting the adoption and scaling of EVORA. For instance, the platform needs access to various ESG data sets, and these are both challenging and expensive, particularly in relation to real estate. Therefore, there is a need to further develop the SIERA software to be able to automate the data collection and integration processes. The development of the open API by the company will make this automation drive much

Table 6.6 ESEP Efficiency Analysis of Evora

ESEP	Inefficiency with the Traditional Approach	How the Tool/Platform/Product Addresses the Inefficiency
Environmental	Traditional real estate investment management strategies do not focus on environmental sustainability, and those that aim to consider this factor do not have the tools to support this drive.	EVORA provides a platform for firms, investors and fund managers who aim to integrate environmental sustainability and ESG factors to do this. This can lead to reduced emissions and wastage. The service that this platform provides has the potential to transform investment management in line with the current global ESG drive. The advisory services ranges from net-zero to green finance, infrastructure, and climate resilience.
Social	With the increasing drive towards ESG, investors and fund managers that are interested in integrating ESG into their investment and funds strategies are unsure of how to go about this. There is also a lot of uncertainty and a lack of information in around this issue.	This platform provides valuable insights and information that investors can utilise to develop their ESG strategies and make more informed decisions.
	Sourcing ESG data and analysing them can be quite cumbersome, as most of the data is fragmented.	ESG data sourcing for real estate is a lot more streamlined and less stressful and cumbersome using this platform.
Economic	Investment decisions such as funds allocation, asset selection etc. take time, and this problem is exacerbated when ESG considerations are involved. Considering that timing is essential in real estate decisions (investment, development, valuation, letting, sales, etc.), market opportunities may be missed, and wrong decisions may be taken.	This platform can provide the necessary data needed to make ESG-related decisions in an investment and this can ensure that investment decisions are not delayed.
	ESG data sourcing and analytics using traditional approaches cost money (personnel, software, equipment etc.).	With this platform, data becomes available using a streamlined process and this can lead to huge savings on time, equipment, labour and other associated resources. Data sets can be downloaded from the platform, making data use easier.
	Sourcing ESG data manually requires a high level of consultation and searching which is time-consuming.	EVORA saves the time that would have been spent collecting ESG data, and the time that is saved can be allocated to other tasks and activities.
Physical/ spatial		

Source: Authors analysis.

quicker. Data quality is also a challenge. The platform reports that only 14% of investors have investment-grade data that leave a huge gap in potential insights that is untapped. As stated in the previous section, poor data quality can lead to poor and costly investment decisions. Although the platform has integrated data validation checks at several stages (validation by the data team during onboarding, and automated data checks in the software), it would be valuable to improve the scope of data collection to ensure that a broader range of insights is integrated into the advisory services provided.

EVORA is currently planning to extend the geographical scope of its services by opening offices in other parts of Europe, the US, and Asia, with plans to also make their data available to a broader group of financial professionals.

Further details on Evora can be accessed from their website using this link: https://evoraglobal.com/.

6.4.4   Case Study 4: Fifth Wall

6.4.4.1   Background, Key Details, and Technologies Utilised

Fifth Wall supports VCs that are focused on technology for the global real estate industry. The company invests in built environment technology companies that are addressing some of the world's pressing challenges such as climate change, ageing buildings and infrastructure, unreliable supply chains, inaccessible housing markets, and the future of work. Fifth Wall is a Certified B Corporation, founded in 2016, and one of the largest venture capital firms focused on technology in real estate. With approximately $3 billion in commitments and capital under management, Fifth Wall connects many of the world's largest owners and operators of real estate with the entrepreneurs who are redefining the future of the Built World. This company is backed by a global mix of more than 90 strategic limited partners (LPs) from more than 15 countries, including BNP Paribas Real Estate, British Land, CBRE, Cushman & Wakefield, Host Hotels & Resorts, Kimco Realty Corporation, Lennar, Lowe's Home Improvement, Marriott International, MetLife Investment Management, Related Companies, Starwood Capital, Toll Brothers, and others.

Fifth Wall raised its first capital in 2016 for built environment technology. It has since expanded beyond the real estate asset class, building a team of operating and advisory professionals, and scaling globally to Europe, Asia, and Latin America. After its first funding round, it announced the close of its $212 million real estate technology fund (Fund I) in May 2017. Co-Founder Wallace said that the firm identified real estate technology as an underserved investment niche because the industry has traditionally used very little technology and was therefore open to disruption. In July 2019, the firm announced the close of its second fund (Fund II), which closed at $503 million. While Fund I was raised from nine investors in the US only, Fund II consisted of a global investor base of over 50 investors from 11 different countries.

In 2020, the firm announced its retail fund of $100 million to back emerging retail concepts and their brick-and-mortar growth. As of February 2020, it manages over $1.1 billion in assets. The firm announced that it would raise $500 million for its "Climate Tech Fund", a sustainability venture fund investing in technologies focused on energy efficiency, water reuse, climate resilience, decarbonisation and risk management. In 2021, Fifth Wall announced that it had raised more than $1.1 billion across its funds and had a global investor base of over 90 investors. In addition, several large real estate players announced their investments in the firm's Climate Tech Fund, including Equity Residential, Hudson Pacific

Properties, Invitation Homes, Ivanhoe Cambridge and others. The company has made over 120 investments: the most recent investment was in June 2022, when WiredScore raised $15 million. Fifth Wall has also made 21 diversity investments. The most recent in June 2022, when $55.5 million was raised. Fifth Wall has had seven exits; the most notable include ClassPass, Opendoor, and Harbor.

6.4.4.2 Key Efficiencies and Solutions the Platform Provides

The solutions that Fifth Wall provides are itemised using the ESEP efficiency framework (Table 6.7).

Table 6.7 ESEP Efficiency Analysis of Fifth Wall

ESEP	Inefficiency with the Traditional Approach	How the Tool/Platform/Product Addresses the Inefficiency
Environmental	Typical VCs focus on raising capital for digitals in the built environment without necessarily focusing on promoting environmental sustainability.	By actively promoting raising funds for tools that can combat the carbon emissions and environmental waste caused by the built environment, Fifth Wall is increasing and promoting environmentally sustainable innovation in the built environment, and the development of tools that can significantly promote environmental sustainability in the built environment.
	Traditional real estate activities continue to increase carbon emissions and environmental wastage.	With more funding made available to PropTech tools that can decarbonise the built environment, carbon emissions that are generated through real estate use, operations and services can be significantly minimised.
Social	With the increasing drive towards ESG, investors and start-ups that are interested in accessing funding to develop their ideas have a lot of uncertainty, particularly as most funds will focus on economic viability rather than environmental sustainability.	Fifth Wall provides access to capital for investors and start-ups that are primarily interested in tackling environmental sustainability in the built environment. This can reduce the anxiety associated with raising capital, particularly in a world where economic benefits are still prioritised over environmental sustainability.
Economic	Raising start-up capital often takes a long period of time, particularly in the case where funds are required for projects that would not necessarily meet the benchmark for investment capital.	Fifth Wall provides specialist funds to support start-ups with the funds needed to fund PropTech tools, particularly those that can enhance the environmental efficiency of the built environment.
	It takes a long time to secure funding for PropTech tools, particularly those in niche areas.	Fifth Wall provides quicker access to funding for niche (environmentally driven digital solutions to the built environment).
Physical/spatial		

Source: Authors analysis.

6.4.4.3   Limitations, Challenges, and Prospects

The effective utilisation and expansion of the Fifth Wall are limited by several factors. Firstly, the legislation on ESG and the built environment is weak, and thus, asset owners and managers do not feel responsible for the emissions generated from their assets. With stronger legislation, asset owners, fund managers and asset managers will gain more awareness of the environmental impact of their activities, and this can begin to make investment in this area more robust. There are also misaligned incentive structures across the industry. Many design and development teams are primarily concerned with new construction costs, especially because operations (and operating costs) end up as another team's responsibility. In too many cases, climate technologies that could quickly pay back their initial costs are omitted in the name of reducing upfront costs.

Fifth Wall's current mission is to focus on the intersection of climate technology and real estate. The company estimates that about $18 trillion is required to get the real estate industry to net zero within the next decade; their current strategy is therefore focused on climate change PropTech is dedicated to decarbonising the built environment. They, therefore, aim to support companies that are best positioned to provide innovative changes to the built environment industry through the deployment of PropTech tools.

Further details on xx Technologies can be accessed from their website using this link: https://fifthwall.com/.

6.4.5   Case Study 5: MSCI

6.4.5.1   Background, Key Details, and Technologies Utilised

Morgan Stanley Capital International (MSCI) is an investment research firm that provides investment data, stock indexes, portfolio risk and performance analytics, and governance tools to institutional investors and hedge funds. It is one of the largest providers of Environmental, Social, and Governance (ESG) indexes with over 1,500 equity and fixed-income ESG indexes designed to help institutional investors to more efficiently benchmark ESG investment performance and manage, measure, and report on ESG mandates. MSCI offers a range of Climate Indexes for investors who seek to incorporate climate risks and opportunities into their investment process.

With over 50 years of expertise in research, data, and technology, MSCI enables clients to understand and analyse key drivers of risks and returns in real estate investments and for them to use these to build effective portfolios and develop effective investment strategies and processes. The platform aims to provide market intelligence and analytics that can provide greater transparency to the financial market and use innovation to drive global economies. MSCI continues to expand its product range, serving global investors with climate change solutions. MSCI's recent acquisition of Carbon Delta, which is a global leader in climate change scenario analysis, strengthens the suite of climate risk capabilities with technology that supports climate scenario analysis and forward-looking assessment of transition and physical risks, as well as extensive company-level analysis of publicly traded companies globally. It contributes to supporting real estate investors to integrate climate, performance, and risk analysis to build more sustainable portfolios.

MSCI has various services, tools, and solutions. One of them is Global Intel. MSCI's data set represents $2 trillion in private real estate assets across 30 countries and over $450 billion in publicly listed real estate. With 150 comparable measures, this powerhouse of tens of thousands of properties globally drives the suite of analytical solutions and enables clients and stakeholders to answer the high-level and granular questions that really matter. The platform also provides income analytics. INCANS powered by Income Analytics uses intuitive dashboards, bond-equivalent rating scores, and a proprietary global tenant grading system, to provide a common, standardised language across pan-regional teams and investment strategies. Furthermore, the platform provides Real Capital Analytics (RCA). This service effectively aggregates timely and reliable transaction data and provides valuable intelligence on market pricing, capital flows, and investment trends in more than 170 countries. This unique insight is used to formulate strategies, source new opportunities, and execute deals. Other products include real estate climate solutions, enterprise analysis, indexes, and benchmarks. Real estate climate solutions are provided through the real estate climate Value-at-Risk (Climate VaR) which is a forward-looking and return-based valuation assessment for individual assets and portfolios. The real estate enterprise analytics provides clients with a common language, and the actionable insight they need to determine future allocation, translate idiosyncratic risk, quickly compare yields against the market, and make better investment decisions. Additionally, the real estate indexes and benchmarks are backed by almost 40 years of benchmarking leadership, and their consistent, comparable, and independent indexes provide the real estate industry with a shared understanding of global markets and set the foundation for effective strategy development and robust performance and risk measurement.

6.4.5.2   Key Efficiencies and Solutions the Platform Provides

The solutions that MSCI provides are itemised using the ESEP efficiency framework (Table 6.8).

Table 6.8 ESEP Efficiency Analysis of MSCI

ESEP	Inefficiency with the Traditional Approach	How the Tool/Platform/Product Addresses the Inefficiency
Environmental	The traditional approach to data sourcing and collection can lead to the extensive use of computing power and energy (e.g., electricity).	The integrated data sourcing and analytics dashboard in MSCI reduces the energy and computing power that would normally be used to access, integrate and process data. By reducing the time and manpower required to source and analyse it, the computing power and energy that would have been used to complete these tasks is significantly minimised.
	Collaboration of real estate professionals requires a high level of travel and commute.	This platform enables collaboration with a significantly reduced commuting level; meetings can be held remotely with reference to the shared dashboard information, leveraging remote meeting tools such as Zoom and Microsoft Teams. The MSCI platform also supports remote working which will further minimise carbon emissions in real estate analytics and investment management.

(Continued)

Table 6.8 ESEP Efficiency Analysis of MSCI

ESEP	Inefficiency with the Traditional Approach	How the Tool/Platform/Product Addresses the Inefficiency
	Traditional investment analysis and reporting require a high volume of stationery.	The MSCI platform is a fully digital system that provides information and data to be shared with various stakeholders in remote locations; teams can therefore access the digital versions without the need to print out the report or output. This can significantly reduce the use of stationery.
Social	There is a high level of uncertainty in real estate investment.	MSCI provides valuable insights and information that investors can utilise to make more informed decisions. Although it does not eliminate uncertainty, it provides information that can be utilised to minimise the uncertainty in investment and advisory services.
	Lack of transparency in real estate markets.	The platform provides a higher level of transparency through access to various forms of data from different markets. MSCI subscribers are provided with access to robust, detailed, and reliable intelligence on a diverse range of data points. This enables the investors and real estate service providers to understand the market they are investing in, and the relevant trends and forecasts in those markets. This further reduces stakeholders' anxiety. The News services also contribute to improving the transparency in real estate markets.
	Data sourcing and analysis is cumbersome and complex using traditional approaches.	The data sourcing process is a lot more streamlined and less stressful. The data from the platform can be accessed and downloaded with relative ease.
	Manual data collection and analysis will require extra effort and expertise to conduct data visualisation.	MSCI provides a range of data analytics that are accompanied by visualisation aids. These make it easier for investors and service users to comprehend the data, make sense of it and draw inferences.
Economic	Investment decisions such as funds allocation and asset selection take time. Considering that timing is essential in real estate decisions (investment, development, valuation, letting, sales, etc.), market opportunities may be missed and wrong decisions may be taken.	The extensive data, update, analytics, and news services provided by the MSCI platform enable the market participants and stakeholders to make well-timed decisions.

(Continued)

Table 6.8 ESEP Efficiency Analysis of MSCI

ESEP	Inefficiency with the Traditional Approach	How the Tool/Platform/Product Addresses the Inefficiency
	In addition to the cost of the time taken for delayed decisions, data sourcing and analytics using traditional approaches cost money (personnel, software, equipment, etc.).	With MSCI, the data collection process is streamlined, leading to huge savings on time, equipment, labour, and other associated resources. Data sets can be downloaded from the platform, making use easier.
	Sourcing data manually requires a high level of consultation and searching which is time-consuming.	The time taken to collect data is significantly minimised, and the time that is saved can be allocated to other tasks and activities.
	Ineffective investment strategies resulting from poor or insufficient insight can lead to loss of revenue and incur unforeseen costs.	The platform improves valuable insights that can be used for investment and for other advisory services. The higher level of predictability and forecasting capabilities can lead to an increase in revenue and a reduction in unforeseen costs. These can also lead to further client retention offering an added benefit for the investment manager. Furthermore, through the data and news provided, portfolio and asset managers can take advantage of investment opportunities and also minimise risks in their portfolios and assets with access to unique market insights.
Physical/spatial	Optimising spatial utility is difficult to accomplish using traditional investment approaches.	MSCI provides important insight that developers, investors, and asset managers can use in their development, marketing and property management strategies to minimise the vacancy rate on properties and thus utilise the space more optimally. It can also provide insights on emerging trends which the portfolio or asset manager can use to modify the space uses to accommodate changing trends e.g., changing work patterns being accommodated with flexible leases or spaces.

Source: Authors analysis.

6.4.5.3   Limitations, Challenges, and Prospects

MSCI provides free access to some of the services and market intelligence; however, there are other services and tools that require a subscription. This requires a financial commitment from firms and service providers which some firms, particularly smaller firms may find challenging. While reducing the use of computing power and energy can offer environmental advantages, the maintenance, and utilisation of the platform itself may offset these benefits due to the high computing power and energy requirements. Additionally, subscribing to cloud services and other data systems leads to a surge in demand for data centres, which, as mentioned in Section Two, contributes significantly to carbon emissions and global energy consumption.

Further details on MSCI can be accessed from their website using this link: https://www.msci.com/.

6.4.6   Other Cases (Exercises for Students)

6.4.6.1   CASAFARI

CASAFARI is an online platform that digitises, centralises, and clarifies all information available on the various real estate markets in the world. Information in real estate markets is chaotic, fragmented, unclear, and in different formats. This makes it unclear to tenants and various built environment stakeholders. This platform was therefore designed to enable real estate professionals to quickly source new properties, find historical data, track their portfolio, be aware of exclusivity breaches, and keep an eye on other agencies' activities. The platform also enables users to download a complete comparative market analysis within a few minutes. Property agents in the network can also be contacted to facilitate transactions; thus, the real estate professional gets access to big capital and closes more transactions in both residential and commercial properties, helping to create and grow a new asset class – single-unit portfolios at large scale.

Further details on CASAFARI can be accessed from their website using this link: https://www.casafari.com/.

6.4.6.2   CROWDPROPERTY

CROWDPROPERTY is a platform that offers various property project funding using crowd-funding. It caters for all real estate finance needs including development, bridging, refurbishment, and auction finance. Investors can invest in a variety of real estate projects for different durations, and in turn, they receive interest. Individuals and businesses can also apply for funding for their projects which may range from residential to commercial and operational real estate assets. Additionally, the platform provides a function referred to as "AUTOINVEST" which utilises an algorithm to enable an investor to build a diversified investment portfolio.

Further details on CROWDPROPERTY can be accessed from their website using this link: https://www.crowdproperty.com/.

6.4.6.3   EGI

EGI is an online platform that provides data intelligence and analytics for real estate markets. The platform provides high-quality data and ESG solutions through insight into the marketplace and how this can generate business for its clients. The platform is built on data power, analytics, critical intelligence, news updates, and industry insights which investors and real estate professionals can use to develop their strategies and make effective investment decisions. The company aims to inspire positive change in the real estate industry through its technology, ESG, well-being, diversity, and inclusion programmes with environmental, social, economic, and physical benefits.

Further details on EGI can be accessed from their website using this link: https://www.eg.co.uk/.

6.4.6.4   JLL Spark

Jones Lang LaSalle (JLL) Spark is a professional services company that specialises in real estate and investment management. The company offers a range of services, including leasing, capital markets, integrated property and facility management, project management, advisory, consulting, valuations, and digital solutions. Its clients include for-profit and not-for-profit entities, public–private partnerships, and governmental entities across a variety of industries. JLL currently has about 100,000 employees worldwide directly in its books. With client annual transaction worth over $208 billion, assets under management (Lasalle Investment Management) estimated at $60 billion, and corporate presence in 80 countries as at the year 2020, JJL as a brand, is valued at $8.8 billion.

JLL Spark, founded in 2017, is the corporate venture fund of JLL, it is responsible for catering to clients by identifying innovative real estate, buying into it, building it, potentially partnering with such innovative companies worldwide and even investing in the so-identified innovations as future game changers. JLL Spark has invested more than $340 million in more than forty early stage PropTech start-ups ranging from IoT sensors to tenant experience services to deployment of AI for real estate solutions.

Further details on JLL Spark can be accessed from their website using this link: https://spark.jllt.com/.

6.4.6.5   Landis

Landis is an online platform that supports individuals to achieve their homeownership goals by providing low-to-moderate-income individuals with financial literacy and providing individualised mortgage coaching. Landis allows clients to choose their dream home and rent it for up to two years while they get their mortgage ready, saving up for a down payment and building credit. The platform connects clients with one of their partner agents who will support them in their home search. Landis will then put in an offer, and handle the appraisals, inspection and all the paperwork. When the client is ready, they can repurchase the house. They also provide a free coaching mobile app, the Landis Homeownership Coach. The app provides a dashboard view of credit, down payment savings, and debt, with insights and actions for clients toward reaching their goal of qualifying for a mortgage.

Further details on Landis can be accessed from their website using this link: https://www.landis.com/.

6.4.6.6   Plentific

Plentific is an online platform that provides end-to-end solutions to the connectivity of real estate investment stakeholders. It connects property owners, operators, service providers, investors, and tenants to a single platform, making operations simpler, faster, and more efficient. By working with clients to streamline operations, unlock revenue, increase tenant satisfaction, and remain compliant, Plentific empowers clients with data-driven insights that drive action. It currently has a network of over 1 million properties and 35,000 service providers worldwide.

Further details on Plentific can be accessed from their website using this link: https://www.plentific.com/en-gb/.

6.5   Chapter Summary

Real estate investment, funding and finance are the underpinnings of the real estate industry. Although this segment of the value chain has been quite resistant to change, it is now experiencing an acceleration as a result of increasing economic, political, and epidemiological shocks in the last few years. The COVID-19 pandemic has particularly contributed significantly to the increase in the digitisation of real estate investment functions, as capital and occupier markets undergo huge transformations and adaptation to the changing times. Investment managers are increasingly expected to measure and forecast real estate market indicators, despite the uncertainty in global economic and political systems; they thus are increasingly looking to various digital tools and systems to enable them to identify opportunities and risks in order to maximise returns and minimise costs.

As advanced data and analytics become commonplace in real estate investment, it can lead to standardisation around data benchmarks, and help to create institutional grade, transparent data platforms to support freely trading assets, or fractions of those assets in capital markets, although this may still take several years to achieve. Short-term and medium-term opportunities for real estate investment include areas around raising capital, onboarding clients, and standardising and digitising existing data to accurately and efficiently measure and then position their portfolios. When data are standardised, it can provide enhanced insight through the introduction of alternative big data sources and shed light on new components of value.

Incorporating advanced analytics into a portfolio is a complex task that requires the collection of ample data to develop algorithms. Manual data scraping for analytical purposes can be costly, and despite the growing number of organisations attempting to leverage the benefits of advanced analytics across various industries, only a limited few are successful in achieving large-scale implementation. The absence of industry-wide data standards gives rise to numerous challenges, and there is currently insufficient understanding of how to optimise data usage. Furthermore, there is little motivation for data sharing and a mismatch between the data being collected. Additionally, a significant amount of capital is invested in legacy software, leading to a reluctance in adopting emerging systems.

Furthermore, as evidenced by the case study analysis sub-section, real estate investment is transcending its traditional economic focus to encompass environmental and ESG considerations. This represents a significant paradigm shift in this segment of the real estate value chain, where economic performance has historically been the core consideration. The industry is now increasingly recognising the importance of sustainability and Environmental, Social, and Governance (ESG) factors. This evolving perspective acknowledges that the value of real estate is not solely determined by financial metrics, but also by its environmental impact and long-term sustainability. Embracing ESG principles in real estate investment demonstrates a forward-thinking approach that not only maximises economic factors but also contributes positively to the environment and society, people, and places.

Notes

1 Cash flow analysis is the evaluation of the potential income and expenses associated with a real state property to determine its net operating income and cash flow; ROI entails calculating the expected return on investment by analysing the cash flows, property appreciation, and potential tax benefits; risk assessment entails the assessment of the risks associated with a real estate investment, including market conditions, property-specific factors, and financing risks.

2 Appraisal and determining a property's value refers to the assessment of a property's value by a qualified valuer using methods such as comparable method, investment method, cost method, and profit and loss; capitalisation rate (Cap Rate) is a metric used to estimate the value of income-generating properties by dividing the net operating income by the property's value or purchase price.
3 Securitisation refers to bundling of real estate loans into investment products and selling them to investors; syndication is forming partnerships or investment groups to pool funds for large-scale real estate projects.
4 Financial modelling refers to creating projections and analysis to evaluate investment feasibility, profitability, and risks; risk mitigation is employing strategies to minimise risks, including diversification, insurance, and hedging; due diligence is conducting thorough research and analysis to assess investment viability and risk.
5 The Fourth Revolution is a term coined by Klaus Schwab, the Founder and Executive Chairman of the World Economic Forum, to describe the ongoing transformation of industries through the integration of automation, data-driven systems, and digital technologies.
6 Disadvantage or discrimination based on prejudice (Becker, 1957).
7 Disadvantage or discrimination based on the level and quality of information available on certain demographic, racial, or gender groups (Phelps, 1972).
8 Disadvantage or discrimination based not directly on prejudice or intention to be prejudiced but based more on prior disadvantaged position, e.g., resulting from lower education, skills, and income (Arrow, 1998; Higgs, 1977).
9 Section 6.3.5 will provide further insight into FinTech Capital and VC (Virtual/Venture capital).
10 This is known as real estate "fractionalisation".
11 It should be noted that real estate investment sourcing is still done using traditional approaches.
12 It should be noted that these stages may vary in terms of investment size and specific names used in different regions or industries. However, the general concept remains consistent: as a company progresses and demonstrates growth potential, it will attract investment at different stages to support its development and expansion.
13 The ESEP efficiency analysis for each case study is not exhaustive; students and readers are encouraged to identify and note down other areas of efficiency which the named tool/platform provides.

References

Altus Group. (2019). *The innovation opportunity in commercial real estate: A shift in PropTech adoption and investment.*
Altus Group. (2020). *Altus Group report reveals CRE industry on verge of significant PropTech consolidation as technology adoption reaches tipping point.* Available at: https://www.globenewswire.com/en/news-release/2020/01/27/1975486/0/en/Altus-Group-Report-Reveals-CRE-Industry-on-Verge-of-Significant-PropTech-Consolidation-as-Technology-Adoption-Reaches-Tipping-Point.html (accessed 5 July 2023).
Arrow, K. (1998). What has economics to say about racial discrimination? *Journal of Economic Perspectives, 12*(2), 91–100.
Baum, A. (2015). *Real estate investment: A strategic approach* (3rd ed.). Routledge.
Becker, G. (1957). *The economics of discrimination.* University of Chicago.
Blackledge, M. (2016). *Introducing property valuation* (2nd ed.). Routledge.
Blundell-Wignall, A., & Atkinson, P. (2010). Thinking beyond Basel III: Necessary solutions for capital and liquidity. *OECD Journal: Financial Market Trends, 1*, 9–33.
Brown, R. (2018). *PropTech: A guide to how property technology is changing how we live, work and invest.* Casametro.
Businesswire. (2018). *VTS announces VTS marketview, first-ever real-time benchmarking and market analytics for commercial real estate.* Available at: https://www.businesswire.com/news/home/20180618005302/en/VTS-Announces-VTS-MarketViewTMFirst-Ever-Real-time-Benchmarking (accessed 5 July 2023).

Cetorelli, N., & Goldberg, L. S. (2012). Banking globalization and monetary transmission. *Journal of Finance*, *67*(5), 1811–1843. https://doi.org/10.1111/j.1540-6261.2012.01773.x.

Chandra, L. (2019). *The PropTech guide: Everything you need to know about the future of real estate.* PropTech Asset Management LTD.

Dermine, J. (2013). Bank regulations after the global financial crisis: Good intentions and unintended evil. *European Financial Management*, *19*(4), 658–674.

Fernando, F., Markevych, M., Schwarz, N., Kao, K., Chen, K., Poh, K., & Jackson, G. (2021). Virtual assets and anti-money laundering and combating the financing of terrorism. *FinTech Notes*, *2021*. https://doi.org/10.5089/9781513593821.063.

Feyen, E., Kawashima, Y., & Mittal, R. (2022). *Crypto-assets activity around the world evolution and macro-financial drivers.* Policy Research Working Paper Series, No. 9962.

FIBREE. (2022). *FIBREE industry report blockchain real estate 2021.*

Giambona, E., Golec, J., & Schwienbacher, A. (2014). Debt capacity of real estate collateral. *Real Estate Economics*, *42*(3), 578–605. https://doi.org/10.1111/1540-6229.12034.

Glickman, E. A. (2014a). Chapter 6 - Capital markets. In E. A. Glickman (Ed.), *An introduction to real estate finance* (pp. 153–188). Academic Press. https://doi.org/10.1016/B978-0-12-378626-5.00006-1.

Glickman, E. A. (2014b). Chapter 7: Property finance: Debt. In E. A. Glickman (Ed.), *An introduction to real estate finance* (pp. 189–218). Academic Press. https://doi.org/10.1016/B978-0-12-378626-5.00007-3.

Higgins, D. (2015). Defining the three Rs of commercial property market performance: Return, risk and ruin. *Journal of Property Investment and Finance*, *33*(6), 481–493. https://doi.org/10.1108/JPIF-08-2014-0054.

Higgins, D. M. (2013). The black swan effect and the impact on Australian property forecasting. *Journal of Financial Management of Property and Construction*, *18*(1), 76–89. https://doi.org/10.1108/13664381311305087.

Higgins, D. M. (2014). Fires, floods and financial meltdowns: Black swan events and property asset management. *Property Management*, *32*(3), 241–255. https://doi.org/10.1108/PM-08-2013-0042.

Higgs, B. Y. R. (1977). Firm-specific evidence on racial wage differentials and workforce segregation. *American Economic Review*, *67*(2), 236–245.

HM Treasury. (2023). *Future financial services regulatory regime for cryptoassets: Consultation and call for evidence.*

Innovate Finance. (2022). *FinTech investment landscape 2022.*

Innovate Finance and Young. (2016). *Capital markets : Innovation and the FinTech landscape.*

IPF. (2018). *Current practices in benchmarking real estate investment performance.*

Jackson, C., & Orr, A. (2011). Real estate stock selection and attribute preferences. *Journal of Property Research*, *28*(4), 317–339. https://doi.org/10.1080/09599916.2011.586469.

JLL. (2020). Why real estate transparency is harder to achieve than ever. *Article.* Available at: https://www.jll.co.uk/en/trends-and-insights/investor/why-real-estate-transparency-is-harder-to-achieve-than-ever (accessed 5 July 2023).

Knight, A. (2022). Why e-conveyancing is less risky than you think. *Property Journal.* Available at: https://ww3.rics.org/uk/en/journals/property-journal/why-e-conveyancing-is-less-risky-than-you-think.html (accessed 4 July 2023).

Laurent, P., Chollet, T., Burke, M., & Seers, T. (2018). The tokenization of assets is disrupting the financial industry. Are you ready? *Inside Magazine*, (19), 1–6.

Law Commission. (2023). *Digital assets : Final report.*

O'Halloran, J. (2020). Microsoft Teams usage growth surpasses Zoom. *Computer Weekly*, 24 June, pp. 5–8.

Oladiran, O., Hallam, P., & Elliot, L. (2023). The COVID-19 pandemic and office space demand dynamics. *International Journal of Strategic Property Management*, *27*(1), 35–49.

Oladiran, O., Sunmoni, A., Ajayi, S., Guo, J., & Abbas, M. A. (2022). What property attributes are important to UK university students in their online accommodation search? *Journal of European Real Estate Research*. https://doi.org/10.1108/JERER-03-2021-0019.

Oxford Future of Real Estate Initiative. (2020). *Tokenisation - The future of real estate investment*. https://doi.org/10.1201/9781003140283-20.

Palmer, D. (2014). Cai-Capital opens up UK property to foreign cryptocurrency investors, 19 June, pp. 1–9.

Phelps, E. (1972). The statistical theory of racism and sexism. *The American Economic Review*, *62*(4), 659–661.

Phillips, G. C. (2009). The paradox of commercial real estate debt. *Cornell International Law Journal*, *42*, 335.

Pi Labs. (2020a). *Technology and the future of real estate investment management*.

Pi Labs. (2020b). *Transparency through technology*.

PriceWaterhouseCooper. (2016). *Blurred lines: How FinTech is shaping financial services, Global FinTech report*.

Renshaw, R. (2017). First bitcoin homes sold in the UK' as crypto investors cash in, 18 December, pp. 1–9.

Saull, A., & Baum, A. (2019). *The future of real estate transactions (report summary)*.

SEI. (2020). *The future of the future of work, the future of real estate investing: 2020 SEI/Preqin survey of real estate managers and investors*. https://doi.org/10.4337/9781786438256.00014.

Smith, J., Vora, M., Benedetti, H. E., Yoshida, K., & Vogel, Z. (2019). *Tokenized securities and commercial real estate, MIT Digital Currency Initiative*. https://doi.org/10.2139/ssrn.3438286.

Souza, L. A., Koroleva, O., Worzala, E., Martin, C., Becker, A., & Derrick, N. (2021). The technological impact on real estate investing: Robots vs humans: new applications for organisational and portfolio strategies. *Journal of Property Investment and Finance*, *39*(2), 170–177. https://doi.org/10.1108/JPIF-12-2020-0137.

Starr, C. W., Saginor, J., & Worzala, E. (2021). The rise of PropTech: Emerging industrial technologies and their impact on real estate. *Journal of Property Investment and Finance*, *39*(2), 157–169. https://doi.org/10.1108/JPIF-08-2020-0090.

Stiglitz, J. E. (1998). The role of the financial system in development. *The Fourth Annual Bank Conference on Development in Latin America and the Caribbean* (Vol. 29, pp. 1–17), The World Bank, San Salvador, El Salvador.

University of Oxford Research. (2020). *Tokenisation: The future of real estate investment?*

US Department of the Treasury. (2022). *Fact sheet: Framework for international engagement on digital assets*. Available at: https://home.treasury.gov/news/press-releases/jy0854.

Chapter 7

Digital Technology and Innovation in Real Estate Valuation and Appraisal[1]

7.1 Overview of Real Estate Valuation and Appraisal

A core element of real estate is value; thus, property valuation and appraisal are essential segments of the real estate value chain. This section provides an overview of real estate valuation with a focus on valuation as a concept and then goes on to review the process and methods involved in valuation.

7.1.1 Real Estate Valuation and Appraisal

Real estate valuation and appraisal are integral processes in the real estate industry that determine the financial worth or value of a real estate asset. To manage the process and as a result of the importance of valuation and appraisal, there are several national and international professional bodies that guide and regulate the valuation practice. These include the Royal Institute of Chartered Surveyors, the International Valuation Standards Council (IVSC), etc. These organisations provide resources, CPD, and accreditation for appraisers, and they also publish appraisal and valuation standards:

- **International Valuations Standards Council (IVSC)**: This council began publishing international valuation standards in the early 1990s, and the latest version was published in 2017.
- **Royal Institution of Chartered Surveyors (RICS)**: The Global Standards (also known as Red Book) was first published in 1983; the most recent version was published in January 2020.
- **Appraisal Foundation**: This foundation first published mass appraisal standards in 1987; the most recent version – Uniform Standards of Professional Appraisal Practice (USPAP) – was published in 2020.

The value of a property is an opinion of either: the most probable price to be paid for an asset in exchange; or the economic benefits of owning an asset. Blackledge (2016) provides a property valuer's definition of value as "…the present price for the rights to receive income and/or capital in the future". In most cases though, the valuer's instruction transcends the mere estimation of the value of a property, to the estimation of the market value. The market value of a property as defined in IVS 104 paragraph 30.1 as the:

> estimated amount for which an asset or liability should exchange on the valuation date between a willing buyer and willing seller in an arm's length transaction, after proper

DOI: 10.1201/9781003262725-9

marketing and where the parties had each acted knowledgeably, prudently and without compulsion.

(RICS, 2021, VPS4 paragraph 4)

Although value, worth, and price can be used interchangeably from a layman's point of view, they have different meanings in the context of real estate valuation. Blackledge (2016) highlights the various descriptions of the different terminologies from a valuer's point of view:

- Value: An estimate of the price that would be achieved if the property were to be sold in an open market.
- Worth: An investor's perspective of the capital that he/she should be prepared to pay or receive for a steam of expected benefits that he/she expects the investment to generate.
- Price: The actual observable exchange price on a property in the open market.

Value in this context has three core elements: time, objectivity, and capital/income:

- Time: Time is an important element of valuation. Values are calculated at a specific date (the valuation date) and are only valid for a limited period of time after that date, depending largely on the market, demand and supply interplay and its volatility. Establishing the date that a valuation was carried out is therefore important.
- Objectivity: Value needs to be assessed objectively without bias or favour to a particular party's viewpoint. Although this process typically entails some level of subjectivity from the valuer, there are principles, approaches, and techniques that require objectivity in order to achieve an accurate estimation of value.
- Capital and (or) income: Depending on the purpose of the valuation, the valuation of a property is usually targeted at estimating the worth of the capital outlay or the periodic income to be obtained from the property. Capital is a one-off lump sum either as outlay (in the case of a purchase) or obtainable (in the case of a sale) on a property.[2] Income refers to the sum receivable at periodic intervals over a period of time and they could come from rent or interest payment.

As stated in Section 6.1.2, valuation and appraisal are the most crucial aspects of real estate investment managers' decision to purchase (or not to purchase) a property. More broadly, they enable buyers, sellers, lenders, investors, and several other real estate stakeholders to make informed decisions regarding real estate transactions, financing, taxation, and portfolio management. Due to the importance of this function, specialist property management valuers and appraisers are employed to ensure that the property being assessed is what it appears to be.

Real estate valuation and appraisal are carried out for several purposes. They include the following:

a Investment analysis: Valuation is carried out to evaluate the potential returns and risks of a real estate investment.
b Finance: Assessing the value of a property for collateral for mortgage loans or for determining LTV (loan-to-value) ratios.
c Real estate transactions: Properties can be valued in order to determine the fair market value for buying, selling, or refinancing purposes.

d Insurance: Valuation is carried out to determine the appropriate insurance premium that will be paid on a property or in the cases where there are insurance claims.
e Taxation and assessment: In order to determine the appropriate taxes and tax liabilities, there is a need to assess the property to ascertain its value.
f Litigation: Valuation is needed to provide expert opinions and evidence in legal disputes, divorce settlements, and eminent domain cases.

Valuation reports may be required by the private sector or public sector organisations. The private sector includes individuals, pension funds, banks, insurance companies, property companies, developers, REITs, etc., while the public sector includes charities, local authorities, central government departments and public authorities. Valuation usually entails commissioning a valuation report, including the collation of comparable evidence and opinions concerning the strength of the local market. It can also entail the assessment of the present value of the expected income from a property, discounted at a rate that captures the risk of the investment while taking other factors into consideration. The use of comparables is important in valuation as it enables the valuer to identify a baseline for which the property price should be adjusted; the reliability of the comparables is important in establishing the correct property value.

The valuation process is summarised below[3]:

1 Context: Understanding the brief, the purpose of the valuation, and the client's requirements.
2 Data collection: Gathering property-specific data such as the location, size, condition, amenities, recent sales data, and comparables. This usually requires an inspection of the property to conduct a physical examination and to evaluate its condition, features, and potential issues.
3 Market analysis: Assessing the macroeconomic and regional factors, and analysing local real estate market conditions, trends, and other factors that affect property value.
4 Valuation process: Applying one or more valuation approaches based on the property type, data availability, and the purpose for the valuation. It may also entail financial analysis where data such as income, expenses, rental rates, vacancy rates, and operating costs are analysed.
5 Reporting: Compiling the data and presenting a professional report where the key information relating to the valuation is presented with a highlight of findings and opinion of value.

Property value is determined by several factors. A core factor is location (hence the common phrase "location, location, location"). This accounts for the desirability and proximity of a property to amenities, schools, transportation, employment centres, and other influential factors. Another is the physical characteristics of the asset including the size, layout, condition, construction quality, age, architectural style, and unique features of the property. The income potential of a property is also an important consideration. This takes into account the rental income, occupancy rates, expenses, and market rental rates for income-producing properties. The market conditions are also important because they account for demand and supply dynamics, economic factors, interest rates, and overall real estate market trends. Additionally, valuation and appraisal usually take into account zoning and regulatory factors such as land use regulations, zoning restrictions, environmental considerations, and legal constraints.

The methods of valuation are given as follows:

Comparison method or market approach: This approach is based on comparisons derived from current market evidence to find rental or capital values directly. It requires a comparison to be made between the subject property and other similar properties that have been recently transacted.

Cost approach: This approach provides an indication of value using the economic principles that a buyer will pay no more for an asset than the cost to obtain another asset of equal utility, whether by purchase or by construction. This is usually used for properties for which there is insufficient direct comparable market evidence and where profits may be insufficient from the accompanying business, making the use of comparison or profits methods impossible.

Investment method or income approach (traditional or discounted cash flow): Traditionally, the investment method involves the capital value of a property investment being derived by multiplying the annual income flow by a multiplier, which is the cumulative present value factor also known as years purchase (YP) through a process referred to as capitalisation. The income approach depends on the discounted cash flow (DCF) accounting principle which explicitly incorporates a number of factors into the calculation, making allowances for inflation, tax liability, irregular receipts, or payments for anticipated rental and capital growth.

Profit method: This method is used to value trade-related properties. Trade-related properties refer to "any type of real property designed for a specific type of business where the property value reflects the trading potential for that business" (International Valuation Standards Council, 2020: 8). These properties include licensed and other "specialised" properties in the leisure, entertainment, sports, tourism, filling stations, hospitality and private care homes.

Residual method: This method is used to calculate the site value where direct market comparable evidence is not available. It is often used to examine the feasibility and viability of a proposed development scheme or for redevelopment and refurbishment purposes.

Due to the complexities of property valuation, the property valuer needs a diverse range of core skills and competencies such as calculation, measurement, research methods, law, management and business finance, information technology, and the knowledge of economics, buildings construction, and environmental sustainability.

7.2   Inefficiencies in Traditional Real Estate Investment, Finance, Appraisal, and Valuation

Due to risk and uncertainty and high capital associated with real estate investment, finance, and development, these financial valuation and appraisal are essential because they provide useful insight into the worth of an interest in a property that is considered for sale, purchase, or holding. There are several challenges and inefficiencies associated with the traditional approach to the valuation and appraisal of real estate. Although the valuation practice is highly regulated and standardised, it has several inefficiencies and problems.

7.2.1   Data Access, Market Transparency, and Time Sensitivity

Data are considered as an essential element of valuation and appraisal; thus, access to high-quality market data is important for its accuracy and reliability. Comparables are obtained and adjusted to determine appropriate rents, yields, and prices for comparative and investment valuation methods; cost and value data are required for residual valuation; market and sales data are required for profits valuation; cost data are required for replacement cost methods of valuation. Using traditional approaches to valuation, access to these data can sometimes be difficult, particularly in opaque or emerging markets. Regardless of market maturity or transparency, valuers often rely on recent and relevant data to estimate property values accurately. However, real estate markets can be volatile and property values can fluctuate over time, and this problem can be exacerbated in areas with limited transaction activity. This can lead to difficulty in obtaining reliable and up-to-date data, and it can further affect the accuracy of valuations.

Traditional real estate appraisal methods rely on historic data such as records of previous transactions. However, these data may not always reflect the current market conditions and may not accurately capture the distinctiveness of the subject property. For instance, the forces at play in a market at the time when a transaction was completed may have changed when that transaction is being used as a comparable, and this can further lead to uncertainty and it can limit the accuracy of the valuation, particularly in rapidly changing markets. This issue underscores the importance of timing in valuation. The volatility of markets, the unpredictability of the black swan events, and even economic and geopolitical shocks can render valuation reports obsolete in some contexts as they may not reflect the market a few years down the line.

The type of data being considered in valuation may also create some inefficiencies. Although core demand and supply indicators are usually integrated into valuation, there are several other factors that traditional valuation models may not be able to capture. Research has shown that property-specific and neighbourhood characteristics can affect property values. Hedonic pricing models are therefore being used (albeit non-traditionally) to estimate the impact of certain tangible and non-tangible property features to estimate value. However, most valuation models still do not sufficiently capture the value of non-economic factors such as environmental sustainability, environmental impact, social dimensions, and community value, which are increasingly being promoted.

7.2.2 Asset Heterogeneity and Information Asymmetry

Each property is unique with specific characteristics that impact its value. Although some buildings may appear to be identical (particularly residential property), they are in fact different in terms of their location – i.e., no two properties occupy the same piece of land, and geospatial coordinates can be obtained for each square inch of the earth. In addition to the locational variation, there are also several other property distinctiveness such as size, layout, conditions, zoning restrictions, amenities, and potential income (for income-producing properties). In fact, there are some specialist properties that require additional expertise and research for their value to be accurately determined. These properties include historic buildings, development sites, and properties with specialised income streams such as operational real estate assets, and traditional approaches to valuation may not produce accurate values for these non-traditional uses.

Information asymmetry and lack of transparency are also issues in real estate valuation. A valuer may inspect a property but fail to pick up on some defects or structural issues either because some of the defects have not manifested or because they may have been concealed (or not disclosed) by the selling party. Examples of these include environmental issues, structural problems, legal encumbrances, etc., and they can significantly affect the accuracy of the valuation.

7.2.3   Subjectivity and Biases

The accuracy and variation in real estate valuations have been the subject of debates in literature (Boyd & Irons, 2002). Traditional approach to real estate appraisal and valuation is subjective and, in many cases, different valuers may arrive at different opinions, leading to discrepancies in property valuations (Klamer et al., 2017; McAllister et al., 2003). It requires the consideration of various factors, including location, property condition, market trends, comparable sales data, income potential, etc.; however, the valuer needs to subject these indicators and data to his/her analysis and ultimately make a judgement call. The valuer's analysis and judgement may also be influenced by his/her biases (Howell & Korver-Glenn, 2018; Perry et al., 2018). Furthermore, market sentiment, information available, stereotyping, data deficiencies, or the motivations of stakeholders involved in the valuation also increase the subjectivity in valuation. Additionally, emotional factors or personal interests may affect the objectivity of the valuation, and clients may place the valuer under pressure to produce a certain preferred valuation, and this can lead to discrepancies in values (Levy & Schuck, 1999, 2005).

There is currently no universally accepted method for valuing real estate. The valuer has to decide on the approach and method to adopt depending on the property type and the purpose of the valuation. These different methods may lead to different values, and this can create discrepancy and inconsistency in valuation, which can further limit comparability and sometimes lead to disputes and disagreements among stakeholders.

7.2.4   Contextual Variation

Property valuations need to comply with legal and statutory regulatory requirements in a given geopolitical context. For instance, valuations for mortgage, taxation, investment, and eminent domain will need to conform with specific standards and guidelines in a country or region. Failure to adhere to these requirements can have legal and financial consequences. This, therefore, in a sense, limits valuation experience and expertise to a specific geopolitical location, potentially increasing the cost of context-specific appraisal, particularly when the client or valuer is overseas

7.2.5   Expertise and Experience

The quality and expertise of a valuer may vary and due to the subjectivity involved in the process, this variation in expertise and experience may lead to inconsistencies in the valuation. This may manifest through insufficient knowledge, inadequate analytical skills, or a high level of bias which can affect the accuracy and reliability of the valuation. For this reason, clients will typically want to verify the competence, experience, and expertise of the valuer to ensure they are qualified to carry out the valuation and to ensure that the valuation output is credible. Furthermore, insufficient local context knowledge can significantly affect the accuracy of the valuation, particularly if the subject property is overseas.

These issues can ultimately lead to several issues and inconsistencies in the valuation, such as smoothing (Edelstein & Quan, 2006; McAllister et al., 2003) and anchoring (Burak & Baycar, 2019; Saddiqi, 2018; Tversky & Kahnemann, 1974).

In some contexts, the demand for real estate appraisals can outpace the availability of qualified and certified valuers, and this can create long turnaround times, delays in completing valuations and potential bottlenecks in the real estate transaction process. Thus, satisfying the demand for timely valuations is a significant challenge, particularly considering that high quality is promoted.

7.2.6   Rigid Processes and Systems

Traditional valuation approaches have existed for several decades, and there is no feedback system in place that can be used to improve them. The feedback from conferences, CPDs, and seminars is fragmented. This is one of the reasons that real estate valuation processes have remained relatively manual (or semi-automated), and the automation of advanced analytics, AI, and automated valuation models (AVMs) have faced significant backlash in the last couple of decades. The real estate industry could benefit from a more robust system of evaluation, feedback, and improvement.

7.2.7   Scale

Scaling valuation can be challenging using traditional valuation approaches. The traditional approaches to valuation may be suitable for the valuation of individual properties at a smaller scale; however, if the valuation is required for thousands of properties, for mortgage applications, for instance, it is difficult to hire thousands of experienced surveyors at short notice to value each property in a mortgage lending portfolio each time a loan book is being sold, or when an updated valuation is required. Due to the labour and time intensity involved in traditional valuation, scaling the process for mass appraisal[4] becomes a major challenge.

7.3   Innovative/Emergent Systems and Services in Real Estate Valuation and Appraisal

As real estate asset investors and developers continue to diversify to various real estate sectors, particularly with the acceleration of investment in operational real estate (Oladiran et al., 2022), the process of valuation and appraisal becomes more complex. The advancement of digital technology across the various segments of the real estate value chain contributes to making valuation more complex. For instance, the use of online mortgage platforms has facilitated the growth of automated valuations, and the growth of online investment platforms and crowdfunding reveals the need to have advanced and automated valuation processes (AVPs). Several digital tools have been utilised in real estate valuation and this has led to the emergence of contemporary digital tools, systems, and platforms. This sub-section provides a review of innovative and emergent digital technological-driven innovations and buzzwords that have emerged in the realm of real estate valuation. Some of these phrases and buzzwords were more or less unheard of two decades ago; some others have been used by the authors due to the absence of a suitable.

7.3.1   Automated Valuation Models (AVMs)

The integration of advanced digitalisation in valuation has led to the emergence of advanced valuation models (commonly known as AVMs). AVMs are well established in several jurisdictions and typically applied to assets in the residential real estate sector. AVMs are becoming widespread and are now used in influencing and informing valuation and transaction-related activities in real estate. A wide range of real estate stakeholders are increasingly using it. They include asset managers, lenders, REITs, tax authorities, accounting firms, regulators, financial supervisors, etc. One of the areas where AVMs are most commonly used is in mortgage lending and property brokerage. There are online tools that can now provide value estimates for prospective lenders, and other property listing platforms that also provide automated valuation estimates for free.

AVMs are sometimes misconstrued as a fully automated valuation process, whereby the valuation process including data collection, analysis, modelling, and reporting is produced using a computer-based system with very minimal human intervention; indeed, that is what the name appears to imply. However, AVMs, in their true practical sense, are much narrower than that. It is therefore important to examine the concept of AVMs their current application and context, and to further identify the boundaries of the current AVM application.

7.3.1.1   The Current Scope and Application of AVM

AVM involves a broad spectrum and a hybrid involving varying degrees of automation, data collection, analytics, and various forms of human involvement and intervention. Most traditional AVMs can be classified as hedonic pricing models following its introduction to real estate by Rosen (1974), with the proposition that a property's value is derived from basic property attributes such as its age, size, location, and other tangible and non-tangible features. AVMs are generally used with widely traded homogenous properties (particularly residential assets) for the purposes of lending, consumer-facing valuations, and mass appraisal. Their performance becomes less accurate with more heterogeneous assets. In the residential sector, for instance, there is a higher probability of identifying properties of similar features which are traded relatively often. Mature markets with higher levels of market transparency and data availability are experiencing an increasing use of AVMs. Markets such as the UK, the US, Canada, the Netherlands, Switzerland, and Australia appear to be leading the way. AVMs will typically be suitable for properties that require a lower degree of human intervention. For instance, given the scale of many social housing portfolios, AVMs are already being utilised to provide estimates of value (Royal Institution of Chartered Surveyors, 2022).

Baum et al. (2021) and Royal Institution of Chartered Surveyors (2022) provide further insight into the various uses of AVMs given as follows:

- Non-performing loans
- Risk-weighted asset calculations
- Residential mortgage lending origination
- Re-mortgaging
- Mortgage-backed securities

- Property tax
- Collateral valuation
- International Financial Reporting Standards (IFRS) requirements
- Loan book valuations by rating agencies (including for CRE assets)
- Open-ended funds

In the residential brokerage/agency market, AVMs are also used to provide clients with an estimate of value for marketing purposes, setting prices and expectations in the emerging iBuyer business model. Central and local government parastatals are also employing AVMs, for periodic mass appraisals and revaluation exercises across both residential and non-domestic properties.

7.3.1.2   Boundaries and Limitations of AVM

Even in residential properties, AVMs cannot be applied where properties don't fulfil the basic criteria of homogeneity and a sufficiently liquid and transparent market. Furthermore, AVMs' use may be restricted based on construction type, for instance, when contemporary construction methods are adopted. Additionally, HMOs (houses in multiple occupations) and buy-to-let properties may not be suitable for AVMs, mainly because they are not as homogenous, and they are relatively thinly traded. Royal Institution of Chartered Surveyors (2022), however, notes that institutional build-to-rent (BRT) is increasingly being valued using AVMs in many markets.

AVMs are not suitable for the various commercial real estate asset classes: office, industrial, retail, etc., and for operational real estate assets such as data centres, retirement housing, PBSAs, co-working, and self-storage. This is primarily a result of the polarised nature of these asset classes, the heterogeneity of the assets, and the fact they are thinly traded (Baum et al., 2021). Furthermore, the valuation of commercial assets is typically based on the income-producing features which are then linked to various attributes of the properties such as the occupier covenant strengths, rental growth projections, lease terms, market yields, void, capital, and operating costs. Thus, the mechanism of AVMs, the data required, the data sources, and the modelling have stark contrasts.

There are semi-automated valuation processes; these should not be classified as "AVMs". This is because they require a high level of input and support from a valuer in order for the values to be accurately derived. For instance, software that is used for valuation modelling is usually limited in terms of its capabilities. In many cases, they just perform the work of a Microsoft Excel spreadsheet, but they do this at a more advanced level. Regardless, this does not qualify them to be termed AVMs.

7.3.1.3   Re-evaluating "AVM"

There is a need to re-evaluate the term AVM in the context of its current applications, boundaries, and limitations as discussed in Sections 6.3.5.2 and 6.3.5.3. As the Royal Institution of Chartered Surveyors (2022) suggests, a broader discussion and response to the use of automation in valuation is needed, with considerations for three key factors: data, models, automation, and processes. The scope of automated valuation must go beyond merely the current limited scope of residential real estate to the broader spectrum of real estate asset classes. Whether or not there are current automated systems and processes in place is not the issue, rather, it is important to have a suitable term that can be used broadly. This book therefore proposes a shift from AVM to AVPs in line with the recommendations in the Royal Institution of Chartered Surveyors (2022).

7.3.2   Automated Valuation Processes (AVPs)

7.3.2.1   Introducing the Concept Automated Valuation Processes (AVPs)

AVP is described as the automation and digitisation of the various aspects of the valuation process for various asset types and classes. Unlike AVM (which is primarily focused on modelling), AVP may include automated data collection, sourcing comparables, property inspection, data analysis, modelling, reporting, delivery, and any other aspect of the valuation process (Figure 7.1). As emphasised earlier, this definition should not be misconstrued to imply that the whole valuation process must be fully automated in order for it to be described as "digital"; rather, it extends the same principles that are being used in other aspects of the real estate value chain where individual areas can be automated and still be seen as a PropTech. This further broadens the scope of valuation automation, and hopefully, real estate students and PropTech enthusiasts can begin to conceptualise and develop solutions beyond the current restrictive and narrow AVM scope. This effectively makes AVM a sub-set of AVP. Furthermore, AVPs do not imply that they challenge the traditional methods of valuation; rather, they "challenge" the processes of carrying out valuation exercises (as they have done to other areas of real estate) and attempt to develop more efficient approaches and systems.

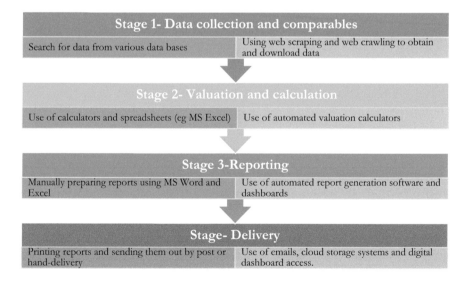

Figure 7.1 An illustration of an automated valuation process.

7.3.2.2   The Demand Drivers for AVPs

The development of automated valuation has been underpinned by the need for improved speed, cost, scale, consistency and accuracy in valuation processes and output.

- **Speed**: The traditional valuation process is time intensive – data collection, market analysis, modelling, reporting, etc., take a lot of time; thus, automation is being explored to reduce the time used to prepare and provide valuation reports. These automated processes provide a time-efficient approach to valuation and reporting.
- **Cost**: Valuation reports are typically prepared by professionals who charge a fee. These are usually in line with the professional scale of practice and are often perceived as being very high. Individuals and businesses would make financial savings if automated valuation models and systems are adopted.
- **Scale**: There are times when a high volume of valuation reports is required for taxation purposes, funds, portfolios, and non-performing loans; it may be practically impossible to execute high-volume instructions of this nature without automation.
- **Consistency**: Valuation is an art and a science; the "art" element is embedded in subjectivity; thus, it is almost impossible for two (professional) valuers to carry out a valuation exercise and arrive at the exact same values. This seeming lack of consistency is often a worry for clients, and automation is often perceived as one of the ways of standardising valuation.
- **Accuracy**: There is hardly an accurate valuation (value); however, accuracy in valuation can be achieved when there are minimal errors and mistakes in the various aspects. For instance, there could be errors in the measurement of the property, in gathering comparables, in the valuation calculation, or even in the reporting. Automated systems are less prone to errors, and when errors are observed, they are more easily managed.

In the past, AVMs were used to double-check traditional methods and approaches rather than being used as a total replacement. In some cases, some funds employed the use of AVMs internally, with a human reviewing the data and output to ensure accuracy and consistency. The COVID-19 pandemic however accelerated the development of this area, particularly as there were significant constraints to travel and by extension, physical inspection of properties.

7.3.2.3   The Scope of AVPs

Several aspects of real estate valuation processes are already automated. For instance, some automated systems are producing capital valuations on the assumption of vacant possession, with others focusing on forecasting market rents. There are also various automated property market indices that capture the movement and forecasting of various market matrices such as rents, yields, and capital values, at sector and subsector levels. At the portfolio level, automated systems are used for commercial real estate, although the accuracy becomes questionable when the models are applied across various segments and individual assets. There are also semi-automated systems where some of the aspects of the valuation process such as data collection and comparables are manually done, and the data are inputted into a software that performs the valuation based on pre-set parameters, and then, the output is generated, and finally, the reports are prepared. Software such as Argus Developer and Argus Enterprise are some examples.

Advanced data sourcing and analytical techniques are increasingly being used leveraging digital technology to enhance the efficiency of data collection, analysis, and utilisation. Non-conventional data such as mobile phone location data, tenant experience surveys, and online reviews in the hospitality industry are now being used to identify localised patterns, complemented by demographic and socio-economic information like crime rates and median age, which contribute to long-term market forecasts (Asaftei et al., 2018). These emerging data sources have the potential to accurately predict specific hyperlocal areas that exhibit significant potential for price changes.

Advanced data analytics using tools such as artificial intelligence (AI) and machine learning are also essential in AVPs. AI-enabled AVMs, for instance, have flexibility in their functional forms which makes them mitigate issues relating to statistical models for within-sample predictions. This is particularly valuable for valuations of property characteristics across various settings. For instance, the features of a home (e.g., number of rooms) and its amenities, e.g., green spaces may be priced differently in various locations. Baum et al. (2021) note that green space may be a more valuable amenity for a house in the city centre than it would be for a house in the countryside and the willingness of buyers to pay for certain property features may vary over time. The power of AI can enable AVPs to account for these variations through analysis and modelling.

Various apps such as Placense, Gyana, and Geoblink now offer aggregated and anonymous real-time location insights. These encompass a range of metrics, including customer visits, visit duration, customer journey paths, gender, age groups, income levels, transportation modes, and other demographic data. These can improve the predictions of retail valuations and valuations of operational real estate assets such as hotels, leisure facilities, and restaurants. For instance, WeWork makes use of diverse data matrices in the selection of office locations. Through collaboration with Factual, a data company that specialises in digital mapping and categorising location information from online sources, WeWork incorporates data on community and governance, local landmarks, transportation infrastructure, healthcare services, business amenities, and other desirable features from approximately 52 countries. This enables them to create an index of essential amenities within the vicinity of potential office locations. WeWork can then assess how well a particular office location aligns with their desired amenities and target tenant demographic. This data set has gained economic value, with companies like Uber, Amazon, Apple, and SquareSpace also leveraging the data for their own purposes.

Property inspection is an important aspect of valuation and appraisal. Property inspection traditionally involved physical on-site inspections; however, advancements in digitisation have made it possible to conduct assessments without the need for a physical inspection. Digital technology is now used to capture, store, and present comparables and relevant data (RICS, 2017). These technologies include 3D imaging software and virtual tours, digital twins, Building Information Modelling (BIM), smart buildings equipped with Internet of Things (IoT) connected sensors, and building passports.

The wider adoption of AVPs can lead to the better integration of emerging trends in markets such as ESG and sustainability which traditional valuation models and processes would typically not account for. There is a growing trend of "brown discounts" in which buildings without sustainability ratings, like LEED or BREEAM, are sold or leased at lower prices (Pi Labs, 2020). WiredScore currently provides a building accreditation based on digital connectivity and network capabilities (Wright, 2020). This development enhances the assessment of ESG benchmarks, including the difficult-to-measure social aspect (S), which can potentially impact property value when quantified (Saull, 2020). Furthermore, these digital systems may lead to the emergence of a new building value centred around occupants' approach, driven by openly accessible satisfaction and productivity benchmarks, similar to those commonly found in the hospitality industry. These benchmarks can be crowd-sourced and enable a more occupant-centric approach to evaluating property value.

There appears to be scepticism among real estate professionals and scholars alike that automated valuation is not feasible and does not appear to be possible. The Royal Institution of Chartered Surveyors (2022) expresses the same sentiment: "...with complex, high-value CRE investment properties, the need for detailed valuation, provided by valuers with an intimate knowledge of the asset and the local market will continue" (12). The Royal Institution of Chartered Surveyors (2022) however, further cautions that automated valuation will continue to be developed and applied to CRE and its use will increase. This is indeed true. The fundamental principle behind the concept of innovation and the concept of innovative destruction was propounded by Joseph Alois Schumpeter who described innovation as replacing today's pareto optimum with tomorrow's different new things, and where new innovations and technologies replace existing economic structures and create new ones.[5] This suggests that innovation is founded on the ability to look beyond the challenges and impediments to developing new systems and processes, and this is what has underpinned the development of PropTech and innovation in real estate in the last decades.

Students and PropTech enthusiasts are therefore challenged to consider ways through which digital technology can be used to produce a fully automated variation process for commercial real estate with minimal human input. Some of the core challenges in this respect would be data sourcing, collection, and input functions; however, tools such as AI are increasingly being used for advanced web scraping (Oladiran et al., 2022), and these can be applied to the data sourcing, collection, inputting and valuation functions. It is expected that innovation will soon lead to the development of tools and capabilities that can be used for the fully automated valuation of office, retail, and industrial, assets, as well as data centres, retirement housing, PBSAs, co-working, self-storage, etc., regardless of their heterogeneity. These AI-enabled tools can be trained to account for the income-producing features of an asset, and then link them to various attributes of the properties such as the occupier covenant strengths, rental growth projections, lease terms, market yields, void, capital, and operating costs, thereby unifying the data sourcing (data scraping and unified APIs), analytics, modelling, and reporting.

7.3.2.4   Risks, Limitation and Challenges of AVP Adoption

The efficacy of AVP (AVM inclusive) is dependent on the data used to develop and operate it. The use of AI for data sourcing and data collection as well as the use of AI-AVMs, requires a high volume of data, first to train the model (machine learning), and next to run and operate them. This will only improve with time, as professionals become more comfortable with the use of these automated systems. In Europe, for instance, data availability and regulatory landscapes play a significant role in the varied adoption of automated valuation. Countries such as Spain, Portugal, Italy, Greece, and Turkey, as well as regions in northern Europe with better data availability, are witnessing increased AVM adoption. In the Asia-Pacific region, AVM adoption has also been observed in Singapore, Hong Kong, China, and Australia, and similar trends can be observed in Brazil and in the Caribbean. Better access to data can therefore improve the adoption of AVPs across the world.

Although data access is a good start, there are several other data-related considerations for AVPs. They include the following:

- **Collection methodology**: This focuses on the data collection methods such as web scraping, web crawling, filtering and selection of data, and data integration with APIs.
- **Consistency**: This is a concern about the assurance that the data will provide a consistent level of efficacy over time.
- **Recency**: This relates to the time the data were collected and the validity of the data relative to the time of collection.
- **Security, ethics, and privacy**: This focuses on the ethics of data collection and storage with respect to legal, ethical, data protection, licencing, and ownership considerations.
- **Scale and range**: This is concerned with the depth and scope of the data in terms of asset type, region, neighbourhood, amenities, and sample size.

In addition to data-related issues, there are also challenges and potential risks with the implementation of AVPs. They include the following:

- There may be a potential conflict of interest, particularly in markets where a small number of professionals have multiple roles in developing and using the tools within the AVP.
- AVM outputs intended for risk assessment purposes are being treated as valuations, which can lead to misinterpretations. The limitations of AVP-based tools or platforms should be clearly stated so that the users are fully aware of the limits to its applications.
- Users in the audit and accountancy sector may perceive AVM outputs as equivalent to traditional valuations.
- Considering the unpredictability of real estate markets, it is not clear how AVPs will exist regarding how AVMs will handle market adjustments and corrections, market uncertainties with limited transactions, extreme market volatility, and their lack of forward-looking capabilities.
- AVMs can potentially contribute to market volatility or even act as market makers rather than merely reflecting market conditions.
- Without regular calibration against reference data like manual valuations, there is a risk of AVMs becoming self-reinforcing and circular in their outputs.
- The lack of standardisation in the development of AVPs, with various models and approaches being used, may negatively affect consistency and hinder the market's ability to assess their effectiveness and application across different markets and asset types.
- In many cases, the input of valuation professionals is limited in the development of some of the existing AVP tools and platforms. Including valuation professionals has the potential to identify pitfalls and propose solutions in advance.

Generally, current automated valuation systems, particularly AVMs, are being used to facilitate lower-risk transactions. As Baum et al. (2021) point out, it is unlikely that a lender would use an AVM-backed valuation to underwrite a newly constructed home for a first-time buyer at a loan-to-value ratio of 95%. This disposition to use the AVMs to facilitate higher-risk transactions is the result of the error distribution (forecast standard deviation) as well as the question of accountability to a set of professional standards that professional chartered surveyors must adhere to.

7.4   Case Study Analysis

Following the review of core areas of digital transformation in real estate valuation, this section analyses specific PropTech platforms and firms within this area of the real estate value chain. Each case study begins with an overview and summary of key details of the platform/company; the solutions it provides are then analysed using the ESEP efficiency framework introduced in Chapter 1.[6] The analysis concludes with a brief overview of the limitations and further prospects.

7.4.1   Case Study 1: Altus Group (ARGUS Enterprise)

7.4.1.1   Background, Key Details and Technologies Utilised

Altus Group is a leading digital real estate valuation software company that offers software, data solutions, and expert services with technology-enabled capabilities to the worldwide commercial real estate sector. With a track record of over three decades, the company has built a reputation for commercial real estate valuation service support. The platform provides comprehensive solutions that cover several aspects of commercial real estate, enabling their clients to obtain valuable insights, mitigate risks, and enhance investment performance. The software can aid in the generation of detailed cash flows using lease-by-lease modelling, run commercial property valuations using DCF and traditional valuation, stress test cash flow and valuation assumptions to manage risk and improve reporting transparency for investors. The company has approximately 7,000 active software clients in more than 100 countries. They also provide services to over 200 Universities and colleges worldwide.

Altus Group provides a range of solutions and tools, namely, Altus valuation, Altus portfolio performance, Altus market insight, Altus property tax, Argus software, data and analytics, and professional services, and these tools and services intersect between valuation, portfolio management and asset management. ARGUS enterprise, for instance, is a commercial real estate specialist software that supports the valuation, appraisal, and general portfolio management of commercial and operational real estate assets, namely, office, commercial, industrial, logistics, retail, multi-family, build-to-rent, and hotels.

The ARGUS Enterprise platform is a cloud-based software with a collaborative Microsoft Environment on an SQL database, supporting multiple properties and portfolios to house properties, templates, and scenarios for multiple funds, internal departments, and clients. The platform, therefore, supports common Microsoft features such as files, filters, and views. Each portfolio can contain various scenarios for thousands of individual asset models representing several sensitivity results and various periods. It is also useful for individual asset cash flow modelling, with various asset details. Assumptions can be adjusted and edited, and results are automatically generated using a single tab click. This platform enables the client to build detailed cash flow forecasts, stress test market and leasing assumptions, and create comprehensive property valuations using discounted cash flow analysis and other valuation methods. ARGUS Enterprise replaced legacy products and solutions such as ARGUS Valuation DCF, Valuation Capitalisation (ARGUS ValCap), and ARGUS Asset Management (DYNA) into one platform. With this centralised solution, clients can build accurate cash flow forecasts, run stress testing on market and leasing assumptions, and create comprehensive commercial property valuations and several other services.

7.4.1.2   Key Efficiencies and Solutions That ARGUS Enterprise Provides

The solutions that ARGUS Enterprise provides are itemised using the ESEP efficiency framework (Table 7.1):

Table 7.1 ESEP Efficiency Analysis of ARGUS Enterprise

ESEP	Inefficiency with the Traditional Approach	How the Tool/Platform/Product Addresses the Inefficiency
	The analogue nature of data analysis and modelling using traditional approaches can lead to more extensive use of computing power and energy (e.g., electricity)	The integrated data analytics and reporting dashboard reduces the energy and computing power that would normally be used to process information. By reducing the time and manpower required to source and analyse data, the computing power and energy that would have been used to complete these tasks are significantly minimised.
	Collaboration with investors and other stakeholders involved in a valuation requires a high level of travel and commute.	Collaboration is done with reduced commuting; meetings can be held remotely with reference to the shared dashboard information, leveraging remote meeting tools. This platform can also support remote working which will further minimise carbon emissions.
	Traditional valuation and reporting require a high volume of stationery.	ARGUS Enterprise provides a platform that can be shared with various stakeholders without the need to print out the report or output. This can significantly reduce the use of stationary.
Social	Cash flow analysis and appraisal are usually tenant-specific, and because leasing terms and market conditions change rapidly, traditional cash flow analysis can be cumbersome, and recent updates can be omitted. The lack of clarity into the future and the difficulty of getting comparable transactions make valuation using this method challenging and inefficient.	Using ARGUS Enterprise, detailed cash flow can be built and updated, and market and leasing assumptions can be stress tested. This can be used to provide comprehensive commercial property valuations using the DCF and other valuation methods.

(Continued)

Table 7.1 (Continued)

ESEP	Inefficiency with the Traditional Approach	How the Tool/Platform/Product Addresses the Inefficiency
	In order to conduct an effective sensitivity analysis and scenario analysis, there is the need to be able to reasonably predict future trends as accurately as possible and to test several scenarios. Doing this using traditional approaches can be cumbersome and inefficient.	ARGUS Enterprise provides a "what-if" analysis feature that enables the valuer to conduct an analysis of individual property or an entire portfolio to measure the impact of various factors and changing trends through the adjustment of the valuation assumptions. Several scenarios can be tested to assess the effect of the asset performance in different market conditions.
	In carrying out investment valuation using traditional approaches, the valuer needs to assess the potential revenue, capital return and costs of an investment, while identifying risks. Depending on the size, tenancy, lease structure and other factors, this process often requires a high level of data analysis which is often cumbersome and challenging.	The ARGUS Enterprise model enables the valuer to build a robust returns model, taking an array of property factors and market assumptions into a holistic view of each property. Data can be imported from the seller with an overlay of in-house assumptions. Regardless of the complexity, this automated system can make worse and best-case scenarios more efficient, and a more comprehensive analysis can be carried out with greater ease.
	Due to the complexities of carrying out valuations and the various aspects that are often taken into consideration, traditional valuation approaches are prone to error, and in implementing corrections manually, there may be inaccuracies and inconsistencies. This can have significant effects on clients' decisions.	The high level of automation and integration of the analytics and reporting systems in Argus Enterprise significantly minimises the changes of errors, inaccuracies and inconsistencies, and when errors are spotted, the corrections can be reflected across the full workflow, making final reports more consistent and accurate. Clients' decisions can be taken with a higher level of accuracy and confidence, and this can make investment strategies more effective and efficient. The software also safeguards against data loss and guarantees the integrity of the calculations.
	Collaboration, updates and discussion with investors and stakeholders is usually more difficult, often requiring a high level of commute.	This platform provides an avenue for collaboration, discussions, and conversations to be conducted remotely, allowing for a higher level of comfort. Feedback and changes to the valuation can also be affected with ease.
Economic	Due to the complexities of carrying out valuations and the various factors to consider, valuation using the traditional approaches and processes often takes a long period of time to complete.	Due to the automated approach to valuation appraisal, scenario analysis, and other analyses using ARGUS Enterprise, the time taken to complete valuations is shorter.

(Continued)

Table 7.1 (Continued)

ESEP	Inefficiency with the Traditional Approach	How the Tool/Platform/Product Addresses the Inefficiency
	Due to the time taken to conduct valuation using traditional approaches, clients often have to wait for several weeks to get their valuation concluded. This can delay investment and other related decisions.	This digital tool significantly reduces the time taken to conduct valuations. Clients can therefore take decisions quicker.
	The delay in completing valuation reports can lead investors to miss out on some investment opportunities or for them to make mistimed decision.	Due to the speed in completing valuations, investors and other clients can make quicker decisions thereby taking advantage of specific investment decisions.
	The delay in getting valuations completed can lead to delays in investment decisions such as funds allocation, and asset selection. Considering that timing is essential in real estate investment, investors may miss out on deals or make lagged investment.	The higher speed in completing valuations can enable investors, investment managers and asset managers to make timely decisions and take advantage of investment opportunities.
Physical/ spatial		

Source: Authors analysis.

7.4.1.3   Limitations, Challenges, and Prospects

The adoption of ARGUS Enterprise has its limitations. First, the software is designed mainly for commercial and operational real estate assets, thus, it cannot be used for residential valuation. The tool is also semi-automated and it requires a high level of human involvement in areas of data entry, assumptions, etc. There are other economic and environmental factors. For instance, for real estate firms to be able to use the platform, they need to pay the requisite fee, and their staff will need to undergo training and upskill to be able to use the tool and to get the best out of the systems. This will require huge financial resources, and it can be a challenge particularly, for smaller fund managers. Although the platform has the potential to attract environmental benefits by minimising high computing power and energy in analysis, it can also be argued that due to the high performance of the platform, high computing and energy usage will still be required to integrate all the systems, and the subscription to the cloud and other data systems will also lead to an increase in the demand for data centres which, as discussed in Section Two, is also significantly contributing to the carbon emission and global energy consumption.

ALTUS group is investing in supporting API integration with other property management and accounting systems, in-house spreadsheets, and other solutions. This has the potential to significantly improve the data sourcing, sharing, and analytical capability of the platform and software.

Further details on Altus Group (and ARGUS Enterprise) can be accessed from their website using this link: https://www.altusgroup.com/.

7.4.2   Case Study 2: ATTOM (ATTOM AVM)

7.4.2.1   Background, Key Details, and Technologies Utilised

ATTOM is an online platform that provides property data with flexible delivery solutions to companies that need data-enabled products and services, as well as companies that are building data analytics platforms. They support various sectors including mortgage, insurance, financial institutions, marketing listing brokers, government, real estate, etc. The platform leverages data such as property data, foreclosure data, ownership data, transactions, mortgages, boundaries, school data, neighbourhood, hazards, risk, and climate change, and these enable the platform to provide a wide range of services and solutions: ATTOM Cloud, bulk data licensing, real estate data API, property reports, marketing lists, real estate market reports, match and append, real estate websites, WordPress plugins, and IDX solutions.

ATTOM AVM is a digital tool that enables lenders to estimate the values of residential properties. The AVM results can be independent or flexible – with other ATTOM data using match and append, and marketing lists. It can be delivered as part of a bulk file, custom list or API. This tool utilises about 80 million homes in its database across the 50 US States and it employs multiple real estate ARM models to choose the modelling approach that is more accurate to the geographical area surrounding the individual property. It requires a minimum of 15 comparable properties, and it may use as many as 100 properties to generate estimated values. After the valuation, each property receives a unique confidence score that represents the precision of the AVM estimate and measures the deviation between the range of values and the point use itself. In displaying the most accurate estimate of the house values, the platform also displays detailed property profile information.

7.4.2.2   Key Efficiencies and Solutions That Argus Provides

The solutions that ATTOM AVM provides are itemised using the ESEP efficiency framework (Table 7.2):

Table 7.2 ESEP Efficiency Analysis of ATTOM AVM

ESEP	Inefficiency with the Traditional Approach	How the Tool/Platform/Product Addresses the Inefficiency
	The analogue nature of data analysis and modelling using traditional approaches can lead to more extensive use of computing power and energy (e.g., electricity).	The integrated data analytics and reporting dashboard reduces the energy and computing power that would normally be used to process information. By reducing the time and manpower required to source and analyse data, the computing power and energy that would have been used to complete these tasks are significantly minimised.
	The collaborative work that goes with traditional valuation may sometimes require commuting and travel.	This automation approach to ATTOM's AVM means that valuation can be completed with very minimal human effort. This means that the commute-related carbon emission associated with this activity will be significantly minimised.

(Continued)

Table 7.2 (Continued)

ESEP	Inefficiency with the Traditional Approach	How the Tool/Platform/Product Addresses the Inefficiency
	Traditional valuation and reporting require a high volume of stationery.	The need for printing the valuation report is minimised as the valuation estimate is made available to the lenders and the clients online, minimising the need to use stationary.
Social	Three is usually a trade-off between the quantum, quality and application of comparables when valuation is conducted using traditional approaches and processes.	ATTOM AVM uses up to 100 comparables which strengthens the quality of the output and its reliability. The confidence score generated by the platform is an added advantage.
	The manual gathering and adjusting of comparables takes a considerable amount of effort and processes which are inefficient.	Gathering and adjusting comparables is automated using ATTOM AVM, and this saves a considerable amount of energy and effort.
	Valuation for insurance purposes is typically to enable the underwriter to assess the risks associated with the house. Although the valuer will attempt to consider all of the risks associated with an asset, this may not always be possible using traditional approaches and processes.	In calculating property values for insurance, ATTOM AVM takes account of the home profile, neighbourhood and other risk information which would be impossible using manual methods and approaches. This is enabled by the platform's access to a wide range of data sets which are integrated for better property value estimation.
	Due to the complexities of carrying out valuations and the various aspects that are often taken into consideration, traditional valuation approaches are prone to error, and in implementing corrections manually, there may be inaccuracies and inconsistencies. This can have significant effects on clients' decisions.	The high level of automation and integration of the analytics and reporting systems in the AVM significantly minimises the changes of errors, inaccuracies, and inconsistencies. Lenders and underwriters can therefore make their decisions with a higher level of accuracy and confidence, and this can help to minimise risks and unforeseen costs.
Economic	Due to the complexities of carrying out valuations and the various factors to consider, valuation using traditional approaches and processes often takes a long period of time to complete.	Due to the automated approach to ATTOM AVM, the valuation and appraisal can be completed in a short time.
	Sourcing comparables using manual approaches takes time.	The automated approach to accessing and analysing comparables using ATTOM AVM saves a considerable amount of time.
	Due to the time taken to conduct valuation using the traditional approaches, potential homeowners often have to wait for several days and or weeks to get their valuation concluded. This can have adverse effects on their mortgages as interest rates may increase and the cost of borrowing may also increase.	This digital tool significantly reduces the time taken to conduct valuations which means that the borrowers can conclude their mortgage transactions in a shorter time, minimising the risks associated with economic and political changes.

(Continued)

Table 7.2 (Continued)

ESEP	Inefficiency with the Traditional Approach	How the Tool/Platform/Product Addresses the Inefficiency
	The delay in getting valuations completed can lead to delays in investment decisions for residential real estate investors. Considering that timing is essential in real estate investment, investors may miss out on deals or make lagged investment.	The higher speed in completing valuations can enable investors, investment managers and asset managers to make timely decisions and take advantage of investment opportunities.
Physical/ spatial		

Source: Authors analysis.

7.4.2.3   Limitations and Challenges

ATTOM AVM is limited by its application to residential real estate assets. This means that other forms of real estate assets cannot be automatically estimated using the model, thus limiting the highlighted benefits to the residential real estate sub-market. It will be valuable to explore the channels through which this technology can be extended to other real estate areas. Also, one of the features of the tools is that it requires a minimum of 15 comparable properties. In the case where a property does not have up to 15 comparables, the tool's application is also limited. Although the platform has the potential to attract environmental benefits by minimising high computing power and energy in analysis, it can also be argued that due to the high performance of the platform, high computing and energy usage will still be required to integrate all the systems, and the subscription to the cloud and other data systems will also lead to an increase in the demand for data centres which, as discussed in Section Two, is also significantly contributing to the carbon emission and global energy consumption.

Further details on ATTOM AVM can be accessed from their website using this link: https://www.attomdata.com/.

7.4.3   Case Study 3: CoreLogic

7.4.3.1   Background, Key Details, and Technologies Utilised

CoreLogic's AVM is a platform that provides real-time estimates of residential asset values. As one of the main custodians of property data in Australia, the platform collects, analyses, and connects billions of data points from thousands of disparate sources, to create a unique and all-rounded view of over 14 million properties in Australia and New Zealand. They support various sectors including real estate professionals and service providers, mortgage brokers, construction professionals, valuers, lenders, financial institutions, insurers, government, etc. The platform utilises advanced digital technologies such as AI-powered analytics and cloud-based computing, real-time and dynamic access to data with a rapid matching of over 10 million properties, accessed through AIP or bulk match and delivery, and a fully integrated machine learning model which ensures rapid and accurate deployment. The AVM-generated valuation and indices are independently audited to ensure best practices are in line with Australia's Prudential Regulatory Authority (APRA) guidelines.

CoreLogic uses its RP Data and ValConnect to provide advanced property market valuation and analytics. This is in response to the demand for remote valuations during the pandemic. The ValConnet has a unique feature known as the "Upload Portal" which enables clients to securely upload service for property images for digital valuations. The RP Data mobile app further provides remote and mobile access to deep market insights. The CoreLogic's AVM provides a snapshot of recent sales over the last 90 days across all the suburbs in selected postcodes and the Home Value Index is a mapping market tool that provides a national overview of each suburb's current median value and quarterly and annual changes in value. The model also takes various factors into consideration including insurance rebuild estimates, dwelling types, occupancy scores, and geographic data, and these are analysed using advanced AI modelling and data science expertise. This ensures rapid and accurate deployment.

7.4.3.2   Key Efficiencies and Solutions That CoreLogic AVM Provides

The solutions that Core Logic AVM provides are itemised using the ESEP efficiency framework (Table 7.3):

Table 7.3 ESEP Efficiency Analysis of Core Logic

ESEP	Inefficiency with the Traditional Approach	How the Tool/Platform/Product Addresses the Inefficiency
	The analogue nature of data analysis and modelling using traditional approaches can lead to more extensive use of computing power and energy (e.g., electricity).	The integrated data analytics and valuation service provided by CoreLogic AVM reduces the energy and computing power that would normally be used to process information that is required for a valuation. By reducing the time and manpower required to source and analyse data, the computing power and energy that would have been used to complete these tasks are significantly minimised.
	The collaborative work that goes with traditional valuation may sometimes require commuting and travel.	The automation approach to CoreLogic AVM means that valuation can be completed with very minimal human effort. This can significantly reduce commute-related carbon emissions associated with this activity such as site visits and data collection.
	Traditional valuation and reporting require a high volume of stationary.	The need for printing the valuation report is minimised as the valuation estimate is made available to the lenders and the clients online, minimising the need to use stationary.
Social	Three is usually a trade-off between the quantum, quality and application of comparables when valuation is conducted using traditional approaches and processes.	CoreLogic AVM uses tends of comparables which strengthens the quality of the output and its reliability. The confidence score generated by the platform is an added advantage.

(Continued)

Table 7.3 (Continued)

ESEP	Inefficiency with the Traditional Approach	How the Tool/Platform/Product Addresses the Inefficiency
	The manual gathering and adjusting of comparables takes a considerable amount of effort and processes which are efficient.	Gathering and adjusting comparables is automated using CoreLogic; this saves a considerable amount of energy and effort.
	Valuation for insurance purposes is typically to enable the underwriter to assess the risks associated with the house. Although the valuer will attempt to consider all the risks associated with an asset, this may not always be possible using traditional approaches and processes.	In calculating property values for insurance, CoreLogic takes account of the home profile, neighbourhood and other risk information which would be impossible using manual methods and approaches. This is enabled by the platform's access to a wide range of data sets which are integrated for better property values estimation.
	Due to the complexities of carrying out valuations and the various aspects that are often taken into consideration, traditional valuation approaches are prone to error, and in implementing corrections manually, there may be inaccuracies and inconsistencies. This can have significant effects on clients' decisions.	The high level of automation and integration of the analytics and reporting systems in the AVM significantly minimises the chances of errors, inaccuracies, and inconsistencies. Lenders and underwriters can therefore make their decisions with a higher level of accuracy and confidence, and this can help to minimise risks and unforeseen costs.
	Due to the complexities of carrying out valuations and the various factors to consider, valuation using traditional approaches and processes often takes a long period of time to complete.	Due to the automated approach of CoreLogic AVM, the valuation and appraisal can be completed in a short time.
	Sourcing comparables using manual approaches takes time.	The automated approach to accessing and analysing comparables using CoreLogic AVM saves a considerable amount of time.
	Due to the time taken to conduct valuation using the traditional approaches, potential homeowners often have to wait for several days and weeks to get their valuation concluded. This can have adverse effects on their mortgages as interest rates may increase and the cost of borrowing may also increase.	This digital tool significantly reduces the time taken to conduct valuations which means that the borrowers can conclude their mortgage transactions in a shorter time, minimising the risks associated with economic and political changes.
	The delay in getting valuations completed can lead to delays in investment decisions for residential real estate investors. Considering that timing is essential in real estate investment, investors may miss out on deals or make lagged investment.	The higher speed in completing valuations can enable investors, investment managers and asset managers to make timely decisions and take advantage of investment opportunities.
Physical/ spatial		

Source: Authors analysis.

7.4.3.3   Limitations and Challenges

CoreLogic AVM is limited by its application to residential real estate assets. This means that other forms of real estate assets cannot be automatically estimated using the model, thus limiting the highlighted benefits to the residential real estate sub-market. It will be valuable to explore the channels through which this technology can be extended to other real estate areas. Also, although the platform has the potential to attract environmental benefits by minimising high computing power and energy in analysis, it may also be argued that due to the high performance of the platform, high computing and energy usage will still be required to integrate all the systems, and the subscription to the cloud and other data systems will also lead to an increase in the demand for data centres which, as discussed in Section Two, is also significantly contributing to the carbon emission and global energy consumption.

Further details on CoreLogic can be accessed from their website using this link: https://www.corelogic.com.au/.

7.4.5   Other Cases (Exercises for Students)

7.4.5.1   Collateral Analytics

Collateral Analytics (CA) is an AVM that uses property data and transaction information to estimate residential property values for more than 95% of the US housing stock, generating instant reports. For each residential property, the model utilises a multiple model approach and neighbourhood-specific comparable selection which ensures that the most recent data and information are considered in the property value estimates. CA's products include CA Value, CA Interactive, CA Commercial, CA Consumer, CA Value Forecast, CA Neighborhood Value Range, CA Rental, and CA Complexity Profiler.

Further details on Collateral Analytics can be accessed from their website using this link: https://collateralanalytics.com/.

7.4.5.2   HouseCanary

HouseCanary has home price indices (HPIs) and AVMs that have been developed by a team of data scientists and backed by advanced machine learning and a robust range of data sources. The company was founded in 2013, and the client base is currently made up of financial institutions, investors, lenders, mortgage investors, consumers, etc. In determining the property values, HouseCanary AVM examines all publicly available data about a property and analyses the information to provide an estimate of the property's value through direct API integration or per-asset via match and append. The Property Explorer tool is an engaging data mapping tool that utilises HouseCanary's vast data set of valuation and rental insights. These insights are available to real estate brokers who use the valuation models and data to support their clients in identifying and transacting on properties more efficiently.

Further details on HouseCanary can be accessed from their website using this link: https://www.housecanary.com/.

7.4.5.3   Redfin Estimate

Redfin Estimate provides property value estimates, using its complete and direct access to multiple listing services (MLSs).[7] The data on recently sold homes on MSLS are used to create the estimate of the property value; this process is not perceived as an appraisal nor is the output perceived as a professional opinion of value; rather, the output is used as a guide to the real estate agent or broker. To produce the estimates, the algorithm considers hundreds of data points about the local market, neighbourhood, and the tangible and non-tangible features of the property. These data are analysed using advanced computing power and machine learning software and stored using cloud-based technology.

   Further details on Redfin Estimate can be accessed from their website using this link: https://www.redfin.com/.

7.4.5.4   VeroVALUE

VeroVALUE is an AVM that combines predictive modelling and technology with quality data and industry expertise to create valuations that are reliable and accurate. It combines multiple predictive technologies with traditional real estate market fundamentals; these are then supplemented with advanced analytics that provide deeper insights into price trend information. The platform also provides information on regions that have been declared disaster areas as a value-added service. It takes account of several factors such as satellite imaging and advanced data analytics to analyse hurricanes, earthquakes, wildfires, windstorms, volcanoes, tornadoes, floods, storm surges, and tsunamis. The input data are sourced from multiple third-party aggregators, county assessors, sales data updates from county clerk/recorder offices, and other publicly available data sources.

   Further details on VeroVALUE Estimate can be accessed from Vero's website using this link: https://www.veros.com/.

7.5   Chapter Summary

This chapter has provided a review of real estate valuation and appraisal and highlighted the inefficiencies associated with its processes and procedures. It has also provided insights into the innovative tools and platforms that have emerged in real estate valuation and appraisal and some of the emerging terminologies. The case study analysis also provided further insights into specific real estate valuation digital tools and platforms, highlighting the enhanced efficiency that they provide. There are several innovations and digitisation in real estate valuation that should be further researched. For instance, it would be valuable to explore the channels through which AVP can be further entrenched in valuation practice, particularly in commercial assets and AI may offer solutions in this respect. There will also be the need to develop innovative approaches to collecting, adjusting, and analysing comparable data, and ultimately integrating this in the valuation models to produce accurate and reliable estimates. Real estate students, scholars, and professionals are encouraged to explore the possibilities that lie in automated valuation across all real estate sectors.

Notes

1 It should be noted that this chapter does not delve into debates and arguments around valuation and appraisal methods; rather, it provides a holistic review and analysis of the processes and approaches to carrying out these functions and providing related services.
2 This may also apply to mortgages.
3 It should be noted that this process may not necessarily be linear.
4 Mass appraisal is the process of valuing a large group of properties at a given date by using standardised methods and common data which are subjected to data analysis.
5 Pareto optimality describes a situation where no further improvements to society's "well-being" can be made through a reallocation of resources that makes at least one person better off without making someone else worse off.
6 The ESEP efficiency analysis for each case study is not exhaustive; students and readers are encouraged to identify and note down other areas of efficiency which the named tool/platform provides.
7 MLSs are the databases that real estate agents use to list their properties in the US.

References

Asaftei, G. M., Doshi, S., Means, J., & Sanghvi, A. (2018). Getting ahead of the market: How big data is transforming real estate. In *Mckinsey & Company* (Issue October). https://www.mckinsey.com/industries/real-estate/our-insights/getting-ahead-of-the-market-how-big-data-is-transforming-real-estate#/.

Baum, A., Graham, L., & Xiong, Q. (2021). *The future of automated real estate valuations (AVMs)*. Oxford Future of Real Estate Initiative. Available online at: https://www.sbs.ox.ac.uk/sites/default/files/2022-03/FoRE%20AVM%202022.pdf (accessed 5 May 2023).

Burak, U., & Baycar, K. (2019). Historical evidence for anchoring bias: The 1875 cadastral survey in Istanbul. *Journal of Economic Psychology, 73*, 1–14.

Blackledge, M. (2016). *Introducing property valuation* (2nd ed.). Routledge.

Boyd, T., & Irons, G. (2002). Valuation variance and negligence: The importance of reasonable care. *Pacific Rim Property Research Journal, 8*(2), 107–126.

Edelstein, R., & Quan, D. (2006). How does appraisal smoothing bias real estate returns measurement? *The Journal of Real Estate Finance and Economics, 32*(1), 41–60.

Howell, J., & Korver-Glenn, E. (2018). Neighborhoods, race, and the twenty-first-century housing appraisal industry. *Sociology of Race and Ethnicity, 4*(4), 473–490.

IVSC. (2020). International valuation standards. In *IVSC valuation standards* (Effective). International Valuation Standards Council.

Klamer, P., Bakker, C., & Gruis, V. (2017). Research bias in judgement bias studies – A systematic review of valuation judgement literature. *Journal of Property Research, 34*(4), 285–304.

Levy, D., & Schuck, E. (1999). The influence of clients on valuations. *Journal of Property Investment and Finance, 17*(4), 380–400.

Levy, D., & Schuck, E. (2005). The influence of clients on valuations: The clients' perspective. *Journal of Property Investment and Finance, 23*(2), 182–201.

McAllister, P., Baum, A., Crosby, N., Gallimore, P., & Gray, A. (2003). Appraiser behaviour and appraisal smoothing: Some qualitative and quantitative evidence. *Journal of Property Research, 20*(3), 261–280.

Oladiran, O., Sunmoni, A., Ajayi, S., Guo, J., & Abbas, M. A. (2022). What property attributes are important to UK university students in their online accommodation search? *Journal of European Real Estate Research*. https://doi.org/10.1108/JERER-03-2021-0019.

Perry, A., Rothwell, J. & Harshbarger, D., 2018. The devaluation of assets in Black neighborhoods, The case of residential property, Brookings Institution. United States of America. Available online at: https://policycommons.net/artifacts/4139385/the-devaluation-of-assets-in-black-neighborhoods-the-case-of-residential-property/4947807/ (accessed 26 Feb 2024).

Pi Labs. (2020). *Technology and the future of real estate investment management.*

RICS. (2017). *The future of valuations: The relevance of real estate valuations for institutional investors and banks – Views from a European expert group* (Issue November). https://www.rics.org/news-insights/research-and-insights/the-future-of-valuations#:~:text=This insight paper examines the developments and changing client expectations.

RICS. (2021). RICS valuation - Global standards 2021 - Red book. In *RICS global standards* (Effective, Issue September). Royal Institution of Chartered Surveyors (RICS).

Rosen, S. (1974). Hedonic prices and implicit markets : Product differentiation in pure competition. *Journal of Political Economy, 82*(1), 34–55.

Royal Institution of Chartered Surveyors. (2022). *Automated valuation models (AVMs): Implications for the profession and their clients* (Issue April). www.rics.org.

Saddiqi, H. (2018). Anchoring-adjusted capital asset pricing mode. *The Journal of Behavioural Finance, 19*(3), 249–270.

Saull, A. (2020). *The future of real estate occupation: Issues.*

Tversky, A., & Kahnemann, D. (1974). Judgment under uncertainty: Heuristics and biases: Biases in judgments reveal some heuristics of thinking under uncertainty. *Science, 185*(457), 1124–1131.

Wright, E. (2020). *WiredScore launches world's first smart building council.* EG Radius. https://www.egi.co.uk/news/wiredscore-launches-worlds-first-smart-building-council/.

Digital Technology and Innovation in Real Estate Brokerage, Agency, Marketing, and Other Allied Services

8.1 Overview of Real Estate Agency, Brokerage, and Marketing

8.1.1 Real Estate Agency and Brokerage

Real estate agency involves the representation of a client (landlord) and his/her interest in the process of buying, selling, renting, leasing, or managing a property.[1] A real estate agent therefore serves as an intermediary between a property owner and prospective buyers, sellers, or tenants. Real estate brokerage is quite similar to the agency as they also serve as intermediaries between buyers and sellers of real estate. Brokers are licensed professionals who provide guidance, oversight, legal compliance, and representation in real estate transactions. Some of the functions and expectations of real estate agents/agencies and brokers include the following:

- **Services**: Real estate agents may be required to provide a range of services such as property marketing,[2] advertising, listing, inspections, negotiation, offer management, valuation, contract preparation, and assistance with legal processes and transactions.
- **Expertise and market knowledge**: Real estate agents and brokers are expected to have expertise in real estate and allied services and to have extensive property market knowledge. They have access to various resources, including databases, market reports, and historical sales data, and they possess in-depth knowledge of property market data, market trends, neighbourhood characteristics, and legal requirements. They use this data access and expertise knowledge to provide guidance and advice to clients enabling them to make informed decisions regarding asset pricing, investment opportunities, and property selection.
- **Client representation**: Real estate agents and brokers typically represent either the buyer/tenant or the seller/landlord. Some agencies may specialise in one type of representation, while others may offer services to both parties. Agents and brokers work on behalf of their clients to protect their interests, negotiate favourable terms, prepare legal documents, and ensure a smooth transaction process.
- **Networking and connections**: Real estate agencies often have extensive networks and connections within the industry. These connections can be valuable for finding suitable properties, reaching a broader pool of potential buyers or tenants, and collaborating with other professionals involved in real estate transactions, such as mortgage lenders, attorneys, and inspectors. The connections can facilitate an information flow and a smooth transaction process.

DOI: 10.1201/9781003262725-10

- **Licensing, compliance, and regulation**: Real estate agencies are subject to regulations and licensing requirements by the relevant governing bodies in their jurisdiction. Agents therefore must adhere to these regulations and operate within the legal framework, standards of professionalism, and ethical conduct to protect the interests of their clients and maintain professionalism.
- **Customer service**: A good real estate agency should prioritise customer service and strive to provide a positive experience for their clients. This includes effective communication, responsiveness, transparency, and personalised attention to client needs throughout the entire transaction process. A real estate brokerage's reputation is built on its track record, client testimonials, and the satisfaction of past clients; thus, providing a high level of customer service and satisfying clients' needs is an important element of real estate agency and brokerage.

It's important to note that the specific services, expertise, and reputation of a real estate agency can vary significantly. Clients are therefore advised to research and evaluate agencies based on their track record, client testimonials, market knowledge, and alignment with their specific needs and goals.

8.1.2   Real Estate Marketing

Although real estate marketing is one of the core aspects of real estate agency and brokerage, this task can also be performed by individuals, landlords, and corporates who may not necessarily consult with property agents. This function involves promoting and advertising properties to attract potential buyers or tenants and it comprises various strategies and tactics that are aimed at creating awareness, generating leads, and facilitating successful property transactions. Some property management functions include the following[3]:

- **Market research and analytics**: Real estate marketing is enhanced by effective market research and data analysis. By monitoring market trends, competitor activities, and consumer behaviour, marketers can refine strategies, identify opportunities, and make evidence-based and data-driven decisions to maximise and enable their clients to achieve their objectives.
- **Defining the target market**: Effective real estate marketing starts with identifying the target market for a specific asset or development. This involves understanding the demographics, preferences, needs, socio-economic status, and socio-cultural characteristics of potential buyers or tenants. By defining the target market, marketing efforts can be tailored to reach and resonate with the right individuals.
- **Advertising**: Properties and developments are often advertised through: print media such as newspapers, magazines, brochures, signage, and billboards; audio-visual channels such as radio and television adverts; social media; and direct mail campaigns. These can be targeted to specific geographic areas or niche markets.
- **Branding**: Having a strong brand presence and positioning is important in real estate marketing, particularly for agencies and brokerages. This involves creating a unique identity and value proposition for the real estate agency. An agency/brokerage's branding efforts should convey the property's unique features, benefits, and lifestyle attributes, and also highlight the expertise and professionalism of the agency.

- **Event marketing and open houses**: Hosting events, such as property launches, open houses, or community engagement activities, enables potential buyers or tenants to experience the property first-hand. These events create opportunities for personal interactions, showcasing property features, and answering questions to generate interest and attract individuals to a property or a development.
- **Networking and partnerships**: Real estate marketing involves building relationships and partnerships within the industry. Collaborating with other professionals, such as mortgage lenders, real estate attorneys, interior designers, and home staging experts, can enhance marketing and provide additional value to clients.
- **Customer service**: Regardless of if marketing is carried out by an agency or the landlord, customer service is important in real estate marketing. For marketers to reach their customers and to retain their interest in a subject property, they need to provide a high level of service and maintain relationships with their prospective clients, current clients, and old clients.[4]

Real estate agency, brokerage, and marketing are an essential segment of the real estate value chain. They serve as a connector between asset owners/agents (developers and investors) and potential users, which effectively means linking the development and investment markets with the users' market. This link is essential as landlords may find it difficult to identify tenants that will occupy their properties, and prospective buyers; space users may also not be able to identify spaces that best satisfy their needs, taste, and preferences.

8.2   Inefficiencies in Traditional Approaches to Real Estate Agency, Brokerage, and Allied Services

8.2.2   Inefficiencies in Real Estate Agency and Brokerage

Traditional real estate agencies and brokerage firms have faced numerous challenges associated with changing consumer behaviours and market dynamics. These challenges include the following:

1 **Information asymmetry**: In traditional real estate practice, agencies held a significant volume of property information, leading to information asymmetry between agents and clients. This made them gatekeepers of property information, making information and data highly fragmented. The search theory provides some insight.[5] It suggests that information asymmetry can exacerbate search friction associated with factors such as time, consumer circumstances, and the quantity and quality of the information in decision-making in high-involvement products (Qiu and Zhao, 2018).
2 **High transaction costs**: Real estate agency and brokerage are accompanied by high fees associated with services provided. The commission-based compensation structure, where agents and brokers earn a percentage of the property's sale or lease price as commission, can result in substantial costs for buyers and sellers. Furthermore, information asymmetry can increase search costs for buyers and sellers (Albrecht et al., 2016; Turnbull and Sirmans, 1993; Williams, 2018), renters, and landlords (Blowers et al., 2014; Kim, 1992; Read, 1993, 1997). These costs may impact affordability, limit market accessibility, and reduce the level of patronage that agents and brokers would get.

3 **Changing consumer expectations**: Real estate needs to continue to change in line with changing demographic, household, and socio-cultural factors, as well as evolving individual tastes and preferences. Furthermore, real estate transactions are evolving, and customers are also demanding improvement in agency and brokerage service delivery in areas such as speed, convenience, and transparency. Traditional real estate practice is unable to meet up with these changes to taste, preferences, demographic shifts, technological advancements, and socio-cultural factors. This slow pace of agency adaption has further led to an equally low rate of adaption in real estate, particularly in the office market (Oladiran et al., 2023).

4 **Limited market coverage**: Traditional agencies often have limited market coverage, focusing on specific geographic areas or property types. This limitation can restrict client choices and reduce opportunities for both buyers and sellers. In an increasingly globalised and interconnected world, clients often seek broader options and specialised expertise that traditional agencies might struggle to provide.

5 **Inefficient processes, systems, and services**: Modern business requires efficiency in service delivery and a high level of convenience, transparency, and personalised experiences. Clients often expect real-time updates, access to information, and seamless transactions. Traditional approaches to real estate brokerage and agency will be unable to meet these evolving expectations, as they are typically driven by manual and analogue processes, paper-based transactions, and physical interaction and contact.

6 **Inefficient marketing and advertisement**: Traditional real estate agencies and brokerage were highly dependent on methods such as newspaper advertisements and brochures which severely limited the reach and scope of the property being listed in the market. This led to information mismatch between space need and availability, further leading to a mismatch between demand and supply. This could lead to high vacancy rates because of the inability of agents to effectively reach space users, while space users also find it difficult to locate or identify suitable spaces for their needs and businesses.

7 **Adhering to regulation and standards**: The regulatory environment governing real estate agencies can also present challenges. Compliance with licensing requirements, professional standards, and legal obligations can impose administrative burdens on agencies. Moreover, regulatory changes, such as stricter data privacy regulations or shifts in property tax laws, may require agencies to adapt their processes and practices, adding complexity to their operations. This should not be misconstrued as a criticism of the regulation in the real estate agency sector; rather, it is a highlight of the inability of traditional approaches to brokerage and agency to provide strict adherence to ethics, legal obligations, legal requirements, regulations, and standards. For this to be achieved through traditional approaches, there will be high operational and administrative costs incurred.

8.3.2   Inefficiencies in Real Estate Marketing

Traditional real estate marketing has been riddled with numerous challenges and problems. These challenges include the following:

1 **Shifting consumer behaviour**: Consumers' behaviour in the real estate market has undergone significant changes. As stated in Section 8.3.2, sellers and buyers are increasingly changing their preferences and tastes resulting from changes in socio-cultural

factors, technological advancement, etc. This has made the use of traditional marketing and advertising media such as print advertisements and physical brochures relatively obsolete as tools for real estate marketing as they hardly capture the attention of the intended customers.

2 **Limited reach and targeting**: Traditional real estate marketing methods often have limited reach and targeting capabilities. Print advertisements, billboards, and direct mail campaigns may only reach a local or limited audience. They also have restricted capabilities to target specific demographic, geographical, and sociocultural strata-important elements in the execution of marketing strategies.

3 **Cost efficiency**: Traditional marketing methods, such as newspaper ads and physical brochures, can be expensive and have limited shelf life.

4 **Measurability and data analysis**: Traditional marketing efforts often lack comprehensive data analysis and measurability. It can be challenging to track the effectiveness of print advertisements or gauge audience response.

5 **Maintaining personalised interactions**: Traditional marketing often thrives on personalised interactions between agents and clients. Face-to-face meetings, property viewings, and negotiations have been central to the real estate sales process. These processes are limited in cases where face-to-face meetings is impossible, and they can often lead to delays in concluding deals and transactions.

8.3   Innovative/Emergent Systems and Services in Real Estate Agency, Brokerage, Marketing, and Other Allied Services

As the property sector evolved over the last decades, various tools, services, products, and platforms have also emerged in an attempt to keep pace with the changes as well as to address customers' and clients' changing and highly varying needs, challenges, preferences, etc. Technological advancement has also enabled the emergence of several tools, services, and systems. This sub-section therefore provides an analysis of some of these emergent systems, services, tools, platforms, and products related to agency, brokerage, and marketing. It should be noted that this sub-section goes beyond emergent systems and services in real estate brokerage and agency. It also covers other allied emergent and innovative real estate areas such as the shared economy and space commoditisation which have also emerged because of changing demographic, sociocultural, and economic shifts and shocks.[6] Some of these allied areas have changed the nature and approach to the real estate agency and brokerage.[7]

8.3.1   Online Property Listing and Marketing

Real estate marketing requires the consideration of a wide range of factors and the use of various strategies and approaches. This and several other enablers have led to the emergence of various real estate property listing and marketing websites, software, and mobile applications. Online property listing platforms are increasingly gaining popularity as go-to resources for property searches across the world. These platforms offer a wealth of information, property listings, advanced search functionalities, user-friendly interfaces and, in some cases, streamlined transaction processes, offering alternative ways for buyers and sellers to connect more efficiently and effectively and for transactions to be completed. They create and maintain a visually appealing and user-friendly platform that showcases properties and offers targeted advertising, search engine optimisation (SEO), social media marketing, pay-per-click (PPC) advertising, and email campaigns that can effectively reach and engage a wider audience. Successful digital property marketing campaigns leverage strong branding, compelling visual content, digital and traditional advertising, market research, and relationship-building techniques to attract and engage potential buyers or tenants.

Saiz (2020) provides valuable insight into the factors that have facilitated the development of online property listing and marketing. The scholar argues that in many market spaces with initial concentration and large network economies, transactional technology has facilitated the role of the broker, and in markets where network economies are absent, disintermediation turned out to be the outcome.[8] As stated in Section 8.2, information asymmetry is one of the major problems of marketing and brokerage, but digital technology has significantly minimised this problem in property search and selection (Palm and Danis, 2002). Studies have shown that the tangible and non-tangible property features of properties are important in individuals' search and selection (Khozaei et al., 2011; Kobue et al., 2017; Magni et al., 2019), thus, access to a wider range of property information can increase the likelihood that a property will be selected by a prospective tenant or buyer (Oladiran et al., 2022).

Online property listing and marketing have been enhanced by several digital tools including computing, data, analytics, mobile technology, internet, virtual reality, etc. AI is particularly a powerful tool that is further transforming property listing and marketing. It is now commonplace for online listing platforms to use AI algorithms to imitate human conversational skills and chatbots used to answer questions that property searchers seek regardless of the time of the day. Online property listing and search have become the default. Although real estate brokers and agents are still relevant, it is difficult to now think about real estate sales, buying, rental or lease without the use of digital technology for property search, listing, advertisement, and general data collection. At the heart of this system and approach to property marketing is the centralisation of real estate markets.

According to Saiz (2020), online property listing and marketing began in the early days with platforms such as Apartments.com or Craigslist and then metamorphosed with the development of Multiple Listing Services (MSL) in the US. Listing of residential properties was quickly adopted by various regional MLS which in previous years were merely databases with standardised property data fields and photographs that the brokers hardly shared with their customers. The franchising of local real estate brokerages amalgamating into local MLS exchanges incentivised the sharing and dissemination of information to their broker bases motivated by the urge to improve sales and business performance. This metamorphosed into the FSBO (for-sale-by-owner) market in the 1990s where sellers listed properties themselves on the internet and had buyers approach them directly. Americans often visit Zillow to review the value of their properties and sometimes view the characteristics of their neighbour's properties. Zillow has a subscription model that facilitates the generation of sales leads or preferential placement by subscribed brokers. LoopNet is also one of several commercial platforms that have also become popular for the search and listing in the US.

In the UK, some of the early online property listing and marketing platforms were Rightmove and Zoopla. It is now typical for individuals searching for properties to buy, sell, let, or rent in the UK to initially engage with these platforms, particularly for residential properties. For individuals who do not have any knowledge of these platforms and thus conduct a Google search for residential properties, Rightmove, Zoopla, OpenRent, and On The Market are some of the first results to appear. With the first two platforms, agents can list their properties on them, and prospective tenants and buyers can contact the agents directly. However, OpenRent directly links tenants to landlords, cutting out all intermediaries.

Online property listing and marketing have also transformed the agency and brokerage. The services in these sectors are important due to the heterogeneity and diversity in space users' needs, specifications, requirements, etc. Mega global brokerage firms such as Jones Lang Lasalle, Knight Frank, Cushman and Wakefield, and CBRE, although not involved in a substantial proportion of office and industrial listing, have been acquiring large competitors (Saiz, 2020). These integrated CRE services companies (ICRESCs) bundle brokerage with other critical services such as PR, research, marketing, planning, financing, asset management, investment, custodial, site selection, consulting, and advisory services. The digital revolution in real estate brokerage may therefore offer new opportunities for ICRESCs.

Several digital tools and platforms primarily dedicated to CRE have emerged in the last 15 years, and this has further decentralised CRE brokerage. LoopNet, for instance (with CoStar as its holding company) has become one of the major repositories of CRE data. Although LoopNet provides a role similar to the MLS in the residential market, its listing and marketing platform are used for both commercial leases and property transactions. CoStar supplements the listing on LoopNet with information from extensive property record databases and CRE brokers pay a premium to have their properties listed. The digital CRE brokerage space is relatively thin and has fewer players compared to other real estate sectors. The relatively small competition in the CRE transaction data space can be linked to the distinctive core competencies.

The emerging operational real estate (ORE) asset class has also benefitted extensively from the online listing model; in fact, it can be argued that this asset class has been driven or underpinned by this listing and marketing system.[9] It has made the marketing, search, reservation, and booking of assets in this sector more efficient and effective for potential customers and operators. One of the major distinguishing features of online listing platforms in ORE that distinguishes it from residential real estate and CRE is that they have a function to automatically book and reserve spaces, and the vast majority of them are for rental. Examples of some of these platforms are student.com, Uniplaces (PBSA), hotel.com, booking.com (hotels), Big Yellow, and Safestore (storage space).

8.3.1.1   Benefits and Advantages of Online Property Listing and Marketing

Online property listing platforms and tools have significantly enhanced real estate search, selection, listing, letting, rental, and sales in the various real estate sectors (residential, CRE, and ORE). Some of the benefits include the following:

Information access and transparency: One of the major benefits of online property listing platforms is that they provide prospective tenants and buyers with a wide range of information on the listed properties. For platforms that have integrated mapping systems, the prospective buyer or renter can use the street views to understand the street and neighbourhood where the property is located and to also assess proximity to services and amenities such as schools, hospitals, and transport links. This significantly minimises the problems that prospective buyers and renters will have when searching for properties, particularly when they are unable to visit the property. The provision of pictures and virtual viewing capabilities on some platforms offer the added advantage of visuals aiding prospective renters or buyers in their search and selection process. The increase in information access that online listing platforms provide is significantly increasing the level of transparency in markets, particularly for platforms where previous property transactions are provided.

Reduced transaction time and cost: Better access to information and disintermediation is significantly reducing transaction costs associated with property letting, sales, and purchases. Digital marketing tools provide an opportunity for brokers, landlords, and property sellers to reach a wide range of people which can increase their reach with minimal costs. Being able to reach out to the same number of people using traditional marketing approaches such as marketing campaigns, brochures, and magazines will incur huge costs. For prospective buyers and renters, digital marketing offers cost-efficient approaches to searching for properties, inspections (virtual tours), selection, and, in some cases, completing transactions. These digital channels provide wider exposure and longer-lasting impact, often at a fraction of the cost of traditional methods. Additionally, the disintermediation that is achieved through platforms that provide direct contact of buyers and sellers/renters and landlords can further reduce agency and brokerage costs.

More personalised experience: Digital marketing offers a cost-effective approach to providing personalised property search and selection. For most online property listing platforms, users can use the filter tools to narrow down the list of displayed properties based on their budget, distance, specified locations, amenities, accommodation details (e.g., number of rooms), etc. This enables prospective renters and buyers to make quicker property selections and enables them to find properties that closely match their preferences. Personalised services are also essential in meeting the expectations of real estate customers and consumers.

Reduced environmental impact: A significant amount of carbon emission is used in commutes relating to property viewings and inspections; the use of online listing platforms significantly reduces transportation-related carbon emissions. Furthermore, digital marketing platforms can minimise the use of paper brochures and other printed media which can support environmental conservation efforts and wider environmental sustainability goals.

Efficient space utilisation: Online property listing platforms provide a more effective and efficient matching mechanism for tenants and landlords, thus enhancing the efficiency of space use. This makes space available to those who need it, and it enables property owners to let (rent) their spaces more speedily. This can minimise the time that a property stays vacant, thereby maximising returns and minimising costs on an asset.

Advanced analytics capabilities: Digital property listing provides an enormous opportunity for data to be collected, stored, and analysed using advanced techniques to enhance empirical analysis and insight on various search, selection, and other patterns that prospective tenants have which would hitherto not have been captured using the traditional approaches. This can enable brokers, agents, and landlords to make more informed decisions for property selection, sales, and rental and enable them to develop and optimise their investment strategies. Furthermore, researchers can also collect data from these platforms either through exclusive accesses provided by the platform, or through the adoption of advanced techniques such as automated web scraping to enhance empirical analysis.

8.3.1.2   Challenges and Limitations of Online Property Listing and Marketing

Although online property listing platforms have attracted several benefits to real estate agencies and brokerage, they also have several challenges and limitations including the following:

Information overload: Online property listing platforms provide an enormous volume of property information and a wide range of properties to select from, which can overwhelm users. Sorting through numerous listings, filtering search results, and identifying relevant properties can sometimes be a time-consuming and daunting task for buyers, particularly in city markets that have a wide range of properties. Information overload may lead to decision fatigue and frustration, potentially affecting user satisfaction.

Privacy and security concerns: Online property listing platforms require users to share personal information, such as contact details, preferences, and search history. Privacy and security concerns arise regarding the methods of data collection, storage, and use. Users are therefore becoming increasingly hesitant to provide private sensitive information, which can limit the effectiveness of personalised services.

Inaccurate or outdated listings: Maintaining accurate property listings can be challenging for online platforms. Properties might be listed as available when they have already been sold or rented, leading to a disappointing user experience. Keeping listings current requires constant monitoring and timely updates from property owners or agents, which might not always happen. Furthermore, because the volume and quality of the information displayed about a property can affect its selection, properties that are listed with insufficient information or poor-quality photos may likely receive less attention. Furthermore, photos of a building can be edited and presented in ways that are not accurate. For instance, rooms may appear to be bigger than they actually are due to the angle of the photos or the editing. This can lead to property selection on an inaccurate premise, and this can affect user satisfaction.

Quality control: Maintaining quality control over property listings can be challenging for online platforms. Some listings lack essential information, have low-quality images, or include misleading details, and it is difficult to monitor and control this. Without proper quality control measures, users may encounter irrelevant or misrepresented properties, hindering their search experiences.

Limited scope for complex transactions: Complex real estate transactions, such as commercial properties or unique investments, might not fit the standardised format of online property listing platforms. These platforms serve properties in the residential and operational real estate sectors better due to their homogenous nature, although more specialised properties in these sectors may face the same challenges that are experienced in the listing of commercial real estate.

Real estate investors, developers, and users, particularly new entrants into the market may therefore benefit from further consultation with real estate professionals when they are taking decisions on property selection, particularly for large-scale projects.

8.3.2   Virtual Viewing

Property viewing and inspections have evolved from physical inspection to the virtual realm, and this approach to viewings and inspections became more widely adopted during the COVID-19 pandemic outbreak. Virtual property viewing refers to the use of various digital tools and technology by potential buyers or renters to virtually tour a property remotely, without physically being present at the location. It leverages various digital tools, such as virtual reality (VR) and 360-degree videos, to create immersive and interactive experiences for users. Virtual viewing has become increasingly popular, offering several benefits to both buyers and sellers. In most cases, virtual viewings are products of pre-recorded views such as videos and photos of existing spaces and properties; in some cases, however, virtual tours can be taken in properties that have not been developed.[10]

Virtual property viewings require the use of specialised cameras to capture high-quality images or videos of the property and its spaces. This may include 360-degree views of the various spaces and amenities in the properties. These images and pictures are then edited and uploaded for virtual access by interested tenants or buyers. These may be accessed by interested parties through a screen display or through the use of VR headsets and VR-compatible devices. Users wearing a VR headset can navigate through the property by moving their heads or using handheld controllers. This creates a fully immersive experience that gives the feeling of being inside the property. Virtual viewing may also incorporate interactive features that allow users to explore the property at their own pace. For instance, users can choose the rooms they wish to view, zoom in on specific details, and navigate through the property with ease. Furthermore, users may also have access to supplementary information about the property, such as floor plans, property specifications, neighbourhood details, and pricing information. This additional context enhances the understanding of the property's features and potential.

The use of high-quality images, videos, and virtual tours is important in real estate marketing and listing. In addition to the detailed property descriptions, providing professional photographs, virtual tours, and video presentations should enable potential buyers or tenants to visualise a property and its features. Optimised property listings on various platforms, including real estate websites, MLS databases, and online classifieds, increase exposure and attract interested parties. Virtual property viewing is enabled and underpinned by the internet, computing, cloud, mobile technology, and VR/AR technologies. Adoption of virtual property viewings therefore requires investments in equipment, software, and training.

The merits of virtual viewings include the following:

- **Time and cost saving**: With virtual viewing, potential buyers and renters can tour multiple properties more quickly and efficiently. They can filter out properties that do not meet their criteria, saving time and effort in the property search process. Sellers and agents can also reduce the costs associated with preparing the property for physical viewings. Additionally, virtual viewings will ensure that only buyers that are serious about renting or buying a property will come in for physical inspection and this can save the time of the agent or the landlord and by extension, the labour costs associated with the viewings.
- **Increased engagement and conversion**: Virtual viewing creates an engaging experience for potential buyers, increasing their interest and emotional connection to the property. This can lead to a higher conversion rate and ultimately speed up the sales or rental process.
- **Enhanced visualisation**: Virtual viewing provides a realistic and immersive experience, enabling potential buyers to visualise the property as if they were physically present. This helps them to better understand the layout, size, and design of the property, leading to more informed decision-making. Virtual viewing allows sellers to showcase

various features of the property effectively. It can highlight unique selling points, such as architectural details, interior design, or outdoor spaces, which may be challenging to convey through static images or descriptions.

- **Convenience and accessibility**: Virtual viewings eliminate the need for physical travel to the property, making it a convenient option for buyers who are unable to visit the location in person. It enables users to explore properties from anywhere in the world at any time, enhancing accessibility and widening the range of potential buyers.
- **Global reach**: Virtual viewing transcends geographical boundaries, allowing international buyers to explore properties in different countries without the need for extensive travel. It opens up new opportunities for cross-border real estate transactions.
- **Safe and contactless experience**: During times of health concerns or restrictions, virtual viewing offers a safe and contactless alternative to physical property visits. It minimises the risk of exposure to viruses and ensures that social distancing protocols are observed.

Virtual property viewing is transforming real estate marketing providing convenience, time efficiency, and an immersive experience for potential renters and buyers. It enables sellers, landlords, and agents to reach a broader audience and showcase properties in a captivating way. As digital technology continues to evolve, virtual viewing is likely to become an even more integral part of the real estate market, further enhancing the property search and transaction process. Failure to adapt to these advancements may result in outdated marketing approaches that are unable to resonate with modern consumers.

8.3.3   Shared Economy

The sharing/shared economy can be described as a movement designed to enable the sharing of real estate and space use through the adoption of digital technology (University of Oxford Research, 2017). Brown (2018) defines it as an economic system in which assets or services are shared between individuals, either for a fee or free of charge, typically through the use of the internet. It is sometimes referred to as 'collaborative consumption' or 'access economy' which represents an economic revolution built around an economic philosophy that utility can be maximised when space and capital goods are shared. The shared economy has secured its roots in real estate use and brokerage due to its ability and potential to improve collaboration and accessibility to products and services. Organisations are increasingly valuing collaboration which involves exchange through peer-to-peer-based platforms of intangible assets such as expertise, skills, innovation, and user experience. Some of the sectors that have been significantly transformed by the shared economy include transportation, labour, and neighbourhood support.

The real estate sector has been significantly transformed by the shared economy. The concept emerged in the early 2000s, and the GFC, through the search for new business models led to the growth of this sector which has redefined core business models. With significant job losses and financial austerity, consumers have changed their pattern of accessing, buying, and using goods and services. Traditional approaches of personalisation, individuality, and ownership appear to be getting obsolete due to their associated challenges such as the burden of expensive storage space, and now making way for more flexible and cheaper alternatives such as shared/temporary access or ownership. Although this model is thought to have emerged because of necessity, the cost-saving advantage has made it attractive to demographic and generational groups' lifestyles, particularly the millennials. Some of the core drivers of the sharing economy in real estate are shown in Figure 8.1 and further discussed below:

Economic: Austerity, crises, and high house prices, among other factors, have changed consumers' perspectives and priorities which has, in turn, changed consumption behaviour in favour of more cost-efficient models. Some of these factors have driven the need for shared spaces, particularly for start-up companies that need spaces but may not be able to afford to take out a long-term lease or large floor plates. The same applies to individuals who seek shared residential spaces in order to meet up with accommodation costs and bills.

Geopolitical: The rise in political instability globally and the promotion of entrepreneurship and the replacement of institutions by global businesses have created a shared channel for idealism and social enterprise. Brown (2018) also argues that low regulation is an advantage and a driver of the shared economy as it improves access to space.

Environmental: With net zero targets and other ESG goals, companies and individuals are getting more conscious of the effect of ownership and reservation of property on the environment. For instance, companies are promoting car sharing as a way of cutting down travel-related carbon emissions. With the ESG goals becoming more prominent and with the World Green Building Council requiring all new buildings to achieve net zero carbon by 2030 and all existing buildings to meet net-zero carbon standards by 2050, real estate space users are increasingly becoming more conscious of their carbon footprint.

Socio-cultural: This can be associated with generational and demographic shifts. For instance, millennials and Gen Z are different from Baby boomers who were more interested in space reservation. For instance, millennials question homeownership as a necessity for security and a fulfilling life. Individuals are becoming more interested in sustainable living and organisations are now trying to be responsible through ESG initiatives. This is causing a sociocultural change in society. Furthermore, there is also a general change in the nature of work, wellness, well-being, ergonomics, etc., becoming a buzz phrase. These factors, particularly after the COVID-19 pandemic have led to an increasing demand for better space quality and collaborative spaces (Oladiran et al., 2023). An increase in the rate of remote working is a major driver for the real estate shared economy. Reducing or repurposing underutilised space can enhance users' well-being through the provision of spaces for amenities such as gyms and break out areas.

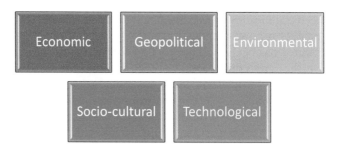

Figure 8.1 Drivers of the sharing economy in real estate.

Source: Authors.

Technological: Although the other factors listed have played a fundamental role in driving innovation, it would be impossible to have a shared economy without the enablement of digital technology. Cloud, mobile, internet, IoT, AI, computing, and other technological advancements have enhanced the development of the shared economy. For instance, consumers who are interested in shared spaces often rely on online listing platforms, depending on the type of shared spaces they are interested in. Other enablers include FinTech platforms such as PayPal which facilitate real estate transactions.

The emergence of space sharing has created several opportunities for more profitable changes to the use of real estate spaces in both residential and commercial sectors. In the commercial sector, it has led to higher intensity of space use on the supply side, changed user expectations on the demand side, and shaped a new economy of flexible jobs. This increased flexibility and mobility of contemporary working continues to deepen the shared working practice in commercial real estate, particularly post-COVID-19.

Some specific emergent areas of the shared economy and related value chain areas are further discussed in the sub-sub-sections.

8.3.3.1   Co-working and Flexible Working

Co-working is one of the models of the emergent shared economy in commercial real estate which has amalgamated open office and hot desking. Co-working and flexible working are terms that are used interchangeably but generally refer to the same concept and system. A flexible workplace is essentially space as a service, i.e., a technology-enabled space with access to wellness and hospitality options, while co-working is the entry-level form of a flexible workplace (Cushman and Wakefield, 2019). Co-working is increasingly becoming widely used and acceptable in the commercial real estate community. A survey conducted by Cushman and Wakefield (2019) revealed that 64% of global organisations were already utilising co-working in their portfolios, and 57% of the majority of corporate real estate executives had a positive/very positive view of co-working. Table 8.1 provides an overview of the evolution of co-working.

Co-working continues to grow aggressively. As of 2019, the global co-working space inventory was approximately 125 million sq. ft, with over 50 million sq. ft located in the United States. Saiz (2020) identifies four factors that have driven co-working. These have been re-constructed (Figure 8.2) as co-working models with an emphasis on their underlying drivers:

- **Adaptation co-working**: This model is driven by space underutilisation probably as a result of architectural design issues or underutilisation of public spaces in the property. It is typically an adaptative measure that is used by firms or landlords to generate cash flow and revenue from unutilised space. This was the case for several companies during the COVID-19 outbreak when there were high levels of unutilised space. For several years (from before the pandemic), Pivotdesk has enabled firms with excess office space capacity to share their existing spaces with other businesses. Other companies such as Liquid-Space also allow current users to sublet or pass on their leases for their current spaces. This enhances transparency and ease of subletting and essentially minimises underutilised office space. From the lens of an economist, it is worth investigating the channels through which these 'recycled' office spaces will compete with regular spaces provided

Table 8.1 Evolution of Co-working (as of 2015)

Year	Event
1995	C-base the predecessor of co-working space was set up in Berlin, Germany.
1999	A space with a flexible desk, established by a software company, opened in New York.
2002	A community centre for entrepreneurs was founded in Vienna, Austria.
2005	The first hub was set up in London UK
2006	Citizen Space, one of the first official co-working spaces was founded in San Francisco, US.
2007	The number of co-working spaces globally reached 75; the concept of co-working was picked up by the media.
2008	The Coworking Visa programme, catering to travelling workers was founded in the US.
2009	The first official co-working space Betahaus, now home to c. 200 entrepreneurs was opened in Germany.
2010	600 co-working spaces worldwide.
2011	The movement nearly doubled every year; large companies started experimenting with co-working spaces.
2012	Open co-working, an organisation promoting cooperation between co-working spaces around the world was formed.
2013	More than 100,000 people were estimated to be using co-working spaces worldwide.
2015	By the end of the year, the number of co-working spaces globally was predicted to reach 7,800.

Source: Authors' illustration (2023); data from Jones Lang LaSalle (2016).

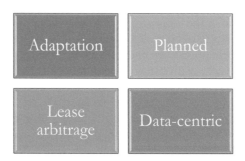

Figure 8.2 Co-working models.

by landlords, and how this can be integrated with office market demand/supply (equilibrium) models as well as rental and pricing mechanisms in the office market.

• **Planned co-working**: This model is based on the planned sharing of office spaces, in which case, the lease would have been prepared to enable this. It is aimed at providing flexible space-time services to corporations or smaller firms such as start-ups. In planned co-working, the space operators tend to be the investor or property owners who have developed or refurbished property with the aim of providing flexible co-working spaces. An example of this is CEG's East-West office property in Nottingham, UK. In addition to the longer-term essential space provided (with traditional leasing terms and arrangements), there are also other flexible co-working provisions such as flex (let-ready office space at any time), custom (bespoke fit-out delivered to customers' needs), and complete (everything covered including furnishing, fit-out, etc.).[11] One of the core advantages of planned co-working spaces is that they are often designed to provide branded private suites with short-term, simplified, all-inclusive leases. These leases allow companies to be flexible as they experience growth and or shrinkage and for them to be able to react accordingly without the baggage of offloading so much space in a short time or paying for space that they do not need. These office spaces include premium amenities such as meeting rooms, kitchens, and fitness centres. The operators in this model absorb the risk associated with the flexibility of space from the established non-real estate companies in exchange for a rental premium. This, however exposes the investor and landlord to risks associated with shorter leases, pricing of financial options, and approach to turn over. Furthermore, tenants' improvement of the property is unlikely as commoditised space-time will have to be taken as it is by the user.

- **Lease arbitrage co-working**: This model is associated with lease arbitrage, similar to securitisation in the mortgage sector. In the co-working case, payment is made for real estate space-time and value is created through the operator taking on large, durable leases and thereafter atomising space-time in a bid to sell it under very short straightforward membership contracts to smaller tenants. This model is premised on the fact that space utilisation per worker by contract firms will be smaller and with a degree of required synchronicity (Wheaton and Krasikov, 2019). If users' requirements are completely asynchronous, firms would simply need a conventional smaller lease and have the workers use the office at different times. These users should however be willing to pay a premium compared to what they will pay in traditional leases. Another opportunity may arise from small firms as the cost of setting up conventional leases may be high relative to their space requirements as they may be willing to pay a premium for avoiding such fixed costs.
- **Data-centric co-working**: Another co-working model is emerging which aims to capture, utilise and potentially monetise office data as firms become more active creators of vibrant workspaces. The large amount of data that can be obtained from operators' large, mobile, and experimental customer base, as well as the use of IoT sensors and other data sources (cameras, key/swipe cards), etc., are valuable for space designers, planners, managers, and operators. This model would be more effective in planned co-working spaces. Operators can experiment with multiple iterations of space layout, space user mix, amenities, environmental conditions, location, lighting, views, decoration, vibe, etc. This can lead to a scientific and data-driven approach to tailor physical office requirements to what new market entrants are seeking. Operators can also create value by designing spaces that increase the productivity of users which in turn justifies higher average rents per square foot and hours and also provides office environments and services to the specific productivity needs of their local base of start-up innovators.

According to Cushman and Wakefield (2019) survey, some of the major benefits of co-working are as follows:

- The ability to quickly ramp up/down a property portfolio.
- Reduced real estate costs.
- Increased employee engagement and satisfaction.
- Networking with other companies and employees.
- Attraction and retention of talents/employees.

- Improved efficiency of employees.
- Improved perception of the company's brand and culture.

Furthermore, the survey reported that the largest corporate concerns when utilising co-working and shared spaces were given as follows:

- Personal privacy.
- Digital security concerns.
- Company culture/cohesion.
- Decreased employee engagement/satisfaction.
- Increased real estate costs.
- Decreased efficiency of employees.

The benefits and concerns of co-working as highlighted in this study appear to be contradictory in some cases. This is an indication that the shared economy in the CRE sector may still possess some disparities in terms of benefits, challenges, and prospects. The variation may also be the result of differences in perceptions and views on co-working across world regions.[12] A study by Leesman (2019) shows that employees' workplace experiences are clustered around three emotional responses: doing, seeing, and feeling; these can be further connected to a complex web of activities, and physical and service features (University of Oxford Research, 2020). Some of these disparities may therefore be driven by the complexities in the web of emotional, physical, and service features in a workplace, and the value, expectations, and satisfaction derived from these factors will vary significantly based on geographical location, demographics, socio-cultural, and economic factors.[13]

One of the major risks that co-working providers and operators need to consider and mitigate is that many start-ups and small businesses are likely to be attracted to this model but having a portfolio where the vast majority of customers (tenants) in this category can be a challenge during economic down turns. During recessions, the first set of companies to fold up or fail tend to be start-ups; furthermore, in a bid to survive, some of these start-ups can easily downsize or outrightly discontinue their contract (depending on the type of contract they have signed). This can affect the cash flow for the investor/landlord. There is usually a mismatch between tenants and landlords' priorities: tenants' will prefer short-term leases which provide more flexibility and landlords prefer long-term leases for their protection and cash flow. This becomes a major risk in investment particularly when potential epidemiological, economic, and political factors negatively affect the economy and real estate markets (Fiorilla, 2018).

8.3.3.2 Space-Time Commoditisation

Space-time commoditisation is one of the contemporary concepts that has emerged as part of the broader shared economy. This concept has been driven by advancements in digital technology, globalisation, and changing consumer behaviour and demographic shifts. The concept encapsulates the transformation of physical and temporal spaces into valuable 'commodities' that can be bought, sold, and traded in the global marketplace. Space-time commoditisation has several dimensions which have emerged in unison with the shared economy which has leveraged digital technology to transform underutilised spaces and idle time into revenue-generating opportunities. This has enabled people and organisations to monetise their workspaces and living spaces, creating a fluid marketplace where space and time become tradable assets.

The convergence of space and time has also enhanced space-time commoditisation. The internet and other digital technologies have collapsed the barriers of distance and time, transforming how we interact, communicate, and conduct business. Space-time commoditisation represents the fusion of physical spaces, such as real estate, transportation, and hospitality, with temporal aspects, such as on-demand services, instant access to information, and real-time communication. On-demand services and instant gratification have further enabled the development of space-time commoditisation. With the rise of on-demand services, consumers now expect instant access to goods and services. From food delivery and ride-hailing to streaming services and e-commerce, space and time have become interchangeable commodities. Businesses that can deliver products and services promptly capitalise on this consumer trend. Digital nomadism and flexible working have further enhanced space-time commoditisation. The concept of digital nomadism has gained popularity as technology allows individuals to work remotely from any location with an internet connection (Leducq et al., 2023). This blurs the boundaries between traditional workspaces and personal spaces, providing the ability to transform various destinations into temporary workplaces. The rise of remote work has transformed how we perceive and utilise physical spaces and the notion of traditional office hours.

The space-time commoditisation has had profound effects on real estate. For instance, virtual viewing technologies, such as 3D tours and virtual reality, enable potential buyers/renters to explore properties remotely, reducing the need for physical visits. Additionally, the popularity of short-term rental platforms has transformed residential properties into investment opportunities, adjusting long-term housing dynamics and rental markets. Space-time commoditisation has assumed new dimensions following the COVID-19 pandemic and Oladiran et al. (2023) provide insight on changes to space quantity and space quality; Saiz (2020) provides insight into a third dimension-cash flow potential.

The first dimension (quantity) examines the potential of digital tools or products to increase or reduce the need for space. It is often argued that innovation can negatively impact the quantity of real estate demanded; it should however be noted that while this may not be true in all cases. E-commerce for instance has led to a decline in retail space demand and the Highstreet, while simultaneously increasing demand for industrial space through warehousing and logistics. The second dimension (quality) considers the changes to the use of the space, highlighting the need for higher quality spaces and the transformation of old spaces. As new requirements emerge for space use, particularly since the COVID-19 outbreak, investment in high-quality spaces may increase. The (Y)OUR SPACE survey by Knight Frank Research Report (2020) revealed that employers are increasingly seeking ways of improving space quality to attract talent and improve their productivity. New spaces are being adopted to account for the general development of technology. For instance, data centres and server farms are increasing, and they need to be equipped with sophisticated cooling systems. The third dimension is the ability of spaces to generate cash flow, which explores the potential for innovation to be monetised through sales, rental, or collateralisation.

While space-time commoditisation offers numerous benefits to real estate, it also raises several challenges. One of these is privacy. Privacy concerns can arise as spaces become more publicly accessible. Additionally, the hyper-connectivity of space and time may lead to burnout and an "always-on" culture, impacting work-life balance. The regulation of digital frontiers is also a challenge. The rapid growth of space-time commoditisation has outpaced regulatory frameworks, leading to legal and ethical questions. Issues such as data privacy, tax compliance, and labour rights are complex when traditional boundaries of space and time are blurred. Striking a balance between leveraging technology's potential and addressing societal and environmental concerns is essential for the navigation of ethical and responsible practices which should be major considerations in the CRE market, and crucial for harnessing the full potential of this paradigm shift.

8.3.3.3   Co-living

Co-living is a term often used to refer to the residential segment of the shared economy. It is an emergent housing trend that offers a unique approach to communal living. It involves shared spaces and amenities and fosters a sense of community and collaboration among residents. Co-living spaces typically consist of private bedrooms and shared common areas, such as kitchens, living rooms, and co-working spaces. Co-living providers often design spaces to promote social interaction, creativity, and a sense of belonging among residents who are often referred to as "co-livers" or "co-residents", who share resources and responsibilities. The properties where co-living takes place are sometimes referred to as houses of multiple occupation (HMO).

Co-living is driven by economic, demographic, socio-cultural, and economic factors. It offers a more cost-effective housing option compared to traditional renting, especially in high-cost urban areas. Shared expenses for utilities, communal resources, and rent result in lower individual living costs. This model is also driven by demographic shifts, tastes, and preferences. For instance, Millennials and GenZs generally have a preference for renting rather than homeownership which is in contrast to older generational priorities. There are several reasons for this. One is that while they increasingly want to live close to the urban centres due to access to social life and proximity to work, these spaces are very expensive and will require high capital down payment for purchase. Even if debt finance is used, the down payment constraint usually remains. In addition, the high cost of purchase will also increase mortgage interest cost, agency fees, taxes, stamp duties, and other costs, typically linked to the property value.

Some of the sociocultural drivers of co-living include community, networking, collaboration, convenience, and flexibility, creating social environments that enhance connectivity, sharing, and meaningful relationships. Due to its communal nature, it fosters a strong sense of community and social connectivity. Residents often engage in communal activities, events, and shared experiences, creating lasting friendships and networking opportunities. Furthermore, co-living arrangements are also designed to provide hassle-free living and fully furnished spaces, inclusive utilities, and flexible lease terms to cater to individuals seeking convenient and transient living solutions. Additionally, co-living environments also attract individuals from diverse backgrounds and professions. This diversity fosters networking and collaboration opportunities, making co-living spaces appealing to entrepreneurs, freelancers, and digital nomads. Several online listing platforms such as Rightmove now typically list shared houses which enable landlords and agents to find tenants for spare rooms or rent entrepreneurs.

The intersection of residential and hospitality sectors has also created another space commoditisation hub known as short let. Platforms such as AirBnB and Homeaway[14] are facilitating the efficient use of what would have been redundant room space for short-term rental. Airbnb for instance, which was launched in 2008, was borne out of necessity.

College graduates, Brian Chesky and Joe Gebbia, who were unable to find jobs took advantage of an industrial design conference that came to their city by taking a few guests who were attending the conference but did not want to pay for expensive hotels. The platform has now grown internationally. According to the University of Oxford Research (2017), it is operating in over 57,000 cities in 192 countries, with a market capitalisation of $30 billion (Zervas et al., 2017). Approximately 2.8 million room nights were booked via Airbnb between September 2014 and August 2015, and this leads to an estimated annual revenue loss of $450 million annually for hotels (Mahmoud, 2016). It is now considered part of the sharing or platform economy where peer-to-peer activities provide access to goods, or services which are enabled by an intermediary digital platform (Schlagwein et al., 2020; Wang et al., 2023).

Co-living has addressed several residential market inefficiencies. For instance, it provides co-livers with the flexibility they need in the case that they change jobs, or other life events occur that require them to change their accommodation. It also provides people with differentiated housing options and particularly provides young professionals with a unique living experience. Furthermore, it is helping to address issues around social isolation and loneliness, especially in urban environments. By providing a built-in community, co-living spaces promote social interactions, mental well-being, and a sense of belonging.

This model however has several challenges and limitations. They include the following:

- **Privacy concerns**: Sharing living spaces with strangers can raise privacy concerns for some individuals. While co-living providers often offer private bedrooms, striking the right balance between social interaction and personal space can be challenging.
- **Compatibility and conflict resolution**: Living in close proximity to others may lead to conflicts, differences in living habits, and potential compatibility issues. Clear guidelines and effective conflict resolution mechanisms are essential for maintaining a harmonious co-living community.
- **Transient nature**: Co-living arrangements may attract residents with transient lifestyles. High turnover rates could impact the stability of the community and its ability to build lasting connections.
- **Legal and regulatory complexities**: In some regions, co-living may encounter legal and regulatory challenges. Zoning laws and local regulations may not approve of the use of residential properties in an area for co-living purposes.

Co-living represents a growing trend in the real estate industry, redefining how people live and interact in urban environments. As this concept evolves, it will be crucial for operators and policymakers to address these challenges while harnessing its potential to create more inclusive and connected living spaces in the future.

8.3.4   Digital Customer Relationship Management (CRM)

Customer relationship management (CRM) is a robust management system for client interactions, which helps to maintain relationships with potential buyers, current clients, and past clients. It allows for targeted communication, follow-ups, and personalised marketing efforts based on individual preferences and needs. Digital CRM has revolutionised the real estate brokerage industry, transforming how agents interact with clients, manage leads, and streamline sales/rental processes.

There are several features and elements of digital CRM in real estate brokerage. They include the following:

- **Lead management**: Digital CRM systems enable real estate agents to efficiently capture and manage leads from various sources, such as websites, social media, and online advertising. Leads can be tracked, categorised, and prioritised based on their potential for conversion.
- **Integration with marketing tools**: Digital CRM platforms are often integrated with marketing tools, enabling seamless execution of marketing campaigns. Agents and brokers can use CRM data to create targeted advertisements, social media posts, and email newsletters to reach potential clients.
- **Customer profiling and segmentation**: Digital CRM tools enable agents and brokers to build comprehensive customer profiles, including preferences, property requirements, and transaction history. This data can be used to segment clients, personalise marketing efforts, and provide bespoke property recommendations.
- **Transaction management**: Digital CRM systems streamline transaction management, enabling agents and brokers to track the progress of deals, manage documents, and automate paperwork. This ensures smooth and efficient transaction processes for clients.
- **Automated communication**: CRM platforms facilitate automated communication through emails, text messages, or personalised notifications. Agents and brokers can set up automated follow-ups, property alerts, and reminders, ensuring consistent and timely interactions with clients.

Digitised CRMs can enhance the customer experience of clients and business performance. They provide effective lead tracking and follow-up which results in increased lead conversion rates. The automated communication service also enhances customer engagement and fosters stronger relationships between agents and clients. Quick responses and relevant property recommendations contribute to improved customer satisfaction and loyalty. This personalised communication, improved customer service, and enhanced transaction management can significantly increase client satisfaction and repeat business. Additionally, improved lead management and data-driven decision-making lead to higher conversion rates and enhanced business efficiency. All of these can lead to increased efficiency, enabling property agents and brokers to concentrate on building relations and closing deals rather than spending time on analogue processes.

The use of digital CRMs in real estate brokerage and the agency also has its drawbacks and limitations. They include the following:

- **Integration and adoption**: Implementing and integrating a digital CRM system into existing real estate brokerage processes can be challenging. Agents may require training and support to effectively use the platform, and resistance to change can hinder successful adoption.
- **Data privacy and security**: Digital real estate CRM systems contain sensitive client data, making data privacy and security paramount. Brokers and agents need to ensure compliance with data protection regulations and implement robust security measures to safeguard client information.

- **Customisation and scalability**: Some CRM systems may lack the flexibility to accommodate unique business processes or scale with the brokerage's growth. Customisation options and the ability to adapt the platform to changing needs are essential considerations.

Although digital customer relationship management systems have streamlined lead management, facilitated personalised communication, enhanced transaction processes, and revolutionised the engagement of agents and clients, the drawbacks are essential considerations, particularly for tools that are being developed. Brokers and agents who leverage these platforms effectively can expect improved customer experience, increased lead conversion, and enhanced business performance, establishing a competitive advantage in the dynamic real estate market.

8.4   Case Study Analysis

Following the review of core areas of digital transformation in real estate brokerage, marketing, and other allied areas, this section analyses specific digital platforms and firms within these areas of the built environment value chain. Each case study begins with an overview and summary of key details of the platform/company; the solutions it provides are then analysed using the ESEP efficiency framework introduced in Chapter 1.[15] The analysis concludes with a brief overview of the limitations and further prospects.

8.4.1   Case Study 1: Convene (Shared Economy)

8.4.1.1   Background, Key Details, and Technologies Utilised

Convene is a digital platform that offers full-service, premium flexible workspaces through the provision of beautifully designed and tech-enabled meetings, events, and workplace locations. The model was inspired by the operations in the hospitality industry, especially hotels. Convene's products and services evolved from physical meeting and event venues to co-working and technology solutions, as a direct response to changing client needs over the years. Convene was founded in 2009 with a mission to help meeting and event planners focus less on the logistics leading up to their events, and instead have more time to spend on programming and ongoing attendee and team needs. It has thereafter expanded its services to offer flexible workplace locations for individuals and teams, as well as virtual and hybrid meeting solutions. This is achieved through a global network of 23 premium locations and advanced digital meeting technology to provide teams with the spaces they need to do their best work.

Convene continues to transform what "space as a service" means in the real estate industry by thoughtfully integrating digital technology into beautiful and inviting spaces, paired with industry expertise in event production, as well as five-star hospitality touchpoints throughout both in-person and virtual experiences, worldwide. At the beginning of 2020, when businesses shifted remotely to respond to the global pandemic, Convene introduced its virtual meetings and events platform which allowed for bespoke registration and event pages, keynotes and breakout sessions, networking, as well as interactive Q&A elements. Each event has dedicated professional support for planning, implementation, and troubleshooting.

Convene has several products and services. Convene Studio, for instance, enables planners to host events in the format that best meets their business needs. They offer an all-in-one solution that combines production services, best-in-class virtual and meeting technology, five-star hospitality, and a network of award-winning properties. Convene Studio events have an option for live broadcasting or pre-recorded programming, with access to film-quality cameras, lighting, and streaming equipment. All Convene venues have meeting and events technology integrated into the spaces to seamlessly bring remote teams together. Each meeting or event is offered its own branded event site for registration and access, with varying levels of security permissions. Virtual production services can host up to 10,000 attendees, and each event comes paired with access to engagement and event performance analytics for the host.

Convene's WorkPlace further delivers mixed digital and physical environments that allow teams to do their best work, whether that's individual, hybrid team meetings, elaborate team retreats, or anything in between. WorkPlace integrates with Convene Studio to be accessible for remote employees, fully equipped with best-in-class technology, support, and hospitality, in award-winning venues at top destinations around the globe.

8.4.1.2   Key Efficiencies and Solutions That Convene Provides

The solutions that Convene provides are itemised using the ESEP efficiency framework (Table 8.2).

Table 8.2 ESEP Efficiency Analysis of Convene

ESEP	Inefficiency with the Traditional Approach	How the Tool/Platform/Product Addresses the Inefficiency
Environmental	Companies traditionally take long leases, and they would need to maintain the space they occupy, regardless of the occupancy levels during the lease. Because space supply relies on occupancy signals in the user market, the development market may respond to the 'false' signals through the provision of more space which will lead to higher emissions and other environmental impacts associated with building and construction.	The flexible workspaces that Convene provide means that firms, particularly start-ups, can take up the space they need per time and adjust this appropriately; this will minimise the 'false' signals being sent from the user to the development market, effectively minimising the building and construction of spaces that would normally have been built This further minimises the associated carbon emissions and other environmental impacts associated with building construction.
	Meetings and team collaborative exercises usually required the physical presence of all team members involved.	With the platform, collaboration is done with reduced commuting; meetings can be held remotely leveraging remote meeting technologies. This can significantly minimise commute-related carbon emissions.
	Searching for office spaces using traditional approaches requires a high level of search-related transportation which leads to high carbon emissions.	With Convene's online office space search, searching for office spaces will require a lower level of search-related transportation which will further minimise carbon emissions.

(Continued)

Table 8.2 (Continued)

ESEP	Inefficiency with the Traditional Approach	How the Tool/Platform/Product Addresses the Inefficiency
	Traditional office-based working models require a high level of commute which causes high carbon emissions.	With the flexible and remote working support that the platform provides, commute-related carbon emissions are minimised.
	High energy consumption within unoccupied spaces and spaces with low occupation.	This platform lowers energy consumption as energy levels can now be based on actual occupation and space usage.
Social	With traditional office-based working models, aesthetics and amenities within the spaces are not prioritised.	This platform provides access to a wide range of workplaces that offer a variety of aesthetically pleasing workspaces with several amenities. This can enhance the physical and mental health of employees.
	Traditional office-based models generally do not provide for social bonding and community among employees.	The quality and design of spaces provided through this platform can increase the level of social interaction, engagement and community for employees.
	Employees within an office block in traditional office settings hardly have the opportunity to interact and engage with employees from other companies and firms.	Through the flexible working options that Convene provides, employees, though affiliated with different companies can work in the same spaces and thus engage and interact more. This can lead to cross-pollination of ideas and enhanced relationships for future collaboration.
	In traditional office-based models where employees feel isolated, uninspired or not valued enough, they will be less productive and efficient, and this can often lead to high staff turnover and low skills retention.	Through the flexible and remote working options powered by Convene, employees can feel more inspired and valued, and this can lead to higher job satisfaction, productivity, and efficiency, leading to higher talent retention.
	Office space search is always a difficult and complex process.	Convene provides a convenient and easier approach to searching for office space.
	It is difficult to collect data using traditional office-based models.	Through the digitisation of office space design, marketing, use and management that Convene provides, data can be collected for improved space usage and management.
Economic	Office space search (from the perspective of the prospective tenants) and space take-up (from the landlord/operators' perspective) takes a long time.	Space search and take-up are significantly reduced through the Convene platform.

(Continued)

Table 8.2 (Continued)

ESEP	Inefficiency with the Traditional Approach	How the Tool/Platform/Product Addresses the Inefficiency
	Companies may be tied up in long leases with space that they do not need or cannot maintain. The high occupancy costs (real estate and maintenance) associated with unused and unneeded spaces can increase the financial hardship for struggling companies.	With the flexible working option that Convene provides, companies can downsize their office spaces when the need arises without having to incur the associated occupancy costs. Similarly, companies can also ramp up their property portfolios when needed.
	With traditional office-based models where staff satisfaction and efficiency are low, a firm's operating costs can be reduced, and this can lead to lower productivity and revenue.	With the flexible, remote working and higher quality of office space that the platform provides, employees' productivity and efficiency may increase and talent attraction and retention will be higher. This can lead to enhanced revenue and minimise personnel costs that may have been accrued through high staff turnover.
Physical/spatial	In traditional office-based work models, there are often high levels of mismatch between space demand and space supply.	With the flexible space provision, and the online support platform that Convene provides, it is easier to use office space more efficiently as the space that is available for use is advertised to a wider range of prospective tenants, and this space can easily and quickly be repurposed when the current use is not more required.

Source: Authors analysis.

8.4.1.3   Limitations, Challenges and Prospects

One of the challenges Convene has is that it's still relatively novel, and although the pandemic has supported a general paradigm shift towards remote working, companies have not been able to establish the level of remote working/office-based working they intend to adopt. Therefore, they may not be willing to invest in the remote working aspect of the platform until they establish their baseline. Regardless, hybrid working is likely to remain relevant, and flexible working may become more widely adopted, particularly with start-ups. The platform therefore has the potential to continue to improve the efficiency of workspace use, planning, and management.

Further details on Convene can be accessed from their website using this link: https://convene.com/.

8.4.2   Case Study 2: HqO

8.4.2.2   Background, Key Details, and Technologies Utilised

HqO is a digital customer-centric workplace platform that offers a wide range of workspaces for people to engage with one another and the building facilities where they work.

The platform creates digital-to-physical experiences that empower people to do their best work. HqO's workplace experience platform creates branded and personalised workplace experiences that accelerate innovative asset strategies for landlords of multi-tenant office buildings, help employers boost talent productivity and retention, and enable data-driven decision-making.

HqO's tools can be used by landlords, operators, tenant companies, and corporate employees. It was founded in 2018 by Chase Garbarino, Kevin McCarthy, Greg Gomer, and Jared Stenquist, with the goal of developing technology that connects people back to their local community and the people around them in a more impactful way. Over time, this concept has evolved into a technology that helps connect landlords with the people who work in, shop, and use their properties every day. In April 2021, the company raised $60 million in Series C funding from leading venture capital and CRE firms, totalling a funding amount of $107 million to date. In October 2021, HqO acquired the leading European tenant and employee engagement platform Office App. The new combined entity is valued at over half a billion dollars. HqO is currently active in 142 cities and 500+ buildings with a footprint of over 200 million sq. ft.

HqO consists of the following:

- **Tenant Experience App**: a mobile app for tenants that supports every aspect of the workplace experience with a flexible digital platform that provides tenants with 24/7, on-demand access to amenities, retail, programming, and property operations.
- **Corporate Employee App**: A mobile app for corporate employees that helps every employee make the best use of their time and get work done—whether they're in the office or working remotely.
- **Member Experience App**: A member experience app that provides solutions for reserving and managing flexible workplaces, also known as flex space. These spaces are similar to co-working spaces but are typically managed by property owners and/or operators. Using this app, property teams can create and manage flex space memberships and invoices, allocate monetary booking credits, and create robust booking rules personalised to each tenant or tenant company.
- **Data & Analytics**: This is a data and analytics offering that provides real-time visibility into property engagement, tenant sentiment, space utilisation, and building population. With easy-to-use dashboards, property teams can increase retention and attract new tenants, evaluate and prioritise amenity investments, and drive a higher NOI.
- **Technology and Service Provider Marketplace**: This is a curated Marketplace that makes it easy to browse and select the best-in-class solutions and services that meet a property's needs. Whether teams are looking for engaging content and events support or building automation technology, they will receive white-glove consultation from technology experts who can enhance their property.

8.4.2.2 Key Efficiencies and Solutions That HQO Provides

The solutions that HQO provides are itemised using the ESEP efficiency framework (Table 8.3).

Table 8.3 ESEP Efficiency Analysis of HQO

ESEP	Inefficiency with the Traditional Approach	How the Tool/Platform/Product Addresses the Inefficiency
Environmental	Companies traditionally take long leases, and they would need to maintain the space they occupy, regardless of the occupancy levels during the lease. Because space supply relies on occupancy signals in the user market, the development market may respond to the 'false' signals through the provision of more space which will lead to higher emissions and other environmental impacts associated with building and construction.	HqO's adaptable workspaces offer companies, especially start-ups, the flexibility to occupy the necessary space according to their current requirements and adjust it as needed. This approach minimises misleading signals in the real estate market, effectively lowering the development of new office spaces that may be unnecessary. Consequently, this practice significantly reduces carbon emissions and other environmental impacts typically associated with construction activities.
	Searching for office spaces using traditional approaches requires a high level of search-related transportation which leads to high carbon emissions.	With HqO's online office space search, searching for office spaces will require a lower level of search-related transportation which will further minimise carbon emissions.
	High energy consumption within unoccupied spaces and spaces with low occupation.	This platform lowers energy consumption as energy levels can now be based on actual occupation and space usage.
Social	With traditional office-based working models, aesthetics and amenities within the spaces are not prioritised.	This platform provides access to a wide range of workplaces that offer a variety of aesthetically pleasing workspaces with several amenities. The platform also provides access to amenities such as onsite restaurants, fitness facilities, electronic vehicle charging, mailboxes, lockers, dry cleaning services, etc. These can enhance the physical and mental health of employees. Furthermore, conference rooms can be booked on demand, using a hot-desk system.
	Traditional office-based models generally do not provide for social bonding and community among employees.	The quality and design of spaces provided through this platform can increase the level of social interaction, engagement and community for employees.
	Employees within an office block in traditional office settings hardly have the opportunity to interact and engage with employees from other companies and firms.	Through the flexible working options that HqO provides, employees, though affiliated with different companies can work in the same spaces and thus engage and interact more. This can lead to cross-pollination of ideas and enhanced relationships for future collaboration.

(Continued)

Table 8.3 (Continued)

ESEP	Inefficiency with the Traditional Approach	How the Tool/Platform/Product Addresses the Inefficiency
	In traditional office-based models where employees feel isolated, uninspired or not valued enough, they will be less productive and efficient, and this can often lead to high staff turnover and low skills retention.	Through the flexible and remote working options powered by HqO, employees can feel more inspired and valued, and this can lead to higher job satisfaction, productivity, and efficiency, leading to higher talent retention.
	Office space search is always a difficult and complex process. It is difficult to collect data using traditional office-based models.	HqO provides a convenient and easier approach to searching for office space. Through the digitisation of office space design, marketing, use, and management that HqO provides, data can be collected for improved space usage and management.
Economic	Office space search (from the perspective of the prospective tenants) and space take up (from the landlord/operators' perspective) takes a long time.	Space search and take up are significantly reduced through the HqO platform.
	Companies may be tied up in long leases with space that they do not need or cannot maintain. The high occupancy costs (real estate and maintenance cost) associated with unused and unneeded spaces can increase the financial hardship that companies that are already struggling to cope with.	With the flexible working option that HqO provides, companies can downsize their office spaces when the need arises without having to incur the associated occupancy costs. Similarly, companies can also ramp up their property portfolios when needed.
	With traditional office-based models where staff satisfaction and efficiency are low, a firm's operating costs can be reduced, and this can lead to lower productivity and revenue.	With the flexible, remote working and higher quality of office space that the platform provides, employees' productivity and efficiency may increase, and talent attraction and retention will be higher. This can lead to enhanced revenue and minimise personnel costs that may have been accrued through high staff turnover.
Physical/spatial	In traditional office-based work models, there are often high levels of mismatch between space demand and space supply.	With the flexible space provision, and the online support platform that Convene provides, it is easier to use office space more efficiently as the space that is available for use is advertised to a wider range of prospective tenants, and this space can easily and quickly be repurposed when the current use is no more required.

Source: Authors analysis.

8.4.2.3   Limitations, Challenges, and Prospects

Flexible working is an emerging area and although the pandemic has supported its transformation of the real estate sector, companies are generally working out the remote working/office-based and flexible working protocol they intend to adopt. Therefore, they may not be willing to invest in the remote working aspect of the platform until they establish their baseline. Flexible working however is likely to remain a core component of real estate use, particularly for start-ups and smaller firms. The platform therefore has the potential to continue to improve the efficiency of workspace use, planning, and management, particularly with the increasing drive to build successful communities in the workplace and companies' increasing ESG awareness and commitment.

Further details on HQO can be accessed from their website using this link: https://www.hqo.com/.

8.4.3   Case Study 3: student.com

8.4.3.1   Background, Key Details, and Technologies Utilised

Student.com is a student listing and marketing online platform that was designed to integrate and display several tangible and non-tangible information on purpose-built student accommodation (PBSA). In addition to displaying the attributes of listed properties, student.com allows students to save their rooms of interest to their wishlists and request for a room to be reserved. The wishlist is a particularly important feature of this platform as it enables students to shortlist properties of interest and save them while continuing their online search. With this platform, students do not need to search on individual PBSA websites but can now access a wide range of properties based on set parameters. The platform has over one million beds in over 1,000 universities across 400 cities worldwide.

Student.com was established in 2011 by Luke Nolan and was initially named Overseas Student Living. Luke worked in China for some years and kept getting requests from his Chinese friends to help them arrange for accommodation in the UK, US, and Australia. The volume of requests was enormous, and Luke realised that this was an opportunity to solve a student accommodation search and selection inefficiency. He began connecting students with professional landlords in various countries, connecting potential students to residential accommodations that would cater for their housing needs while in school away from home. Shakil Khan and John-Paul Jones joined the company in 2014, and by 2015, the company rebranded itself as Student.com to help it focus on a more global reach while recruiting locals from various countries as booking agents to make the bookings more professional and better managed. Student.com assigns the potential student to a booking consultant who follows up on their preferences and works with the student to meet their accommodation needs in the best way possible. Criteria such as student status, budget, special needs, and other sundry features are worked on, and students are guided to make the best decision for the choice of accommodation. Access is provided free of charge to students, but a commission is payable from successful bookings from the user.

8.4.3.2   Key Efficiencies and Solutions That student.com Provides

The solutions that student.com provides are itemised using the ESEP efficiency framework (Table 8.4).

Table 8.4 ESEP Efficiency Analysis of student.com

ESEP	Inefficiency with the Traditional Approach	How the Tool/Platform/Product Addresses the Inefficiency
Environmental	A significant amount of carbon is emitted relating to commute-related property searches and viewings for student accommodation.	With student.com, searching for student accommodation will require a lower level of search-related transportation which will further minimise carbon emissions.
Social	When searching for properties, students are usually frustrated, anxious, and have fewer options to select from. This problem is of high importance for international students.	Using this platform, there is a wider variety of rooms to review, and this can significantly minimise anxiety and frustration, particularly as there are pictures and videos of the rooms on some properties.
	Streamlining the search and selection process for students can be difficult in traditional approaches to student accommodation search.	The platform has a filter option that enables students to narrow down their search criteria such that they are able to have a reasonable number of properties that they can choose from. Factors such as distance to the city centre, type of accommodation, amenities, etc. are typically displayed.
	The physical search for student accommodation can be physically and mentally daunting.	Using this online platform, student accommodation will provide a less daunting and more seamless process for student search and selection.
Economic	Searching for student accommodation requires viewings and by extension, transportation-related costs.	Using student.com, student accommodation search-related commute costs can be significantly reduced.
	Searching for student accommodation using traditional approaches takes a considerable time.	Using student.com, student accommodation search-related commute time can be significantly reduced.
Physical/spatial	The occupancy and take up rates on student accommodation properties may be reduced due to an insufficient demand/supply mismatch arising from inefficient and ineffective property marketing strategies.	Student.com provides a more effective and efficient matching mechanism for students and accommodation operators. This makes space available to students who need it and enables property operators to let out their rooms more speedily. This can minimise the time that a property stays vacant, thereby maximising returns and minimising asset costs.

Source: Authors analysis.

8.4.3.3   Limitations, Challenges, and Prospects

The student.com platform currently does not have an automated booking system (similar to the way hotels are booked). The system requires interested students to send an enquiry. The use of advanced digital tools such as blockchain technology can be used to take students' bookings in real time. This way, students who prefer to book or reserve a room right away can do so without having to wait long periods of time.

Further details on student.com can be accessed from their website using this link: https://www.student.com/.

8.4.4 Case Study 3: VTS

8.4.4.2 Background, Key Details, and Technologies Utilised

VTS (View the Space Inc.) is commercial real estate's leading leasing, marketing, asset management, and tenant experience platform enabling brokers, agents, and landlords to make deals that are underpinned by real-time data. The platform captures a vast first-party data source in the industry which delivers real-time insights that fuel faster, more informed decision making and connections throughout the deal and asset lifecycle. It is utilised by leaders across the office, retail, and industrial sectors including Landsec, The Crown Estate, British Land, Europa, Oxford Properties, CBRE GI, GLP, Blackstone, and Brookfield. Agency businesses across commercial real estate also partner with VTS including CBRE, Colliers, BNP Paribas, Knight Frank, and C&W.

VTS was founded in 2012 by Nick Romito, Ryan Masiello, and Karl Baum, and in 2016, the company raised $55 million in Series C funding. In 2017, VTS merged with Hightower, and the company's co-founders Brandon Weber, Niall Smart, and Donald Desantis joined ranks with Romito, Masiello, and Baum. In 2019, VTS raised another $90 million in funding, which put the company's valuation at a reported $1 billion, and in October 2019, it acquired PropertyCapsule for a reported $20 million to expand its retail offering to customers and launched VTS Retail in January 2020.

In June 2020, amid the COVID-19 pandemic, VTS accelerated the launch of VTS Market, commercial real estate's first integrated online marketing platform enabling landlords to market and lease their space remotely at a time when touring spaces in-person wasn't possible. In October 2020, VTS announced the launch of VTS Data to give landlords access to the real-time insights they needed to drive investment and asset strategies in a volatile and uncertain market. VTS launched its VTS Office Demand Index (VODI) – a monthly report as an additional free resource to the industry in 2020. The VODI is a measure of office leasing demand in the US, capturing 99% of all unique tenant searches in the seven US gateway markets – New York City, Washington D.C., Los Angeles, Chicago, Boston, Seattle, and San Francisco. In March 2021, VTS acquired Rise Buildings for a reported $100 million to enter the tenant experience space and continue to ensure its landlord customers have the technology they need to operate as efficiently as possible during COVID-19 and beyond. Rise Buildings CEO and Co-Founder Prasan Kale joined VTS as Managing Director of Rise. VTS re-launched Rise Buildings under the name VTS Rise while announcing massive market adoption of the product in July 2021.

VTS acquired Lane, the workplace experience platform, for a reported $200 million in October 2021. The acquisition strengthens VTS' existing market-leasing tenant experience solution VTS Rise and enables both Rise and Lane customers to benefit seamlessly from the combined technology. VTS has a massive global customer base for tenant experience with approximately 3 million users across 1,400 buildings in 13 countries. There are over 12 billion sq. ft of commercial real estate managed on the platform and over $200 billion in lease. VTS has over 45,000 corporate users and over 650 employees across six offices in North America and Europe.

There are several products and tools on the platform.

- VTS Lease is an online leasing product that powers over $200 billion in leasing transactions per year, enabling visibility into leases and deals across a company's portfolio, in addition to aiding visibility in performance compared to competitors and unlocking insights to inform leasing and asset management strategy.
- VTS Market and Marketplace is a digital marketing product creating the industry's first integrated online marketing solution. It enables landlords, brokers, and tenants to access unparalleled visibility into real-time market information and the direct connectivity to execute deals with greater speed and intelligence at every point in the planning, marketing, leasing, and asset management cycle.
- VTS Rise is the tenant experience platform for commercial and residential property, offering the most comprehensive and immersive tenant experience solution that enables asset and property managers to power their portfolio-wide operations and provide a premium experience to tenants, building operators, and visitors.
- VTS Data is the market data set offering real-time market data for asset and investment strategy. Capturing active tenant demand and projecting future supply fluctuations, the product is powered by the $200 billion in leasing transactions managed on the VTS Leasing Platform.

8.4.4.2   Key Efficiencies and Solutions That VTS Provides

The solutions that VTS provides are itemised using the ESEP efficiency framework (Table 8.5).

Table 8.5 ESEP Efficiency Analysis of VTS

ESEP	Inefficiency with the Traditional Approach	How the Tool/Platform/Product Addresses the Inefficiency
Environmental	Companies traditionally take long leases, and they would need to maintain the space they occupy, regardless of the occupancy levels during the lease. Because space supply relies on occupancy signals in the user market, the development market may respond to the 'false' signals through the provision of more space which will lead to higher emissions and other environmental impacts associated with building and construction.	VTS' adaptable workspaces offer firms, especially start-ups, the opportunity to utilise the required space as needed and adjust accordingly. This improves the accuracy of the signals sent from the user market to the development market, leading to a reduction in the construction of surplus spaces. Consequently, this practice significantly decreases carbon emissions and other environmental impacts associated with construction activities.
	Meetings and team collaborative exercises usually require the physical presence of all team members involved.	With the platform, collaboration is done with reduced commuting; meetings can be held remotely leveraging remote meeting technologies. This can significantly minimise commute-related carbon emissions.

(Continued)

Table 8.5 (Continued)

ESEP	Inefficiency with the Traditional Approach	How the Tool/Platform/Product Addresses the Inefficiency
	Searching for office spaces using traditional approaches requires a high level of search-related transportation which leads to high carbon emissions.	With VTS' online and mobile office space search, searching for office spaces will require a lower level of search-related transportation which will further minimise carbon emissions.
	Traditional office-based working models require a high level of commute which causes high carbon emissions.	With the flexible working support that the platform provides, employees may not necessarily need to converge at a centralised office. For instance, they can work from spaces that are closer to them, and commute-related carbon emissions can be minimised.
	High energy consumption within unoccupied spaces and spaces with low occupation.	This platform lowers energy consumption as energy levels can now be based on actual occupation and space usage.
Social	With traditional office-based working models, aesthetics and amenities within the spaces are not prioritised.	This platform provides access to a wide range of workplaces that offer a variety of aesthetically pleasing workspaces with several amenities. The platform also provides access to amenities such as onsite restaurants, fitness facilities, electronic vehicle charging, mailboxes, lockers, and dry cleaning services. These can enhance the physical and mental health of employees. Furthermore, conference rooms can be booked on demand, using a hot-desk system.
	Traditional office-based models generally do not provide for social bonding and community among employees.	The quality and design of spaces provided through this platform can increase the level of social interaction, engagement, and community for employees.
	Employees within an office block in traditional office settings hardly have the opportunity to interact and engage with employees from other companies and firms.	Through the flexible working options that VTS provides, employees, though affiliated with different companies can work in the same spaces and thus engage and interact more. This can lead to cross-pollination of ideas and enhanced relationships for future collaboration.
	In traditional office-based models where employees feel isolated, uninspired, or not valued enough, they will be less productive and efficient, and this can often lead to high staff turnover and low skills retention.	Through the flexible and remote working options powered by VTS, employees can feel more inspired and valued, and this can lead to higher job satisfaction, productivity, and efficiency, leading to higher talent retention.

(Continued)

Table 8.5 (Continued)

ESEP	Inefficiency with the Traditional Approach	How the Tool/Platform/Product Addresses the Inefficiency
	Office space search is always a difficult and complex process.	VTS provides a convenient and easier approach to searching for office space.
	It is difficult to collect data using traditional office-based models.	Through the digitisation of office space design, marketing, use and management that VTS provides, data can be collected for improved space usage and management.
Economic	Office space search (from the perspective of the prospective tenants) and space take up (from the landlord/operators' perspective) takes a long time.	Space search and take up are significantly reduced through the VTS platform.
	Companies may be tied up in long leases with space that they do not need or cannot maintain. The high occupancy costs (real estate and maintenance costs) associated with unused and unneeded spaces can increase the financial hardship for already struggling companies.	With the flexible working option that VTS provides, companies can downsize their office spaces when the need arises without having to incur the associated occupancy costs. Similarly, companies can also ramp up their property portfolios when needed.
	With traditional office-based models where staff satisfaction and efficiency are low, a firm's operating costs can be reduced, and this can lead to lower productivity and revenue.	With the flexible, remote working and higher quality of office space that the platform provides, employees' productivity and efficiency may increase, and talent attraction and retention will be higher. This can lead to enhanced revenue and minimise personnel costs that may have been accrued through high staff turnover.
Physical/ spatial	In traditional office-based work models, there are often high levels of mismatch between space demand and space supply.	With the flexible space provision, and the online support platform that VTS provides, it is easier to use office space more efficiently as the space that is available for use is advertised to a wider range of prospective tenants, and this space can easily and quickly be repurposed when the current use is no more required.

Source: Authors analysis.

8.4.3.3   Limitations, Challenges and Prospects

As was noted in previous case studies, firms are working out the remote working/office-based and flexible working protocol they intend to adopt. Therefore, they may not be willing to invest in the remote working aspect of the platform until they establish their baseline. Flexible working however is likely to remain a core component of real estate use, particularly for start-ups and smaller firms. VTS has the potential to continue to improve the efficiency of workspace use, planning, and management, particularly with the increasing drive to build successful communities in the workplace and an increasing ESG awareness and commitment by companies.

Further details on HQO can be accessed from their website using this link: https://www.vts.com/uk.

8.4.5   Case Study 5: Yardi

8.4.5.2   Background, Key Details, and Technologies Utilised

Yardi is an asset management, marketing, and CRM solution provider that makes life easier for prospects, residents, and staff by creating a streamlined digital journey, which includes online leasing, rent payments, maintenance requests, and other services, which are mainly targeted at the commercial and operational real estate markets. The Yardi RentCafe Suite, part of the Yardi Residential Suite, is designed for the residential build-to-rent market. It provides a cloud-based solution that helps promote brands and communities, drives leads and leases, and delivers convenience for residents that enhances the experience, all through a single connected solution. The RentCafe Suite is the ideal tool for clients to create content that represents their brand and make dynamic marketing websites with convenient self-service options. Furthermore, it helps to generate and nurture leads and includes a branded resident app called MyCafe.

The RentCafe Suite is an integrated system that helps improve marketing return on investment, converts more leads to leases, enhances resident services, and streamlines property management and accounting operations. A few highlights of the product include the following:

- Marketing websites

  Marketing websites provide clients with the ability to create an appealing online experience while improving search performance with faster, flexible websites. With a marketing website, clients can create a seamless user experience including tours, applications, and leasing, enabling teams to stay connected with prospects when they are offline. RentCafe marketing websites are highly configurable to meet the visual and functional needs of clients and to promote a client's individual brand, communities, available units, and services.

- Resident app

  The resident app offers a solution for residents where the client can communicate a variety of information to tenants and make their lives easier, such as making and managing payments, raising maintenance requests or logging issues that may affect their well-being at the properties. The resident app can also be used for mental health awareness and advertising events for building community culture. The resident app also includes "deep links" to other providers in the area such as food delivery places and local businesses, which can help promote local businesses and events. This means residents can do everything from one place without leaving the resident app, helping to save time and provide a seamless experience for the tenant.

- CRM solution

  Yardi RentCafe CRM Flex enables leasing teams to automate their day-to-day tasks, which allows them to focus on applicants and residents. Clients can simplify the leasing process and help maximise marketing return on investment with the mobile CRM software. With instant access to resident, property, and lease data, clients can offer prompt assistance and increase residents' satisfaction. Furthermore, clients can manage multiple leasing workflows to accommodate traditional leasing, student housing, corporate leasing, short-term rentals, and co-living spaces in a single connected system.

8.4.5.2 Key Efficiencies and Solutions That Yardi Provides

The solutions that Yardi provides are itemised using the ESEP efficiency framework (Table 8.6).

Table 8.6 ESEP Efficiency Analysis of Yardi

ESEP	Inefficiency with the Traditional Approach	How the Tool/Platform/Product Addresses the Inefficiency
Environmental	A significant amount of carbon is emitted relating to commute-related property searches and viewings.	With VTS' online and mobile office space search, searching for office spaces will require a lower level of search-related transportation which will further minimise carbon emissions.
	CRM traditionally entails manual processes including meetings, collaboration, and site visits which have a high carbon footprint.	With the platform, collaboration is done with reduced commuting; meetings can be held remotely leveraging remote meeting technologies. This can significantly minimise commute-related carbon emissions.
	A high volume of paper is consumed using traditional property marking and CRM activities.	This platform digitises several CRM and marketing activities which means makes it possible for the agents or brokers to send reports to the landlords and investors. Furthermore, communication between the agents, operators and customers is also digitised.
Social	It is difficult for estate agents to capture and manage leads from various sources using traditional and manual processing methods.	Yardi's CRM systems enable real estate agents to efficiently capture and manage leads from various sources, such as websites, social media, and online advertising. These leads can be tracked, categorised, and prioritised based on their potential for conversion.
	With traditional marketing and client management, marketing campaigns are usually generic, and the reach of these campaigns is usually limited.	The CRM system is integrated with marketing tools which enable the seamless execution of marketing campaigns. Agents and brokers can use the CRM data to create targeted advertisements, social media posts, and email newsletters to reach potential clients and to provide bespoke property recommendations. Furthermore, the fact that clients can take virtual tours is also an added advantage.

(Continued)

Table 8.6 (Continued)

ESEP	Inefficiency with the Traditional Approach	How the Tool/Platform/Product Addresses the Inefficiency
	Communication using traditional marketing and customer relationship management is complex. Follow-up, reminders, and other related tasks are cumbersome, inconsistent, and generally inefficient.	CRM and management systems facilitate automated communication through emails, text messages, or personalised notifications. This enables brokers to set up automated follow-ups, property alerts, and reminders, ensuring consistent and more convenient interactions with clients.
	Transactions using traditional approaches are difficult to track and manage and there can often be mistakes in the paperwork.	Yardi's CRM systems streamline transaction management, enabling agents and brokers to track the progress of deals, manage documents, and automate paperwork. This ensures smooth and efficient transaction processes for clients.
Economic	Property marketing, campaigns, and CRM systems traditionally consume time. The time that it would take to manually capture data and information, analyse them, and use them to develop marketing campaign strategies can be enormous.	Yardi's digital marketing and CRM systems provide a time-efficient approach to these tasks with higher conversion and retention rates. Because the system is automated, collecting data from various sources such as websites, social media, and online advertising can be completed and analysed; leads can also be tracked, categorised, and prioritised based on their potential for conversion within a few minutes. The fact that marketing campaigns can be customised using the platform is an added advantage.
	Traditional property marketing, campaigns, and CRM often require a substantial amount of funds. The funds would be used for labour, marketing material, adverts, events, etc.	Yardi's digital marketing and CRM systems provide a cost-efficient approach to these tasks with higher conversion and retention rates. Furthermore, better access to data and information can significantly improve real estate asset and portfolio performance.
	The reach and conversion rates of brokers and agents are severely limited using traditional approaches. For instance, it is difficult to reach an international audience through traditional media channels.	The platform's improved lead management and data-driven decision-making can increase the conversion rates of brokers and agents, and by extension, increase the profitability of real estate agents and brokers.
Physical/ spatial	Space take-up rate can be lower and vacancy rates can be higher due to a demand/supply mismatch, or inefficient marketing strategies.	Yardi's marketing and CRM systems provide a more effective and efficient matching mechanism for tenants and landlords, thus enhancing the efficiency of space use. This makes space available to those who need it, and it enables property owners to get their spaces let more speedily. This can minimise the time that a property stays vacant, thereby maximising returns and minimising costs on an asset.

Source: Authors analysis.

8.4.5.3 Limitations, Challenges, and Prospects

The subscription required for this tool requires a financial commitment from firms and service providers; smaller firms and start-ups may find this challenging.

Further details on Yardi can be accessed from their website using this link: https://www.yardi.co.uk/.

8.4.5 Other Cases (Exercises for Students)

CoworkIntel

CoworkIntel gathers data on global co-working and flexible workspaces to empower operators, landlords, and more with the insights to optimise their businesses. The platform provides accurate and granular data that empower the optimal strategy for co-working and flexible workspace operators and landlords to improve their bottom line. Data will form the basis for informed decisions that will optimise revenue for landlords and operators alike and CoworkIntel is bringing these tools, forged from those of well-established industries with experience of dynamic demand and products, such as hotels and gyms. They enable the exploration of average prices and occupancy of competitors, plus social and website performance, to give operators, landlords, and brokers a competitive advantage.

Further details on CoworkIntel can be accessed from their website using this link: https://www.coworkintel.com/.

Hubble-Co-working

Founded in 2016, Hubble was originally a marketplace for businesses to find flexible office space in London, from serviced office providers and sublets. During the pandemic, Hubble saw a demand in the market for even more flexibility. From January 2021 to December 2021, Hubble expanded globally and grew from just having a presence in London to allowing businesses to access space in over 26 countries worldwide. One of Hubble's solutions is HubbleHQ – an online marketplace for renting flexible office space, with over 5,000 offices live on the platform. Through a combination of the tech platform and property advisors, they help find businesses their new HQ office on a flexible contract. Hubble Pass provides flexible membership that enables businesses to give their employees on-demand access to a global network of top-quality workspaces, whenever they need it. Teams use it to meet up or do individual work, and they also offer day passes, private offices, meeting rooms and event space. The third is Hubble Perks – a product that enables teams to export their company culture to employees' homes, through everything from virtual experiences to better workplace equipment—ensuring that remote teams have productive and happy workspaces.

Further details on Hubble can be accessed from their website using this link: https://hubblehq.com/.

Reapit

Reapit is an estate agency business platform provider that has supported residential estate agencies in the UK, Australia, and New Zealand to grow with a variety of continuous innovative, market-leading products. Reapit customers can access all the PropTech solutions they need to deliver a seamless customer journey, all in an integrated system. Globally, the Reapit Group provides software solutions to over 71,350 users in more than 7,000 offices and supports in excess of 225,000 tenancies. The company is headquartered in London and has offices across the United Kingdom and Australia.

With the backing of investors AKKR, Reapit recently acquired AgentPoint – a real estate agency website and lead generation technology in Australia (March 2021); and Mindworking – a real estate agency websites and lead generation technology in Australia (August 2021). Reapit's product suite includes the AgencyCloud CRM in the UK and AgentBox CRM in Australia, which deliver a fully integrated range of dynamic estate agency solutions including Sales, Lettings, Property Management, Client/Trust Accounts, and Analytics. Reapit also offers four powerful digital marketing products, including websites, digital proposals and brochures, a lead gen tool, and a property lifecycle app, which significantly enhance the marketing capacity and capability of estate agents. Beyond innovative products, Reapit also offers a powerful service and training model, as part of their commitment to being a partner in growth for their customers.

Further details on Reapit can be accessed from their website using this link: https://www.reapit.com/.

8.5 Chapter Summary

Traditional real estate agencies and brokerage service companies face numerous challenges in a rapidly changing landscape. High transaction costs, information asymmetry, limited market coverage, technological disruption, evolving consumer expectations, a lack of innovation, and competition from online platforms all pose significant challenges. However, by embracing technology, adapting their business models, enhancing customer experiences, and staying attuned to regulatory changes, traditional agencies are overcoming some of the current inefficiencies associated with their practice and service delivery.

Digital technology and innovation have contributed significantly to the real estate agency, brokerage, and other allied services. Online listing, marketing platforms, and virtual viewing platforms are improving data access and transparency, reducing transaction cost and time, providing a more personalised experience, reducing environmental impact, improving the efficiency of space utilisation, and advancing analytics capabilities among other benefits. The shared economy is also emerging with systems such as co-working, space-time commoditisation, and co-living yielding benefits such as reduced real estate cost, increased employee engagement and satisfaction, improved networking opportunities, attraction and retention of talents/employees, improved employee efficiency, and improved perception of a company's brand. Additionally, digital customer relationship management is yielding benefits such as enhanced customer experience, effective lead tracking and follow-up, and increased lead conversion rates.

Although these innovations and digital technologies have created several benefits, there are fears around integration and adoption, data privacy, security, scalability, compatibility and conflict, and legal and regulatory complexities. Considering the future benefits that innovation and digital technology can further attract to these real estate value chain segments, there is a need to address the challenges and limitations. To achieve this, real estate brokers and agents must embrace digital contemporary strategies, leverage data analytics, prioritise cost-efficient channels, and strike a balance between personalisation and digital engagement. By leveraging the evolving digital landscape, traditional real estate brokers, agents, and marketers can effectively engage modern consumers and take advantage of emerging opportunities for innovation and growth to drive successful property promotions, rental, and sales.

Notes

1 Although property management is a core component of real estate agency, it has been discussed in Chapter 5; it will therefore not form part of the core discussion in this chapter, although a some areas will be addressed.
2 The next sub-section will focus on real estate marketing.
3 Some of the points listed take cognisance of some digital tools simply because some tools have become an integral aspect of real estate marketing a long time ago.
4 Old clients that have had a good experience with a property marketer are likely to want to deal with that marketer again.
5 The theory provides a lens to understand individual' optimal strategies while choosing from a series of potential opportunities of varying quality with a combination of characteristics which require trade-offs among set of provisions or services (Diamond, 1984; Jovanovic, 1979; McCall, 1970).
6 This should not be misconstrued to mean that the shared economy and space commoditisation in real estate is only in terms of agency, brokerage and management.
7 For instance, the way a traditional residential property would have been let and managed 20 years ago is quite different from the way the same property will be let and managed today if it is used for short let purposes (e.g., Airbnb).
8 Disintermediation refers to the cutting out of middlemen/agents from real estate transactions. In this case, potential renters will contact the landlord directly.
9 Some of the assets in this asset class include hotels, Purpose-built Student Accommodation (PBSA), built to rent (BTR), storage etc.
10 This has been covered in Chapters 1 and 5.
11 This information was valid as of September 2020.
12 Although we are not able to examine this further due to the nature of the data presented in the publication, this is an area that would benefit from further research.
13 For instance, parking may be a concern for employees in city centre offices where parking is expensive and, in some case, unavailable; however, this may not be the case for employees working in offices that are on secondary campuses where there is ample free parking spaces.
14 HomeAway is now Vrbo. Expedia Group recently rebranded its HomeAway and VRBO brands as a single brand, Vrbo.
15 The ESEP efficiency analysis for each case study is not exhaustive; students and readers are encouraged to identify and note down other areas of efficiency which the named tool/platform provides.

References

Albrecht, J., Gautier, P. A., & Vroman, S. (2016). Directed search in the housing market. *Review of Economic Dynamics, 19*, 218–231. https://doi.org/10.1016/j.red.2015.05.002.

Blowers, A., Hamnett, C., & Sarre, P. (2014). The future of cities. *The Future of Cities, April*, 1–319. https://doi.org/10.4324/9780203717172.

Brown, R. (2018). *PropTech: A guide to how property technology is changing how we live, work and invest*. Casametro.

Cushman and Wakefield. (2019). *Cre executive perspectives on coworking how and why the flexible workplace matters.*

Diamond, P. (1984). Money in search equilibrium. *Econometrica, 52*(1), 1–20.

Fiorilla, P. (2018). *Shared space : Coworking's rising star.* https://www.commercialsearch.com/news/shared-space-coworkings-rising-star/.

Jones Lang LaSalle. (2016). *A new era of coworking*.

Jovanovic, B. (1979). Firm-specific capital and turnover. *Journal of Political Economy*, 87(6), 1246–1260.

Khozaei, F., Hassan, A. S., & Razak, N. A. (2011). Development and validation of the student accommodation preferences instrument (SAPI). *Journal of Building Appraisal*, 6(3–4), 299–313. https://doi.org/10.1057/jba.2011.7.

Kim, S. (1992). Search, hedonic prices and housing demand. *The Review of Economics Anc Statistics*, 74(3), 503–508.

Knight Frank Research Report. (2020). *12 dynamics of the post-COVID-19 workplace* (Issue July). https://content.knightfrank.com/research/2033/documents/en/12-dynamics-of-the-post-covid-19-workplace-july-7337.pdf.

Kobue, T., Oke, A., & Aigbavboa, C. (2017). Understanding the determinants of students' choice of occupancy for creative construction. *Procedia Engineering*, 196(June), 423–428. https://doi.org/10.1016/j.proeng.2017.07.219.

Leducq, D., Demazière, C., Abdo, É. B., & Ananian, P. (2023). Digital nomads and coworking spaces: Reshaped perspectives? The paris mega city-region after covid-19. In I. Mariotti, M. Di Marino, & P. Bednář (Eds.), *The COVID-19 pandemic and the future of working spaces* (1st ed., pp. 109–121). https://doi.org/10.4324/9781003181163-11.

Leesman. (2019). *Part 2: Do new workplaces work*. https://www.leesmanindex.com/media/Leesman-EwX-P2-Second-Digital-Edition.pdf.

Magni, M., Pescaroli, G., & Bartolucci, A. (2019). Factors influencing university students' accommodation choices: Risk perception, safety, and vulnerability. *Journal of Housing and the Built Environment*, 34(3), 791–805. https://doi.org/10.1007/s10901-019-09675-x.

Mahmoud, A. (2016). *The impact of AirBnb on hotel and hospitality industry*. Hospitalitynet. https://www.hospitalitynet.org/external/4074708.html.

McCall, J. (1970). Economics of information and job search. *The Qaurterly Journal of Economics*, 84(1), 113–126.

Oladiran, O., Hallam, P., & Elliot, L. (2023). The COVID-19 pandemic and office space demand dynamics. *International Journal of Strategic Property Management*, 27(1), 35–49. https://journals.vilniustech.lt/index.php/IJSPM/article/view/18003.

Oladiran, O., Sunmoni, A., Ajayi, S., Guo, J., & Abbas, M. A. (2022). What property attributes are important to UK university students in their online accommodation search? *Journal of European Real Estate Research*. https://doi.org/10.1108/JERER-03-2021-0019.

Palm, R., & Danis, M. (2002). The internet and home purchase. *Tijdschrift Voor Economische En Sociale Geografie*, 93(5), 537–547. https://doi.org/10.1111/1467-9663.00224.

Qiu, L., & Zhao, D. (2018). Information and housing search: theory and evidence from China market. *Applied Economics*, 50(46), 4954–4967. https://doi.org/10.1080/00036846.2018.1464644.

Read, C. (1993). Tenants' housing search and vacancies in rental. *Regional Science and Urban Economics*, 23(1), 171–193. http://www.journals.uchicago.edu/doi/abs/10.1086/260169.

Read, C. (1997). Vacancies and rent dispersion in a stochastic search model with generalized tenant demand. *Journal of Real Estate Finance and Economics*, 15(3), 223–237. https://doi.org/10.1023/A:1007724006255.

Saiz, A. (2020). Bricks, mortar, and proptech: The economics of IT in brokerage, space utilization and commercial real estate finance. *Journal of Property Investment and Finance*, 38(4), 327–347. https://doi.org/10.1108/JPIF-10-2019-0139.

Schlagwein, D., Schoder, D., & Spindeldreher, K. (2020). Consolidated, systemic conceptualization, and definition of the "sharing economy." *Journal of the Association for Information Science and Technology*, 71(7), 817–838. https://doi.org/10.1002/asi.24300.

Turnbull, G. K., & Sirmans, C. F. (1993). Information, search, and house prices. *Regional Science and Urban Economics*, 23(4), 545–557. https://doi.org/10.1016/0166-0462(93)90046-H.

University of Oxford Research. (2017). *PrepTech 3.0: The future of real estate.*

University of Oxford Research. (2020). *The future of real estate occupation: Issues.*

Wang, Y., Livingston, M., McArthur, D. P., & Bailey, N. (2023). The challenges of measuring the short term rental market an analysis of open data on Airbnb activity.pdf. *Housing Studies.* https://doi.org/10.1080/02673037.2023.2176829%0D.

Wheaton, W. C., & Krasikov, A. (2019). Will coworking work? *SSRN Electronic Journal* (SSRN-id3406049). https://doi.org/10.2139/ssrn.3784792.

Williams, J. (2018). Housing markets with endogenous search: Theory and implications. *Journal of Urban Economics, 105*(December 2017), 107–120. https://doi.org/10.1016/j.jue.2017.12.001.

Zervas, G., Proserpio, D., & Byers, J. W. (2017). The rise of the sharing economy: Estimating the impact of airbnb on the hotel industry. *Journal of Marketing Research, 54*(5), 687–705. https://doi.org/10.1509/jmr.15.0204.

Section Three

Prospects and the Future

PropTech and real estate innovations are still emerging at very much in a growth phase with enormous prospects, opportunities, and unchartered territories. Furthermore, the capital that this emerging area has attracted in the last decades is a testament to the positive sentiment and expectations in this area. Section One introduced the principles and concepts that underpin PropTech and real estate innovations and provided a broad overview of the PropTech evolution. Section Two provided an in-depth review and analysis of the various inefficiencies in real estate and the various innovative and digital tools, platforms, and systems that have emerged across the real estate value chain. This section concludes this book by providing a range of insights, discussions, and perspectives on the current state of real estate innovation, its prospects, and its future trajectory.

As demonstrated in Sections One and Two, the rate of growth and change in the real estate sector has been fast-paced, and it is difficult to accurately predict what the future holds. This section, therefore, does not attempt to predict the future of PropTech and innovation in real estate; rather, it attempts to identify aspects of the real estate innovation and digitisation evolution, with an emphasis on core areas of growth, enablers, and support systems that can further enhance the growth potential in this area. The section also identifies relevant stakeholders, systems, processes, and support systems and provides a reflection on their roles in further deepening the innovation and digitisation drive in real estate.

Chapter 9 provides a review of the key enablers and stakeholders that can further drive the development and growth of innovation and digitisation in real estate, and Chapter 10 provides a summary, reflection, and conclusion of the book.

Learning Outcomes (LO)

At the end of this section, students should be able to:

LO1: Understand the roles of key stakeholders, enablers, and systems that can support the further development of innovation and digitisation in real estate use, operations, and services.

LO2: Understand the roles of education, standards, regulation, and ethics in further enhancing innovation and digitisation in real estate use, operations, and service delivery.

LO3: Identify the core emerging areas of real estate innovation and the digital tools and systems that are likely to define the next phase of innovation and digitisation in real estate.

DOI: 10.1201/9781003262725-11

Chapter 9

Core Stakeholders, Enablers, and Systems of Growth and Development in Real Estate Digital Technology and Innovation[1]

9.1 PropTech Education Development

One of the channels through which digitisation and innovation in real estate can be further promoted is education. Built environment and PropTech educators are therefore core stakeholders and enablers of PropTech growth and further development. Although the various chapters of this book offer a knowledge base and reference material for educators to develop and deliver modules and courses on this subject, it is important to further support them through the provision of education and pedagogical strategies. This section is therefore primarily targeted at real estate and PropTech educators. It begins with relevant concepts and ideas around education, with an emphasis on teaching, learning support, and curriculum development. This sets the stage for a review of PropTech education integration efforts and potential areas of development.

9.1.1 The Concept of Education, Teaching, and Learning Support

Education involves teaching delivery and learning support, and the aim is usually to enable students to gain knowledge and understanding, and the ability to develop the skills that they require to complete certain tasks or provide specified services. From a more technical point of view, education can be conceptualised as a teleological practice – a practice constituted by a "telos", i.e., the Greek word for the purpose of a practice (Carr, 2003). The concept of purpose in education has been elaborated in more recent studies. The question of purpose in education is multidimensional because education typically functions relative to a number of dimensions: qualification, socialisation, and subjectification (Biesta, 2015):

Qualification: This is the acquisition and transmission of knowledge, skills, and dispositions that allow students to "do" something and qualify them. This may be vocational or professional education and can be conceived more generally as a preparatory activity for life in a complex modern society.

Socialisation: This dimension conceptualises education as a representation and initiation of students in traditions and ways of being and doing, emphasising the cultural, professional, and traditional approaches to an area or field. This dimension captures elements of education where existing structure, belief systems, views, perspectives, and potential division and inequalities are perpetuated.

Subjectification: This captures the positive and negative impacts of education on the student as a person. It reflects the ways through which students exist as subjects of initiative and responsibility rather than as objects of the actions of others.

DOI: 10.1201/9781003262725-12

These three dimensions are core elements of education; thus, educators need to be conscious of their roles in defining students' skills development as well as the culture, belief systems, and views that the students hold; by extension, educators shape the future of a profession. Education can also be perceived as the key channel through which the next generation of professionals in a field are trained or endowed with the requisite skills for that profession. Higher education[2] refers to post-secondary education which typically prepares and qualifies individuals (students) for work in a profession or vocational area. Ideally, higher education should contain an emancipatory element that can support the freeing of the mind which can be achieved through critical self-reflection by the students, open learning, interdisciplinarity, and inclusion of philosophical and sociological perspectives in the curriculum (Ronald, 1990). The quality and effectiveness of higher education therefore depend on the curriculum for that course of study, pedagogical considerations, and teaching methods. These will be reviewed in the next sub-sections.

9.1.2   Curriculum Development in Higher Education

To integrate a new or emerging specialisation (PropTech) into an existing and well-established knowledge domain within higher education (real estate), it is essential to undertake curriculum development. The curriculum serves as a fundamental building block in higher education, and its role is pivotal in ensuring the effectiveness of educational programs. Traditional education has focused on knowledge acquisition, and thus, traditional curriculum development would focus on knowledge-based learning. However, in recent times, there has been a growing emphasis on moving away from traditional narrow perspectives in curriculum development. For example, (Fish, 2013) proposes a shift from the conventional focus on skills and specialised knowledge towards a more balanced approach that incorporates teaching, learning, and scholarship to foster intellectual growth and moral behaviour. The study suggests that curriculum management should not only encompass technical skills but also focus on nurturing learners as complete individuals. This perspective is further supported by (Khan and Law, 2015), who argue that curriculum development involves designing comprehensive study plans, teaching strategies, resource allocation, specific lessons, assessment methods, and faculty development. Moreover, they advocate for higher education institutions to adapt positively to changing environments by adopting learner-centred and competency-based curricula. This approach aims to produce graduates equipped with sufficient knowledge, skills, and adaptability to thrive in evolving contexts.

Data has become an essential aspect of curriculum development, gaining increasing recognition in decision-making within educational advancement. Wolf (2007) sheds light on the significance of data-informed educational development, outlining a three-stage process. The initial stage involved curriculum visioning, entailing a comprehensive evaluation through extensive consultation. This assessment involves curriculum evaluation, programme objectives, development or redevelopment, and programme focus. The second stage in the curriculum development process centred on mapping, developing, or redesigning the structure of a programme or course of study. This step involved aligning foundational content and programme objectives with existing and future courses, while also addressing any gaps, redundancies, and opportunities. The third stage focused on alignment, coordination, and further development. It encompassed aligning programme and course objectives, foundational content, and the learning experiences offered by a programme or course. These stages can be followed up with activities such as meetings and workshops to ensure the effectiveness, sufficiency, and appropriateness of the development plan.

9.1.3   Pedagogical Paradigms

Pedagogy refers to systematic ways of teaching in a logical manner, and it is important that it is anchored on interactive activities where the teachers and learners are active participants (Oladiran and Nanda, 2021). Khan and Law (2015) classify pedagogy into two categories: formal/systematic and informal (Figure 9.1).

Teaching may be carried out through a variety of ways (as shown in Figure 9.1), and these pedagogical approaches can be enhanced by varying approaches and social interaction between the teacher and students. For pedagogy to be effective, it needs to be dynamic, diverse, and interesting, and these must be based on a learner-centred approach (Loo, 2004; Oladiran and Nanda, 2021).

9.1.3.1   Learner-Based Paradigm

The learner-based paradigm highlights the connection between curriculum development and the learning processes of students. Students' learning processes, programmes, and plans are important in shaping curriculum development, and this is achieved through a

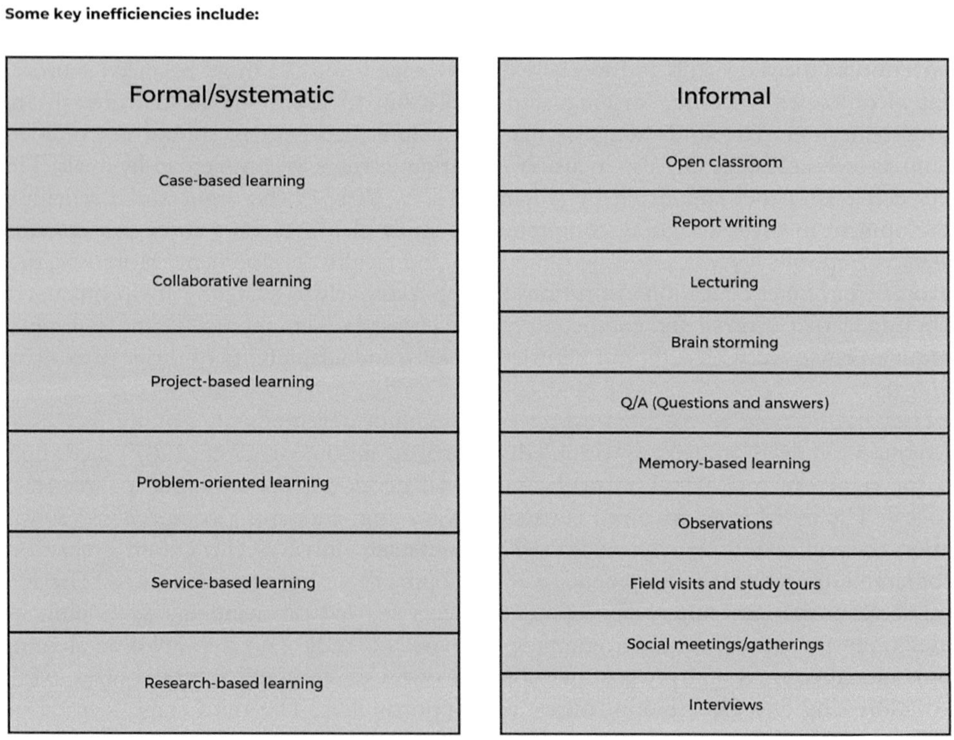

Some key inefficiencies include:

Figure 9.1 Pedagogical classification.

combination of synchronous and asynchronous activities (Nygaard et al., 2008). The three key areas addressed are knowledge (content), skills (both soft and hard skills), and competencies (the practical application of knowledge and skills to enhance performance, make an impact, and solve problems).

It is recommended that higher education institutions re-evaluate their learning requirements to adequately prepare and support students for a world characterised by evolving roles, technologies, and diverse competencies. This preparation is essential for equipping students with the ability to navigate flexible job opportunities in dynamic markets, societies, businesses, and technologies. Such readiness becomes particularly crucial in the current context of rapid change, knowledge-driven environments, and data-centric atmospheres. These insights are relevant to curriculum development strategies.

9.1.3.2   Taxonomy of Educational Objectives

The taxonomy of educational objectives is a pedagogical framework designed to categorise students' learning goals. This framework was developed by Bloom in 1956 and has since provided a means of measurement and it has become established a common language for discussing learning objectives, facilitating effective communication across different individuals, subjects, and grade levels. It also serves as a foundation for defining the specific meaning of broad educational goals and ensuring coherence among educational objectives, activities, and assessments within a unit, course, or curriculum.

The taxonomy's structure established a hierarchy of categories within the cognitive domain: remembering, understanding, applying, analysing, evaluating, and creating (as illustrated in Figure 9.2), and this classification remains applicable in contemporary higher education. The underlying idea is that pedagogical objectives should aim to guide learners' progress from the foundational domain of knowledge (at the base of the inverted pyramid) to the domain of production and creativity, which represents the pinnacle of learning attainment (Krathwohl, 2002). It is the responsibility of educators to organise the teaching and learning environment in a manner that provides all learners with the opportunity to engage in higher-order learning processes. This can be achieved through careful alignment of all components of learning, ensuring that the learning objectives clearly express the level of understanding expected from students. Furthermore, the teaching context should motivate students to undertake the learning activities that will enhance those understandings (Biggs, 2012). Similarly, assessment tasks should align with the appropriate stage of the learning order and clearly specify what is expected of students and the level of attainment needed to meet the objectives successfully.

9.1.4   The Integration of PropTech in Real Estate Education

Sections 9.1.2–9.1.3 have provided valuable insight into core concepts in pedagogy in general. These concepts lay the foundation for the effective and efficient integration of innovation and digitisation in the existing real estate curriculum, teaching, and learning support structures. Unfortunately, the integration of PropTech and innovation in real estate education has moved at a slow pace. One of the reasons for this may be the fragmented nature of the PropTech domain, i.e., torn between real estate and technology, and another reason may be the low volume of research in PropTech education.

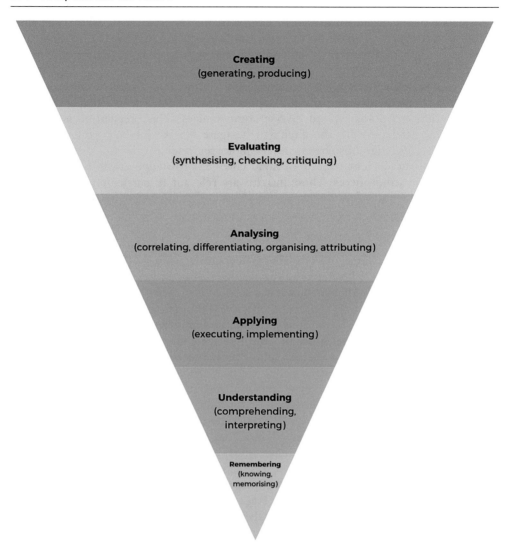

Figure 9.2 Structure of Bloom's taxonomy in the cognitive domain.

The higher education system is a crucial platform for and avenue for the development of PropTech education. This is because, as demonstrated in Chapter 1 and subsequent chapters in this book, PropTech fundamentally links to all areas of the real estate value chain. Integrating PropTech in real estate higher education is therefore an effective and efficient approach to enhancing its development and for the development and grooming of skills and talents needed to further enhance the efficiency of real estate use, services, and operations to become more efficient. Oladiran and Nanda (2021) therefore developed a PropTech Education Integration Framework (PEIF). This framework was inspired by a mix of market and research-led pedagogy. The data collection entailed ten interviews, two focus groups with five participants each and 70 survey respondents, and the respondents included real stakeholders (educators and students), stake watchers (real estate professionals and PropTech practitioners, and stake keepers (professional bodies and institutions). The data from this study were analysed using thematic and content analysis and the data analysis served as the basis for the integration framework.

Some of the key findings from the study are listed as follows:

1 Formally integrating PropTech into the real estate curriculum will open up possibilities for enhanced innovation and creativity within the real estate sector. As a result, the efficiency and effectiveness of real estate use, operations, and services will be significantly enhanced over time.
2 Alongside technology, PropTech education integration should encompass various essential soft skills such as professionalism, creativity, innovation, entrepreneurship, business acumen, problem solving, and communication. These soft skills have played a crucial role in propelling technological advancements within the real estate sector in recent years; they should therefore be promoted in conjunction with technological expertise.
3 Incorporating PropTech into existing real estate courses and programmes in higher education is a positive and commendable initiative that has significant value and prospects.
4 Integrating PropTech in real estate education does not necessarily require real estate students to be data scientists or analytics experts; instead, the primary objective is to enhance their awareness, knowledge, and comprehension of the broader aspects of data, analytics, and digital tools as they relate to real estate.
5 The essential principles of real estate practice and operations, including economics, law, agency, valuation, and other core aspects, should not be disregarded. These foundational elements remain integral to the functioning and practice of real estate; therefore, the incorporation of innovation and digital technology into the real estate curriculum should not be misconstrued as a replacement for these fundamentals.
6 While certain real estate innovations might not fall directly under the category of "technology" (e.g., coworking and co-living), these innovations have been enabled by digital technology and they should therefore be included in the PropTech education and innovation integration plans.
7 The primary goal of the PropTech education integration is harnessing the potential of digitisation and innovation in improving the economic, environmental, social, and physical efficiency of real estate operations and practices.
8 To bolster the drive for PropTech education integration, PropTech and real estate firms should strengthen their collaborative efforts and partnerships with real estate departments/schools. These alliances can take various forms, such as offering guest lectures, organising webinars and events, providing internships, and awarding scholarships.
9 The assessment methods for PropTech modules may be diverse: coursework, case studies, projects, presentations, and conceptual product development. Given the hands-on and applied nature of PropTech, traditional exams may not be appropriate for testing students' knowledge and comprehension in this area.

The results from the survey carried out revealed that the vast majority of respondents are of the opinion that PropTech education integration in real estate higher education is important (Chart 9.1). The results indicate that 97% of respondents agreed that PropTech education integration is important, with a vast majority (78%) believing that it is very important. It is worth noting, though, that the 3% of respondents that indicated that PropTech education integration were academics (as shown in Chart 9.2). This may be a

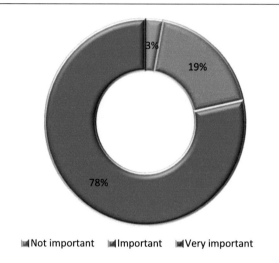

Chart 9.1 Rating the importance of PropTech education in real estate higher education.

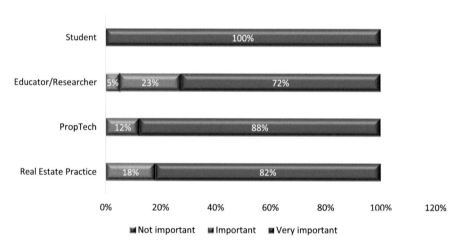

Chart 9.2 Importance of PropTech education in real estate higher education (by respondents' roles/profession).

reflection of the pessimism that some educators and researchers hold about PropTech and its emergence.

The general lower level of optimism and interest in PropTech education integration observed in the educators/researchers should be considered when integration plans are being developed or reviewed.

9.1.4.1   PropTech Education Integration Approaches

Based on these findings, the PEIF proposed two primary approaches to PropTech education integration, namely, the concentrated and dispersal, and a secondary approach, which is referred to as hybrid/blended (Figure 9.3).

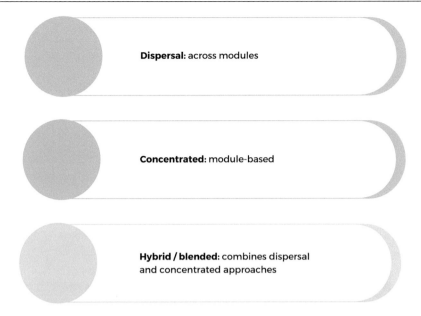

Figure 9.3 PropTech education integration approaches.

9.1.4.1.1   DISPERSAL APPROACH

The integration of PropTech and innovation into real estate higher education can be achieved through its dispersal across various real estate modules or units of learning. Given the diversity of the real estate sector in terms of different sectors, asset classes, operations, and practices, digital and IT applications and innovations can be effectively integrated into existing real estate course/program structures in alignment with core real estate practices and operational areas. This integration can take various forms. For instance, one approach is the asset-class dispersal approach which involves integrating digital and IT applications based on the specific real estate asset classes, namely residential, commercial, and operational real estate (as depicted in Figure 9.4). Although this approach is reasonable given that real estate practice and specialisations often revolve around these asset classes, it will be difficult to implement from a real estate pedagogical point of view due to the significant overlap among the various real estate asset classes. To successfully adopt this method of dispersal, effective cross-module collaboration would be necessary to ensure its efficiency and efficacy. An alternative to this is the module-based dispersal approach where PropTech and innovation are integrated with the core real estate modules (Figure 9.5).

As depicted in Figure 9.5, the modules covering real estate planning, land management, construction, finance, valuation/appraisal, investment, portfolio management, asset/property management, agency/brokerage, and law can serve as starting points for PropTech integration (as demonstrated from Chapters 4 to 8). These changes may require adjustment to existing module aim, objectives, delivery, outcomes, and assessments.

9.1.4.1.2   CONCENTRATED APPROACH

A second approach to integrating PropTech and innovation into real estate higher education is the concentrated approach (depicted in Figure 9.6). This approach involves

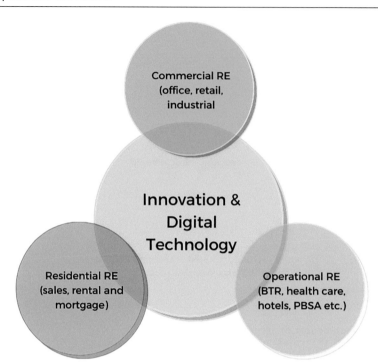

Figure 9.4 An asset-class dispersal approach to PropTech education integration.

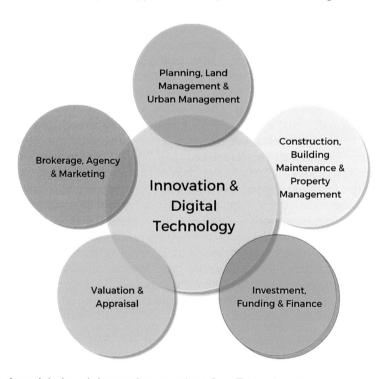

Figure 9.5 A module-based dispersal approach to PropTech education integration.

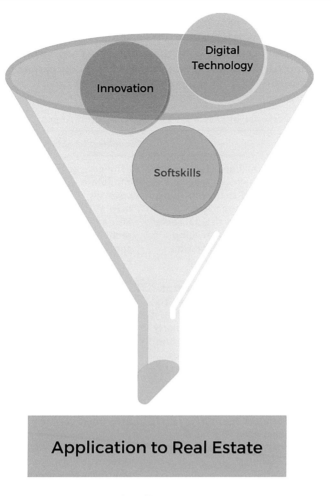

Figure 9.6 A concentrated approach to PropTech education integration.

developing and introducing a PropTech and innovation module in real estate and built environment programmes. The module will introduce students to the significance of innovation and technology in the real estate sector and the aim may be to enable students to understand and appreciate the value that digitisation and innovation bring to real estate use, operations, and service delivery.

The concentrated module approach may provide a comprehensive introduction to IT and digital technology, covering topics like data, analytics, computing, the internet, cloud computing, IoT, blockchain, AI, drones, and more (see Chapter 2). In conjunction with these hard skills, the module should also emphasise the importance of soft skills such as creativity, communication, strategic thinking, ethics, entrepreneurship, and problem-solving. Having provided an understanding of digitisation in real estate, the module could provide insight into the application of digitisation and innovation across the real estate value chain (as demonstrated in Chapters 4–8). The assessment component of a PropTech module may range from essays, coursework, reports, case study analysis, etc. For instance, students may be asked to review an existing PropTech tool or platform, providing a critical analysis of the tools. Furthermore, as part of the soft skills development, students may be asked to work in groups to develop digital tools and make presentations at the end of this. This can strengthen their teamwork, communication, and problem-solving skills.

These primary approaches to integration (dispersal and concentrated) have several advantages and disadvantages that are summarised in Table 9.1.

Table 9.1 Merits and Demerits of Dispersal and Concentrated Approaches to Integration

Dispersal approach	Concentrated approach
Panel A: Merits of the dispersal approach relative to the concentrated approach	
Students are introduced to a wider range of technologies that can expand their knowledge and understanding of various practical applications of innovation and digital IT tools. This can enable them to further a develop broader scope of application of innovation and digital technologies to the different components of real estate.	A module-based approach, though covering several IT-related topics may have a limited scope of application. Students may thus be confined to a few technological tools which can limit the application to different components of real estate.
The dispersal approach may be easier to adapt as it is less likely to require major changes to the current course structures. Furthermore, it is more subtle and less disruptive, particularly when compared to introducing a new module/ suite of modules (see Section 5.1.2). Heads of schools/departments and course directors may encounter less bureaucratic hurdles if this approach is adopted.	The concentrated approach may be more complex in implementation and may require several bureaucratic and procedural challenges (e.g., approval from the RICS and relevant University committees and boards).
With the right level of interest and support from various module leaders, the dispersal approach can be implemented in a much shorter time, in comparison to introducing a new module/suite of modules (see Section 5.1.2).	Developing a module may require long bureaucratic and administrative procedures which may take a lengthy time. Planning and implementing the integration using the concentrated approach may therefore be time-consuming.
The dispersal approach eliminates the need to create a new module. Thus, there is a lesser knowledge/skills trade-off (in terms of getting rid of already existing content and assessments that are embedded in the current course structures) which would be the case if an existing module is to be replaced with a PropTech and innovation module.	Having a new module may be practically challenging for programmes and courses with short duration or special structures (such as postgraduate courses). A major problem is the trade-off between currently existing modules and a new PropTech and innovation module – i.e., which module will be discontinued to accommodate the new module?
Panel B: Demerits of dispersal approach-relative to concentrated approach	
Considering that the integration (dispersal) will be implemented within several modules, it may be more challenging for the heads/leaders of the programmes/courses to coordinate the activities of all the module leaders and appraise the progress and development of the integration.	With the concentrated module, the progress and appraisal of the module (after initial set-up and approvals) can be coordinated more effectively by the heads/leaders of the programme. This can further improve the review and re-development of the module and the PropTech education integration in general.

(Continued)

Table 9.1 (Continued)

Dispersal approach	Concentrated approach
With the dispersal approach, it will be more challenging to monitor students' performance and engagement with PropTech and innovation; this can further make it more challenging to appraise and monitor the progress of the integration plan.	A module-based approach improves the monitoring of students' performance, interest and engagement in PropTech and innovation, thus enhancing the general integration appraisal.
Students without prior knowledge and exposure to PropTech may struggle to appreciate the need and relevance for PropTech and innovation as emergent paradigms if this is dispersed across the different modules.	The concentrated approach offers an opportunity for students to gain foundational knowledge and understanding of the context of PropTech and innovation and their importance to real estate operations and practice in general.

Source: Oladiran and Nanda (2021).

9.1.4.1.3   HYBRID/BLENDED APPROACH

Due to the merits and demerits of separately adopting the dispersal and concentrated approaches (Table 9.1), the PEIF recommended that a hybrid approach to PropTech integration is the most effective and efficient method of PropTech education integration. This approach will entail the integration of a module that will introduce students to property technology and innovation (the concentrated approach); this module will run simultaneously with other modules that will also have elements of digital technology and innovation (dispersal). By combining these two approaches, a balanced integration can be achieved, effectively addressing the weaknesses of each of the primary approaches while maximising their individual potential benefits. This dual approach also enhances the flexibility of integration planning and execution, allowing for a more seamless and efficient integration process.

In the survey that followed the PEIF development, respondents were asked about their preferred method (of the three integration approaches) and the results reveal that the hybrid/blended approach is the most preferred choice (in Chart 9.3). Respondents were further asked to rank their priorities by order preference with respect to respondents' opinion of the most effective and efficient approaches to integration and the least using a ranking mechanism.[3] The results show that the hybrid approach was the most preferred integration method (Chart 9.4), and this was consistent across the various real stakeholder, stake watcher, and stake keeper groups (Chart 9.5). It should be noted that although the hybrid approach is consistently the most preferred, it is not overwhelmingly the preferred option. The preferences for integration could therefore be assessed in further studies.

Further analysis of the various respondent groups (Chart 9.5) shows that the hybrid approach was preferred across the real stakeholders and stake watcher groups, while the concentrated approach was the next preferred approach for all the groups, apart from the educators who had a second preference for the dispersal approach. In general, the responses support the adoption of a more blended integration approach to PropTech education integration.

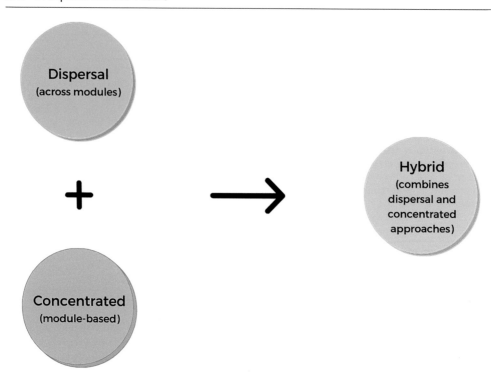

Figure 9.7 A blended approach to PropTech education integration.

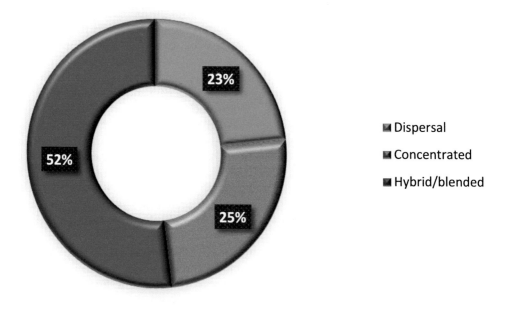

Chart 9.3 Most preferred (first choice selection) integration approaches by respondents.

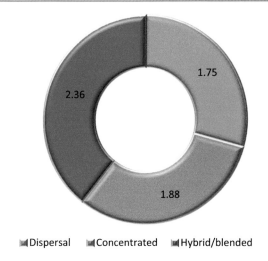

Dispersal   Concentrated   Hybrid/blended

Chart 9.4 Ranking scores of the three PropTech education integration approaches.

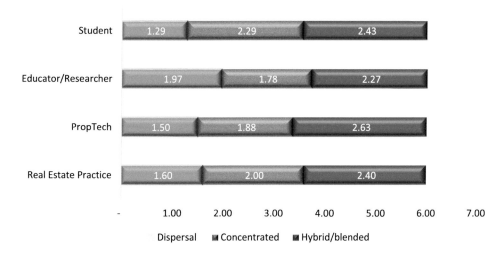

Dispersal   Concentrated   Hybrid/blended

Chart 9.5 Ranking scores of the three PropTech education integration approaches (by respondents' roles/profession).

9.1.5 Reviewing PropTech Education Integration

One of the reasons that PropTech education has been slow in adapting to the changing real estate landscape is that the real estate curriculum has not been agile. It is therefore important that regardless of the integration approach taken, the integration of digitisation and innovation in real estate education is flexible and reviewed periodically. Oladiran and Nanda (2021) therefore propose a three-yearly review plan using the APPIR cyclical model (Figure 9.8), and the idea is that a PropTech education integration plan must culminate in an integration review and a new or updated action plan.

Figure 9.8 The APPIR cyclical model for the PropTech education integration.

The stages of the APPIR cyclical model are summarised below

- **Stage 1 – Assessment**: The step involves evaluating the current course/program structure, identifying areas of needs and challenges, and conducting an assessment of current trends in the real estate industry and the demands of employability.
- **Stage 2 – Prioritise**: The subsequent phase of integration involves identifying the priorities of the course/programme. It is important to bear in mind that various real estate courses may be housed within different departments and schools, such as the Business School or Planning and Environmental Departments, and this can potentially lead to differences in priorities and course administration approaches. Therefore, it is crucial to examine the course/programme priorities in relation to the needs identified in the initial stage to formulate a practical and effective integration plan.
- **Stage 3 – Plan**: Developing and documenting an integration plan is important for effective implementation and evaluation. This plan should consider the specific needs and priorities of both the institution and the individual courses. It should outline the integration objectives (what), the actions and processes involved (how), the timeline for execution and review (when), and the adopted pedagogical strategy. Conducting a SWOT analysis will be beneficial in identifying the plan's strengths and associated opportunities. Additionally, the SWOT analysis will help strategically manage any weaknesses and mitigate potential threats to ensure a well-rounded and successful integration process.
- **Stage 4 – Implement**: Sequel to the development of the integration plan, the integration process can be implemented, with systems put in place for monitoring, feedback, and records. The feedback and observations collected during this stage will serve as the foundation for further review and refinement.

- **Stage 5 – The Review/Reflect**: This marks the conclusion of the initial integration cycle. It involves introspecting on the first cycle of integration, recognising areas of success, and identifying areas that need enhancement. Educators can thus develop their own understanding and insights from their experiences, which can then be utilised to improve current practices and advance knowledge. Incorporating feedback from students is crucial during the review process, ensuring an objective and equitable evaluation. The review outcomes will serve as the basis for the subsequent integration cycles, guiding the continuous improvement and evolution of the integration process.

The review component of the APPIR cyclical model holds significant importance as PropTech education integration plans must consider various dynamic factors. These include the rapid developments in the PropTech domain, changes in programme and course priorities and structures, changes in staff expertise and research interests, as well as advancements in technology and pedagogy, to ensure continued effectiveness. Although the PEIF proposes a three-year APPIR cycle, institutions are encouraged to adopt a cycle period that best suits their specific needs and feasibility.

9.1.6   *Limitations and Challenges in PropTech Education Integration*

The following challenges and limitations can inhibit the integration of PropTech in real estate higher integration:

- PropTech encompasses an extensive range of knowledge and applications, leading to potential complexities and challenges in its integration in higher education.
- In the case of the concentrated approach, there is a scarcity of academic articles and textbooks available for staff to develop module content, which can hinder the development of lecture materials and learning exercises in the case of the dispersal approach as well. An increase in learning resources and academic support materials will significantly boost the integration of PropTech in real estate higher education.
- Real estate innovation and PropTech are highly dynamic and evolving rapidly. This indicates that certain current systems and innovations may become outdated within a short span of time. Consequently, real estate educators must remain receptive to new ideas and stay abreast of changing trends in real estate operations and practice. Establishing close connections with industry and professional bodies can help mitigate this potential challenge.
- The concentrated and hybrid approaches may necessitate the development of new module(s), which, in turn, could lead to significant modifications in existing real estate courses and programs. This may entail substituting an already existing module within the real estate course structure. Balancing this integration with the need for students to acquire a comprehensive understanding of core fundamentals in real estate theory and application poses a challenge and it requires careful consideration to determine the optimal trade-off in terms of core knowledge areas and skills.
- Some real estate departments or schools may not have academic staff with a skillset, interest, or expertise in PropTech, and this can pose a major challenge for all three integration approaches. This challenge can be more pronounced in institutions considering the concentrated approach. Therefore, when considering its adoption, department heads, course leaders, and programme directors should ensure that module leaders possess the necessary interest, skills, knowledge, and understanding of current real estate innovative trends and digital technologies.

9.1.7 Further Recommendations for PropTech Education Development

- The concentrated module can be delivered with the assistance of IT/Computing academics with experience in digital and IT systems. Real estate lecturers can thus gain valuable experience with the guidance and support of experienced lecturers who specialise in teaching IT/digital technology.
- Having students go for placement or internships in PropTech firms can enhance students' enthusiasm for real estate digital technological systems and other innovations. As a result, real estate departments should consider integrating internships or placements into their programs to cater to students interested in PropTech.
- Alternative or supplementary variations to the concentrated approach can be considered. For instance, course induction sessions for new students or regular industry seminars and events can concentrate on diverse aspects of PropTech and innovation. This can complement the integration plan using a dispersal approach in cases where creating or implementing a new PropTech and innovation module becomes challenging or unfeasible.
- Due to varying rates of development in different parts of the world, introducing PropTech elements through different competency levels, such as introductory, intermediate, and advanced modules/content, could prove beneficial. For instance, in regions where PropTech is advancing rapidly, advanced PropTech modules could complement the introductory-level concentrated module. Additionally, an elective PropTech (advanced) module may be developed to build upon either the concentrated or dispersal efforts, allowing students with a particular interest in PropTech to delve deeper into the subject and explore innovations further.
- Flexibility has been a key driver of innovation, particularly in the real estate sector. Likewise, pedagogical flexibility has been recognised as a crucial aspect of curriculum development. Therefore, institutions should exhibit adaptability in adopting and adapting the integration of PropTech to align with their specific realities and dynamic environments.
- PropTech firms and professional organisations, like RICS, the UK PropTech Association, and other PropTech Associations, should increase their collaboration with higher education institutions to foster the integration of real estate digital/innovation. Such collaboration can also mitigate potential challenges that educators may encounter in a rapidly evolving PropTech landscape.
- In the concentrated module and the various core modules of integration (in the dispersal approach), guest speakers from the PropTech space and entrepreneurs can be invited to deliver some sessions.
- Module/course objectives and learning outcomes should be modified to reflect the changing technological climate on a more comprehensive scale than currently practised. These modifications should be guided by both scholarly and market-based pedagogical considerations.

Education is an important service, and the value of this service is maximised through high-quality teaching, student support, and student engagement. To align with the current growth in real estate digitisation, contemporary real estate education must facilitate active student engagement. Higher institutions are encouraged to approach PropTech education integration with flexibility and an effective review mechanism to adapt to the ever-changing landscape of real estate operations and service delivery.

Further research can explore more effective PropTech enhancement approaches and tools to extend the reach and impact of PropTech in real estate education.

9.2   PropTech Skills and Employability

9.2.1   Skills and Employability

Skills development is essential for employability in any profession, particularly PropTech. Skills are increasingly highly valued in today's world; PropTech relies on a divergent pool of talent, and this is essential in enabling the industry to make further progress. Some of the core skills required in general terms are problem-solving, self-management, team-work, analytics, and the use of technology use. Table 9.2 gives insight into the skills that organisations consider to be core in their recruitment globally. It reveals that the top 5 sought-after skills are soft skills, and further classification of these shows that the top 10 skills are generally cognitive, self-efficacy, and technology skills. This underscores the role and importance of soft skills in tech-related and PropTech roles (more specifically).

Employers and recruiters in the real estate industry and PropTech sphere are also increasingly acknowledging the significance of innovation and technology in achieving operational and business success. Professionals in the field are also adapting by incorporating the technical skills necessary to enhance their expertise and proficiency in contemporary innovations and digital applications within the real estate domain. Thus, real estate students who can integrate their core knowledge in real estate along with other soft skills and digital technology are more likely to be distinguished and have enhanced employability prospects.

In the survey conducted by Oladiran and Nanda (2021), 99% of the respondents supported the inclusion of soft skills in PropTech education integration. Similar to the

Table 9.2 Top 20 Skills by Employers in 2023

Ranking	Skill
1	Analytical thinking
2	Creative thinking
3	Resilience, flexibility, and agility
4	Motivation and self-awareness
5	Curiosity and lifelong learning
6	Technology literacy
7	Dependability and attention to detail
8	Empathy and active listening
9	Leadership and social influence
10	Quality control
11	Systems thinking
12	Talent management
13	Service orientation and customer service
14	Resource management and operations
15	AI and big data
16	Reading, writing, and mathematics
17	Design and user experience
18	Multi-linguicism
19	Teaching and mentoring
20	Programming

Source: World Economic Forum (2023).

observed in Charts 9.1 and 9.2, the 1% of respondents that did not support the integration of soft skills with PropTech education are academics (Chart 9.6). This further indicates that educators may not fully appreciate the importance of soft skills.

Data and analytic skills are also essential for employability in PropTech. The vast majority of the respondents (88%) to the survey that accompanied the PEIF indicated that students (or individuals) without basic data and analytics skills are likely to struggle in real estate practice (Chart 9.7). Chart 9.8 further reveals that all real estate professionals and

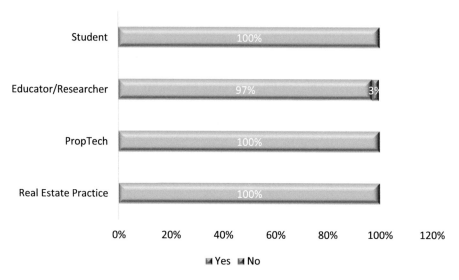

Chart 9.6 Should soft skills be incorporated in real estate higher education? (by respondents' roles/profession)

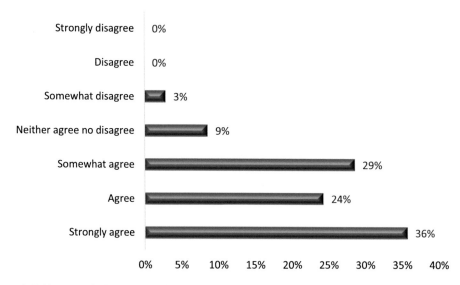

Chart 9.7 How much do you agree/disagree that real estate students who do not have the basic skills and appreciation for data and analytics are likely to struggle in professional practice?

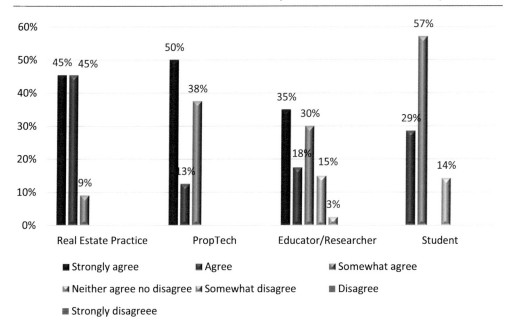

Chart 9.8 How much do you agree/disagree that real estate students who do not have the basic skills and appreciation for data and analytics are likely to struggle in professional practice? (by respondents' roles/profession)

PropTech practitioners agreed that as in previous cases, some academics (18%) did not feel that the lack of skills in data and analytics could jeopardise students' job prospects. It is also interesting to note a similar proportion of students as the academics are also not convinced that the lack of these skills can negatively affect their employability. These results demonstrate that educators and students may not completely understand the role of PropTech and other soft skills in enhancing real estate employability and professional practice in the near future.

9.2.2   Job Satisfaction and Remuneration

An article by LMRE and Fifth Wall (2022) provides valuable insight into the PropTech job setting and also provides an industry salary and employee experience benchmark. The survey was conducted in June 2022, and it included 500 PropTech professionals from a variety of backgrounds and settings such as world region, educational qualification, company size, seniority level, and workplace model.[4]

The LMRE and Fifth Wall (2022) report reveals that the majority of individuals who work in the PropTech feel valued and are happier than when they worked in other sectors, including working in other types of digital-technology-related companies. Overall, 73% feel they are valued, 78% liked their company's culture and 77% believe their company's leadership was innovative. This reflects a high job satisfaction rate, and although the data report does not provide an overall rate, the indicators mentioned (being valued, liking the company's culture, and believing that the company's leadership is innovative) are core components of general job satisfaction, and the data for these are generally consistent with the global job satisfaction rate which according to World Economic Forum (2020) was 74% in 2020. This

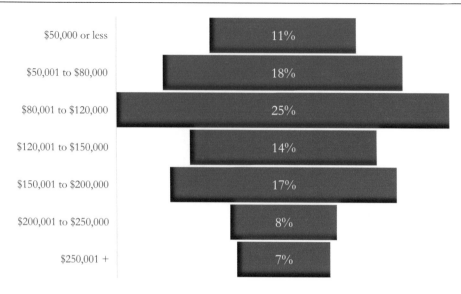

Chart 9.9 Salary range for PropTech employees (2022).

is particularly intriguing because the survey shows that the majority of PropTech employees are relatively new in the sector: 65% of respondents joined the industry less than five years before the survey, and only 9% had more than ten years of experience in the sector. Taking up employment in an unfamiliar sector, particularly when the sector is relatively new, emerging, and quite fluid, is theoretically supposed to be challenging; however, employees in this sector generally have high satisfaction levels. It would be valuable to study some of these issues and to understand some of the drivers of job satisfaction levels in the PropTech sector using empirical evidence. Some of these factors are briefly discussed further.

Remuneration is an essential part of job satisfaction and the (LMRE and Fifth Wall, 2022) report also provides insight into this. The report reveals that those employed in the PropTech sector are generally well paid (Chart 9.9), although this varies by job function and location.[5] The results showed that 33% of those surveyed earned over $150,000 annually and that proportion increased to 45% for directors. More than one-third of C-Suite respondents earn approximately $200,000 a year in contrast to the general data which showed that 15.5% of the respondents earn up to $200,000. Beyond the salaries, most employees in the PropTech sector earn a bonus. The data shows that 67% of the respondents had a bonus or commission with earnings ranging from $30,000–$50,000 a year. Some PropTech firms also provide equity offerings, and the data shows that 40% of respondents have an equity stake in their company and this increases to 46% for directors. Additionally, 92% of C-suites have stock options. The report revealed that alternative compensation options such as bonuses and equity offerings are usually provided to attract talents from other non-PropTech industries where salaries are higher such as banking and commercial real estate.

There are also country-level variations. For instance, on average, the US and Asia-Pacific companies pay higher salaries than their counterparts in Europe. This may be partly due to the higher amount of funding that these regions receive (as shown in Chapter 3. In the US, salaries have increased by 20%–30%, and with real estate investment firms also reeling out their tech and innovation branches, experienced real estate professionals are now being attracted to this sector and they may need to be paid sufficiently to match their experience and the pay in their previous employment for their move to be worth it. Singapore is also a growing PropTech market in Asia-Pacific with the growth expected to increase by 30% by 2025. This may be driven by Singapore's net-zero economy goal and the associated decarbonisation agenda creating an ecosystem for PropTech companies.

Although remuneration and benefits are some of the key drivers of employee satisfaction in the PropTech sector, it is not the only factor. Employees in the survey reported that they like their work locations and general environment, and they also believe in company leadership. The data showed that 59% of employees work hybrid while 31% are remote with only 10% totally in the office. The respondents attributed this to the high investment of the PropTech companies in their workspaces and investment in the creation of spaces where their employees would want to work.

PropTech companies however face challenges. One is the high job mobility in the sector: 52% of the respondents indicated that they were actively searching for new jobs, although most of them were seeking higher remuneration, better benefits, and often more freedom to define their work patterns in terms of location and time. Some of them also revealed that although the pay may be attractive, the benefits may not be as attractive. These include healthcare and retirement packages, raising the bar beyond what emerging companies have to offer to attract and retain a highly skilled and talented labour force. PropTech start-ups are therefore advised to consider these factors as part of their recruitment strategies in order for them to attract and retain a highly skilled, talented, and motivated workforce.

9.3   Professional Bodies, Institutions, and Associations

Professional bodies have played crucial roles in strengthening and supporting the growth of disciplines and sectors. Some of these are summarised in Figure 9.9.

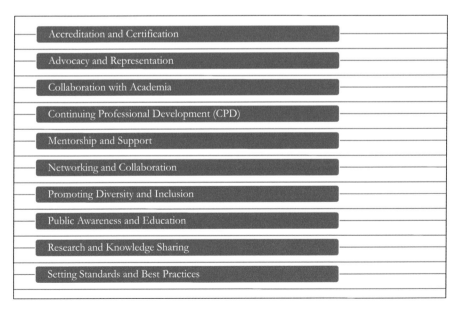

Figure 9.9 **Roles of professional bodies institutions and associations in supporting the growth of sectors and disciplines.**

As shown in Figure 9.9, professional bodies and institutions provide accreditation and certification programmes that validate the skills and knowledge of professionals in a field or sector, and this can boost the credibility of practitioners within the sector. They can also advocate for the interests and needs of a sector to policymakers and industry stakeholders, particularly for emerging disciplines or sectors. They can serve as a collective voice of professionals and promote policies that support growth and development. Professional bodies also facilitate the collaboration of industry with academia. They often achieve this through their work with academic institutions to develop relevant curricula, research collaborations, and opportunities for students to gain practical experience in the developing discipline or sector. Furthermore, they offer continued professional development programmes that can help professionals to stay abreast with the latest development, trends, technologies, and innovations.

Professional bodies also provide mentoring support through programmes aimed at early-career practitioners within an area of discipline. This guidance can accelerate their growth and success within a sector. They can also facilitate networking opportunities, enabling professionals to connect, share ideas, and collaborate. This fosters innovation and the exchange of best practices. Professional bodies also serve as advocates for diversity and inclusion within a discipline or sector, promoting a more equitable and representative workforce. They also raise public awareness about the importance and value of an emerging discipline or sector, and this can attract new talent and investment to the field. Furthermore, they encourage research and knowledge-sharing initiatives that contribute to the advancement of a discipline or sector, and this can generate new insights and solutions. Additionally, professional bodies can establish and promote standards and best practices within a sector, helping to maintain consistency and quality across the industry, and building trust among stakeholders.

This section therefore provides a review of the various professional bodies, institutions, and organisations that are associated with real estate innovation and digital technology, highlighting their roles so far and their prospects. Furthermore, the section contains an overview of events, engagement, and activities that have further enhanced the growth and development of PropTech and real estate innovations. To deepen the insight into the roles and contributions of professional bodies and associations in driving the growth of PropTech, some stakeholders in these were identified using purposive and snowballing sampling techniques, and interviews were conducted. While various parts of this section were inspired by the views and perspectives of the respondents, it should be noted that the content and inferences drawn by the authors do not necessarily reflect their views.

9.3.1 The Royal Institution of Chartered Surveyors (RICS): Innovation and Digitisation in the Built Environment

The RICS dates back to 1792 when the Surveyors Club was formed with a membership of 20 surveyors. The Royal Institution of Chartered Surveyors as it is now known, aims to promote and enforce the professional qualifications and standards in the development and management of land, real estate, construction, and infrastructure (RICS, 2023; Royal Institution of Chartered Surveyors, 2022). With a vast network of more than 134,000 skilled trainees and professionals and offices in key financial markets, the RICS is strategically positioned to impact policy and integrate standards into local market environments, safeguarding the interests of consumers and businesses.

Much of the PropTech evolution has taken place outside the remit of the RICS. For several years, several digital tools, platforms, and software were developed beyond the traditional real estate sector, and by extension, professional bodies such as the RICS, and although it has been increasingly acknowledged that innovation is constantly needed in AEC (Architecture, Engineering and Construction) (see Cao et al., 2015), the RICS did not do enough to promote and drive the innovation and changes. This narrative however began to change from around the mid-2010s. The following is an outline of some key RICS initiatives that have been aimed at promoting data, innovation, and digitisation in the built environment in the last six years:

- **2018 – Pathways and competencies review**: In August 2018, the RICS introduced new pathways and competencies relating to Prop Tech, such as big data, open data, smart cities, and intelligent buildings.
- **2019 – RICS-accreditation**: On 1 January 2019, the RICS published a new global accreditation policy which prescribed that when developing a course for accreditation, the panel would assess ten areas, including data management. Furthermore, some pathways now include some advanced digital technologies. For instance, the Land and Resources pathway includes Smart Cities and Intelligent Buildings as an optional competency.
- **2020 – RICS reports, programmes, and standards**: Following extensive consultation, the 2020 Futures Report – a follow-up to the first Futures Report in 2015 was published (RICS, 2020b). This again highlighted the role of data and technology in the built environment. The RICS Tech Partner Programme was also set up to serve as a collaborative platform for innovative data and technology across the globe producing thought leadership, content, and market insight for the profession (see RICS, 2021b). The RICS Data Standards (see RICS, 2020a) were also developed in 2020.
- **2021 RICS initiatives**: In February 2021, RICS' Governing Council initiated a review of their strategic direction and priorities. This started with Defining our Future (see RICS, 2021a) and an extensive programme of consultation and engagement.

The authors sought further insight on the RICS' digital technology and innovation agenda through an open-question interview with Andrew Knight – the Global Data and Tech Lead at the RICS.[6] Some of the key insights from the interview are:

- Education and qualifications should reflect an understanding that data and technology are important in real estate practice and service delivery; skill set requirements are now beginning to align with these areas.
- There is a growing interest in additional technical areas such as big data, and understanding of data and technology – this is also increasingly being valued by large property firms.
- The RICS is actively developing strategies to improve real estate operations and services through active integration of data and digital technology. In education, they are also engaging in conversations around this integration, although the outcome of this initiative is still unknown.
- There is an increase in the drive for diversity, and inclusion within the profession is growing in terms of gender, demographics, etc. To drive this, there are alternatives to

the more traditional university route into the built environment profession through apprenticeship and apprenticeship degrees and vocational courses.

- It is important that PropTech enthusiasts understand and appreciate the built and natural environment as the baseline for what they do. They should also understand the key domain knowledge from an educational, and training point of view with respect to what they need to do.
- Those who are coming from a purely data and tech perspective need to have the domain knowledge (or seek to have this area covered) due to the complexity, idiosyncrasy, and heterogeneity of the real estate asset class and markets.

When probed about RICS' agenda on PropTech education, soft skills, and their integration into real estate education, Andrew Knight states that the RICS is cognisant of the increasing focus, particularly of employers in the real estate industry to have soft skills, and this predates the prominence of PropTech. He also points out that the competency framework has attempted to address this in terms, so the mandatory skills that are required for professional registration. There may however be the need to integrate areas such as creativity and innovation which the current assessment framework currently does not explicitly account for. Furthermore, the RICS should also consider accounting for interrelationships as soft skills although this may be difficult to measure and may be complex from an EDI and cultural perspective as personality types differ. The RICS is also actively engaging in consultation on data, automation, and ESG, as part of the ongoing review by their EQS (Education, Qualification and Standards) team although the outcome of this conversation is unclear. It should however be noted that at the moment, the RICS does not intend to mandate accredited real estate courses to actively take on PropTech or tech-related curriculum review as a perquisite for accreditation. Although no reason is provided for this, it may be a result of some of the challenges of PropTech education integration listed in Section 9.1.6.

When probed about the fears that automation will reduce the jobs available in the built environment, Andrew stated:

> I genuinely believe this data and tech is augmenting what surveyors do, it's allowing them to almost focus on higher value skills, and avoid some of the kind of lower tech as it were more manual tasks, like many pieces of technology.

The roles, tasks, and functions of professionals are likely to subtly change as automation becomes more widespread, and this will enable professionals to engage in higher-value tasks. For instance, the sophisticated data and analytics provided by several platforms (particularly as demonstrated in Chapters 4–7) still require human interpretation for land acquisition, planning purposes, investment, asset management, etc. These need to be consolidated, analysed, and used strategically in a way that can support decision-making and efficiency. This is indeed a valid point. It should however be noted that automation may already be replacing the roles of professionals in some cases. For instance, automated valuation models are replacing the thousands of valuations that would traditionally have been conducted manually by professionals. As stated in Chapter 7, this is particularly dominant in mortgage origination and mortgaging. However, there are still several roles that valuers still play despite the increasing adoption of AVMs because of the limited scope of application by AVMs or the need for some condition reports that automated systems are currently unable to perform.

Another example that can be cited in response to concerns that automation can cause job losses and redundancies in the real estate industry is related to the use of online property marking and listing platforms. There may have been concerns that these online platforms would replace real estate brokers and agents in the property search and selection process, but it does not appear that this has been the case. Rather, platforms such as Rightmove and Zoopla have made property search and selection a lot more efficient for agents/landlords and prospective tenants. Rather than take away the jobs of the local agents or brokers, these platforms have provided them with a wider reach and a more efficient connection to prospective tenants and provided the prospective tenants with a wider range of information (as highlighted in Chapter 8). Additionally, in many cases, when individuals who are novices to the real estate sector want to buy or rent a property, particularly if it is in a market that they are unfamiliar with, they still value contact with local agents.

It can be argued that advancements in AI can significantly cause the human element of real estate to be diminished. Indeed, AI development is likely to be key to the further transformation of the real estate sector (this will be discussed further in Chapter 10); however, AI still relies on human support in various forms. AI is only as intelligent as the training it receives, and this training is based on several methods, approaches, and data which may have limitations. The human element is therefore important in training an AI-enabled system. For instance, in setting up an automated valuation model, the model has to be trained to identify the correct variables and their weighting, and if this is not done by a professional or not well thought through, it can lead to inaccuracies and inconsistencies. The training that an AI system has been given can create and perpetuate biases and inaccuracies, and they will therefore need to be evaluated and retained based on expertise.

The central message from the interview with Andrew Knight is that the RICS is not oblivious of the changes that are occurring in the sector and there have been concrete steps taken to support data and technology transformation. Through the competencies, potential pathways, and other initiatives, such as having dedicated job roles for data and technology development within the profession, the RICS will continue to promote this agenda in the future.

Another area that the RICS should also consider is changing the data management mandatory competence level from the current level 1 to level 2 as a way of motivating professionals to seek further knowledge. An advanced and more effective approach may be the introduction of PropTech-related pathways or competencies as part of RICS qualifications. These are essentially bundles of competencies. For instance, one of the pathways is commercial real estate, and within this, there are several competencies that APC candidates are assessed against to gain membership. These competencies are measured on three levels (1, 2, 3) including technical and soft skills. In the commercial real estate pathway, for instance, there are specific competencies that would be assessed. Of the two approaches (pathway-based and competencies-based), the latter may be more effective as new competencies can be added to existing pathways with less disruption, and individuals who have interest and experiences to evidence within the existing pathway can do so. This way, many aspiring professionals or technology enthusiasts can pursue their interest in technology in the built environment without having to narrow down their scope of interest.

9.3.2 *PropTech Country and Regional Associations*

The wave of innovation and digitisation in real estate has been accompanied by the development of several country-level PropTech associations, particularly in the last half-decade. These associations support members in their network through industry insights, networking, and collaboration opportunities. It should be noted that these organisations are set up to run as associations rather than professional bodies; thus, they do not provide regulations, standards, or accreditations. This sub-section provides a list of some PropTech associations and specifically explores the UK PropTech Association, its roles, and potential for further development and contribution in driving real estate innovation and digitisation.

A list of some of the world's PropTech Associations is provided below (Table 9.3).[7] There are several countries that currently do not have active associations and organisations that facilitate networking, collaboration, and support; PropTech enthusiasts in these countries are therefore encouraged to establish their associations.

9.3.2.1 *The UK PropTech Association*

The UK PropTech Association, established in 2018, is a non-profit membership organisation that aims to drive digital transformation in the property sector by facilitating collaboration between PropTech and real estate businesses and companies. It aims to create

Table 9.3 List of Some PropTech Associations/Organisations

Countries	Associations
Australia	PropTech Association Australia
Belgium	PropTech Lab Belgium
Bulgaria	PropTech Bulgaria
Canada	PropTech Canada
Switzerland	Swiss PropTech
Denmark	PropTech Denmark
Finland	PropTech Finland
France	French PropTech
Ireland	PropTech Ireland
Italy	Italian PropTech Network
Japan	PropTech Japan
Kenya	PropTech Kenya
Malaysia	Malaysia PropTech Association
Nigeria	Nigeria PropTech Association
Netherlands	Holland ConTech and PropTech
Norway	PropTech Norway
Poland	PropTech Poland
Romania	PropTech Romania
Russian Federation	PropTech Russia
Singapore	PropTech Association Singapore
Slovak Republic	Czech & Slovak PropTech Association
South Africa	SA PropTech
Spain	PropTech in Spain
Sweden	PropTech Sweden
Uganda	PropTech Uganda
United Kingdom	UK PropTech Association

Source: Authors.

a conducive environment that can enhance real estate innovation by leveraging digital technologies. The UKPA effectively advocates for and represents the interests of diverse stakeholders in the PropTech ecosystem such as investors, venture capitalists, technological experts, start-ups, consultants, property firms, and professional service providers. The association maintains robust connections and networks with governmental agencies, regulatory bodies, and prominent industry leaders, thereby affording its members the opportunity to collectively address shared challenges while facilitating growth and advancement within the sector. The UKPA membership may be individual or corporate (company).

The UKPA has four core objectives (The UK PropTech Association, 2023):

1 To provide opportunities for engagement with property professionals and technology investors.
2 To represent the interest of the industry in regard to policy and legislation and industry-wide issues and challenges.
3 To provide an ecosystem of support for PropTech businesses through partnerships with professional service providers.
4 To promote and support the whole of the UK as the best place to start or grow property technology businesses.

The UKPA hosts and organises several events including an annual awards event from 2019 and they also provide market insights that are aimed at supporting their members to access information and networks for their development.

The authors sought further insight into the UKPA's thoughts on the development of PropTech in the UK and its future trajectory. Sammy Pahal, the Director General of the UKPA, highlights three key areas (three S) where real estate companies are prioritising their focus in PropTech:

- **Smart**: This involves integrating the appropriate technology to enhance health, well-being, safety, and automation.
- **Sustainable**: Companies are looking at environmental sustainability across the entire property lifecycle, from construction to materials and management.
- **Services**: Providing more flexibility and better user experiences by balancing technology and human interaction.

Furthermore, Sammy Pahal reiterated the UKPA's agenda of facilitating engagement between PropTech, real estate companies, and technology investors, providing support services, representing members in policy and regulations, and promoting the growth of the UK PropTech industry through education and partnerships. Sammy stated:

> …there's been a greater focus towards integrations and partnerships….. companies are realising that they need to be able to integrate with other technology providers and mainstream technologies such as the CRMs, the finance systems, the systems within the properties themselves, the BMS systems….so it is now easier to implement these technologies for real estate companies.

She, however, highlights that for the benefits of this growing integration to be fully maximised, the data flow across various data systems needs to be improved.

The UKPA's current drive is centred around the following activities and strategies:

Collaboration, integration, and partnerships: This is centred around an increasing focus on the PropTech industry. Companies are realising the need to collaborate with other technology providers and mainstream tech, like CRMs and finance systems, to implement technology effectively and efficiently. The UK PropTech Association collaborates extensively with other industry bodies, both nationally and internationally. They partner with organisations such as the British Property Federation and the Investment Property Forum to address challenges and provide opportunities for real estate companies.

Data and analytics: PropTech companies are emphasising the importance of granular data and insights for their customers. The UKPA acknowledges this and aims to facilitate the provision of data to real estate companies with comprehensive information about their portfolios, enabling better decision-making and understanding of opportunities and risks. They also acknowledge the value of AI-enabled solutions in real estate and the integration of these AI-enabled solutions in real estate operations and services.

Public sector engagement and collaboration: The UKPA is supporting the UK's digital transformation reforms and also working with the government to promote the use of the unit property reference number (UPRN). They also aim to coordinate and contribute to the PropTech policy development in the UK.

Education and skills development: The UKPA acknowledges that addressing the talent and skills aspect of the PropTech industry is important. They are therefore looking to coordinate internships and educational opportunities for students and graduates to encourage their involvement in PropTech and real estate innovation.

Overall, the interview with the UK PropTech Association's Director General highlights the importance of smart technology, environmental sustainability, customisation, collaboration, data, analytics, public sector collaboration, education and skills development, and the association's role in facilitating collaboration, addressing challenges, and supporting the growth of PropTech companies in the UK.

In April 2023, the UKPA announced its intention to merge with the British Property Federation to drive innovation in the property sector. As of September 2023, the nature and other details of this merger were not yet available.

As demonstrated in this subsection, professional bodies and associations are essential parties and partners in the PropTech evolution and the enhancement of innovation and digitisation in real estate. It would therefore be valuable for more professional bodies (globally, regionally, and at the country level) to make more active contributions to these emerging areas, and it would be even more valuable for PropTech Associations to be set-up in countries where they currently do not exist. These associations and organisations can partner with educational institutions and corporate bodies to lead or support debates, competitions, and other education-related activities that can increase students. However, it is important to ensure that these professional bodies, associations, and organisations do not become gatekeepers as this can significantly undermine the rate of innovation and creativity in the industry.

9.4   Standards, Regulations, and Ethics

As highlighted, all through this book, PropTech is synonymous with the use and production of data. Chapter 2 specifically reveals that a high volume is generated every year, and this is accelerating even more with an increase in the use of IoT devices. Furthermore, the data produced in and by the built environment is also accelerating at a similar pace with an increasing need for data sharing and collaboration throughout the real estate value chain with systems such as smart cities and BIM; it is, therefore, important that these data are sourced, stored, analysed, and utilised responsibly and ethically. It is also important to ensure that as the property sector derives benefits from increased use and access to data, there are also systems in place that can ensure that some of the risks that this evolution presents are minimised through the better efficiency of connectivity of data and ethical use and other activities (RED Foundation, 2020).

There are several debates around the standardisation of data. As stated in Chapter 2, it is generated and stored in various forms, and as it is increasingly used across international borders, the varying nature of its generation, use, and storage may be limiting its usability and value. It is often argued that although a large volume of data is currently being generated in the built environment, a lot of this is "siloed". Keeping data in "silos" refers to the practice of storing and managing data in isolated and separate systems or repositories within an organisation. In this context, "silos" represent individual and distinct data storage structures that are not integrated or interconnected with other systems.

Data silos can arise as a result of departments or teams within an organisation using their different management systems or databases that are not designed to communicate with each other. Consequently, it may not be easily accessible or shareable with other parts of the organisation, leading to inefficiencies and hindering the ability to gain valuable insights. These inefficiencies and challenges include

- **Reduced data accessibility**: When data are confined to specific silos, it becomes difficult for different teams or departments to access and utilise the information effectively.
- **Limited data insights**: Data stored in silos may not be combined or analysed holistically and this can limit an organisation's ability to gain comprehensive insights from the data.
- **Inconsistent data quality**: The lack of centralised control and standardised data management practices can result in varying quality across different silos.
- **Data duplication**: Different silos may store redundant or overlapping data and this may lead to unnecessary duplication and potentially conflicting information.
- **Compliance and security risks**: Managing data in isolated silos can raise challenges in terms of data protection, security, and regulatory compliance.

Many companies and organisations are in possession of data that dates to several decades ago. By organising these in a structured form, they can be used to make better decisions. Furthermore, breaking down the silos and promoting data integration and sharing across organisations within a sector can lead to improving collaboration, decision-making, and overall sector efficiency.

There are also several concerns relating to data protection. The General Data Protection Regulation (GDPR) is one of the responses of the European Union to concerns about data protection and regulation. It was developed and adopted in April 2016, and it became enforceable in May 2018 and therefore became prominent across Europe by the end of the last decade. The GDPR has seven core principles that underpin data protection and privacy rights for individuals within the EU:

1 **Lawfulness, fairness, and transparency**: The processing of personal data must be done lawfully, fairly, and with transparency. Data subjects should be informed about the collection and use of their data in a clear and easily understandable manner.
2 **Purpose limitation**: Personal data should be collected for specified, explicit, and legitimate purposes, and they should not be processed in ways that are incompatible with those purposes.
3 **Data minimisation**: Data controllers should only collect and retain personal data that are necessary for the intended purposes. The data collected should be relevant, adequate, and limited to what is required.
4 **Accuracy**: Personal data should be accurate and kept up to date. Inaccurate data should be rectified or erased without delay.
5 **Storage limitation**: Personal data should not be stored for longer than necessary for specified purposes. Data should be retained in a form that allows identification of individuals only for the necessary duration.
6 **Integrity and confidentiality**: Personal data should be processed in a manner that ensures its security, including protection against unauthorised or unlawful processing, loss, destruction, or damage.
7 **Accountability**: Data controllers are responsible for complying with the principles of the GDPR and must demonstrate compliance by maintaining records of data processing activities and implementing appropriate measures to protect personal data.

These principles can serve as a baseline for data use, storage, and analysis when real estate and PropTech tools are being conceptualised, and they can also be used by firms in handling, using, storing, and analysing data.

Several standards have been set in some areas of the built environment such as International Property Measurement Standards, International Cost Management Standards, International Land Measurement Standards, and International Valuation Standards set to provide trust, and ethics and minimise risks. The RED Foundation is one organisations focused on addressing the data-associated concerns in real estate, particularly in light of the recent digitisation in the sector. RED Foundation (2020) suggests that professional bodies, industry groups, and market leaders should develop data standards particularly, as digitisation increases. The report points out that although some standards exist in relation to the creation, use, and management of data, the application of these standards to the emerging PropTech realm may be limited. It further calls for improved discoverability and connectivity which will ensure that the roles of actors in the sector are clarified in the data use. The RICS Tech Partner Programme is however a valuable development as it provides a collaboration platform for innovative data and technology across the globe with the potential to provide thought leadership, content, and market insight for the real estate sector.

9.5   Public Sector Participation

One of the most recent and significant developments in the real estate sphere is the increasing interest and participation of the public sector. At the 2023 London PropTech Show, the level of participation from the public sector was interestingly higher (2.6%) than the attendance by those in academia (2.2%) (London PropTech Show, 2023). Although this does not constitute a significant proportion of the participants, it is notable, since the public sector has had very limited (or almost non-existent) involvement in PropTech and digitisation in the built environment until very recently. Quite remarkably, the 2023 London PropTech show had in attendance the Honourable Minister of Public Works and Infrastructure of South Africa (Honourable Patricia De Lille), the Deputy Minister for Regional Development of the Czech Republic (Honourable Radim Srsen), and the Head of Digital Citizen Engagement at the UK Department for Levelling Up, Housing and Communities (Bridget Wilkins). These public officers also made presentations, highlighting their PropTech initiatives and prospects. These demonstrate the interest of the public sector in supporting the drive towards the digitisation of the built environment.

The authors further sought insight on the UK's public sector PropTech strategy and its future trajectory. Recently, the UK Department for Levelling Up, Housing, and Communities launched a PropTech Innovation Fund to help drive the adoption of digital tools by local authorities and showcase the value that new solutions can bring. They funded 64 pilots in the first 19 months, have had over 600 local authority members attend their events, and have enabled insights from a range of technology entrepreneurs to be fed directly into policymakers. Jessica Williamson, the PropTech Strategist at the UK Department for Levelling Up, Housing and Communities[8] and Bridget Wilkins, the Head of Digital Citizen Engagement in the same department[9] provided some valuable insight. Both respondents highlight the importance of the public sector involvement in PropTech and the impact opportunities the roles provide.

Jessica Williamson stated:

> I've really appreciated the chance to participate in the ecosystem from a new perspective…. Highlights have included launching a fund, organising Ministerial Roundtables, and being able to create new roles for fabulous team members to expand our impact.

Bridget Wilkinson further commended:

> …we are already seeing how adopting digital tools can enable local authorities to generate an increase in the quantity and quality of engagement, tap into more positive community sentiment towards change and create time efficiencies for planners.

Based on these insights, three core aspects of public sector participation in PropTech and real estate innovation in the UK are further discussed: goals and strategies, examples and future prospects.

9.5.1   Goals and Strategies

The use of digital technology, particularly PropTech, within the UK public sector is underpinned by a set of multifaceted goals that encompass diverse aspects of governance, policymaking, infrastructure, and societal inclusivity. Through a content and thematic analysis of the stated goals, several key issues emerged, each carrying significant implications for the broader integration of digital technology and innovation with the built environment.

Strategies around engagement are an important starting point. The ambition to expand innovative approaches and engagement strategies beyond planning into areas such as social housing, infrastructure delivery, and regional development reflects a systemic shift toward a more holistic and inclusive approach to urban development. This goal reflects the recognition of the potential of PropTech and innovation to enhance the planning process and address broader societal challenges. This implies that PropTech has the capacity to transcend its immediate applications and contribute to the overarching policy objectives of equitable urban development and social well-being (as discussed in Chapter 4).

Engaging with innovators and leveraging the UK PropTech sector aligns with the broader trend of public-private partnerships in advancing technological adoption within governmental contexts. This strategy acknowledges the private sector's agility and capacity for innovation and underscores the importance of collaboration between governmental entities and industry stakeholders. This is important because the public sector can harness the momentum of private sector advancements to enhance its services and fulfil its mandate more efficiently, while simultaneously driving technological innovation that aligns with policy objectives.

The assertion that innovative approaches can inform various departmental aims underscores the versatility of PropTech solutions. The integration of innovative technologies and methodologies across domains ranging from community engagement to sustainability implies the potential for a unified and interconnected digital ecosystem that addresses diverse policy areas. This interconnection suggests a paradigm shift from isolated sector-specific initiatives to more integrated policymaking and service delivery systems.

The government's role in fostering relationships with the private sector resonates with the imperative to bridge the gap between technological expertise and market education. This approach not only facilitates the growth of existing PropTech companies but also aligns with broader objectives of capacity-building and market development. The implication is that the public sector's involvement can serve as a catalyst for a well-coordinated and harmonious growth trajectory within the PropTech sector.

The emphasis on unlocking and utilising data holds profound implications for the evolving landscape of data-driven decision-making in governance. The respondents acknowledged that data access and sharing reflect a commitment to open data principles, and this can stimulate innovation, transparency, and collaboration. This objective intersects with wider ambitions of digital transformation within the public sector and sets the stage for transformative advancements that go beyond PropTech applications. Furthermore, recognising the diversity of goals within different parts of the public sector underscores the need for tailored strategies and contextual understanding. This acknowledgement reflects the nuanced nature of public governance and the inherent variability of technological needs across departments. This further reflects the successful integration of PropTech which has unique challenges, objectives, and applications in each sector.

The articulated goals for PropTech utilisation in the UK public sector demonstrate a strategic and comprehensive approach to digital innovation. These goals signify a departure from siloed applications to a more integrated and holistic approach that spans multiple policy domains. The implications of these goals extend beyond immediate technological adoption, touching upon aspects of inclusivity, collaboration, data-driven governance, and sector-specific adaptation. Embracing PropTech as a tool for societal betterment can potentially empower parts of the UK public sector, particularly the built environment to become more transparent, inclusive, and efficient.

9.5.2   Programmes

9.5.2.1   PropTech Engagement Fund

The deployment of digital technology, particularly the PropTech Engagement Fund (launched in August 2021), within the UK public sector, exemplifies a dynamic and multifaceted approach to leveraging digital engagement tools for enhancing the planning process and fostering democratic accountability. The Planning for the Future White Paper sets forth a transformative vision to amplify the utilisation of digital engagement in the planning process, aiming to augment accountability and democratic inclusivity by offering alternative avenues for public participation. This aspiration underscores the evolving nature of public engagement, signalling a departure from traditional approaches towards more accessible, participatory, and technology-enabled forms of civic involvement in the planning process. The implication here is that digital tools have the potential to reshape the very foundations of the planning process, making it more responsive and inclusive.

The launch of the PropTech Engagement Fund also reflects a proactive stance by the UK government to drive the widespread adoption of digital citizen engagement tools within the planning domain. This council-led approach not only emphasises collaborative partnerships but also aims to uncover the barriers inhibiting local authorities from embracing digital engagement at scale. The focus on understanding these barriers hints at a commitment to address implementation challenges and create an environment conducive to technology integration.

DLUHC (Department for Levelling Up, Housing and Communities)'s role as a market facilitator and collaborator between demand (Local Planning Authorities) and supply (the digital citizen engagement sector) highlights the government's pivotal role in catalysing a symbiotic relationship between technology providers and regulatory bodies. By actively engaging with both sides of the equation, DLUHC seeks to ascertain the viability and sustainability of digital engagement tools, while simultaneously challenging established assumptions and stimulating innovation in this domain. The government's role as a catalyst implies a dynamic and active approach to shaping the trajectory of technology adoption within the public sector.

The noteworthy accomplishment of collaborating with 45 Local Authorities underscores the scale and impact of the PropTech Engagement Fund, positioning it as the largest UK Government PropTech Programme to date. This achievement signals the potential for cross-sector collaboration and serves as a model for how governments can effectively work with industry players and local government entities to leverage technology for public benefit. The implication here is that successful collaborations can drive transformative change that reverberates beyond the national level, potentially inspiring similar initiatives globally. The international recognition from the Organisation for Economic Co-operation and Development (OECD) and inclusion in their report on Embracing Innovation in Government: Global Trends 2023 amplifies the global significance of this initiative. This recognition showcases how an effectively designed and executed PropTech initiative can resonate with global trends and foster innovation that transcends regional boundaries. This reinforces the notion that innovative digital solutions within the public sector have the capacity to set benchmarks and influence practices on a global scale.

The outcomes of these initiatives underscore the potential of digital engagement tools to democratise and enhance the planning process. The increase in community sequel to the launch of these initiatives signifies a shift towards a more participatory and citizen-centric approach, aligning with contemporary aspirations for transparent governance. The implication here is that digital engagement can bridge the gap between citizens and policymakers, fostering a sense of ownership and empowerment in the decision-making process.

The PropTech Engagement Fund exemplifies a comprehensive and strategic approach to integrating digital technology within the public sector, specifically in the realm of planning. The initiatives highlighted within this example encompass elements of collaboration, innovation, transparency, and inclusive governance. The broader implications extend to fostering a culture of technological innovation within the government, promoting global recognition for successful initiatives, and fundamentally transforming the nature of citizen participation in governance. This example serves as a blueprint for how digital technology can be harnessed to shape a more responsive, democratic, and technology-enabled public sector.

9.5.2.2 DLUHC Housing Delivery Pilot with Urban Intelligence

Land supply and availability are essential aspects of housing delivery, and this is well recognised by the DLUHC. It is important for them to have the knowledge and sufficient information on where land is available, the ownership structure and the actual (and potential) utility that are inherent in the land. This knowledge and insight can aid the central government and local authorities in making more strategic decisions about land use to support growth and development.

The department ran a pilot to map public sector land use across Oxford-Cambridge (Ox-Cam) and the West Midlands. This pilot worked with land-owning departments and Urban Intelligence to pilot a digital tool for site assessment that could be used to identify suitable land for the development of new homes. The initiative provided the DLUHC with a useful platform to understand how PropTech can be used to support policy development and existing programmes. Mapping available land, understanding ownership, and assessing land utilisation aligns with the broader trend of data-driven decision-making within the public sector. The potential of technology to consolidate and analyse diverse data sources offers a means to enhance the accuracy and efficiency of strategic land use decisions.

The emphasis on understanding how PropTech can support policy development and existing programs within the department highlights the strategic role of technology in shaping governance practices. By utilising the insights generated from the pilot, the government is positioning itself to optimise its policies and programs. This implies that technology can be utilised as an enabler for evidence-based policy development, facilitating a dynamic and adaptive approach to addressing complex issues. It serves as a practical example of how PropTech can be applied to real-world challenges, demonstrating the feasibility of harnessing technology to enhance policy implementation and decision-making.

9.5.2.3 Events and Public Engagement

One of the aspects of the UK public sector is by raising awareness through events and public engagement to promote the adoption of digital technology in the built environment. Examples of such events are the PropTech Roundtable and Public Sector PropTech Summit (both in 2022). These sessions brought together entrepreneurs, investors, and industry leaders to enable the PropTech sector to better understand the UK government's levelling up agenda and discuss areas of support and impact. This reflects the strategy to foster knowledge exchange between different stakeholders within the PropTech ecosystem. By bringing together entrepreneurs, investors, industry leaders, and government officials, these events facilitate a convergence of insights, experiences, and solutions.

The DLUHC has also partnered with TechUK and the UK PropTech Association on industry events to support the PropTech ecosystem, including a Digital Planning and Citizen Engagement Innovation Showcase that brought together several digital engagement companies (approximately 650 attendees drawn from several organisations such as Homes England, Planning Advisory Service, New London Architecture and the Royal Town Planning Institute). This served as a platform for showcasing emerging technologies, fostering networking opportunities, and addressing industry challenges. Collaborative events such as these can lead to the identification of solutions to barriers, ultimately driving the growth and development of the PropTech sector.

At the time of this interview, the respondents had spoken at over 30 industry events in the 12 months prior, advocating for the adoption of digital citizen engagement tools, highlighting the government's commitment to a comprehensive and sustained approach to raising awareness. This advocacy extends beyond policymaking, contributing to the broader discourse around digital transformation in governance. Digital toolkits and best practice guidance have also been developed and they will continue to test with the industry to ensure it meets user needs. This active role in knowledge sharing positions the public sector as a thought leader, contributing to the ongoing evolution of digital toolkits and best practice guidelines.

9.5.3   Next Steps

The provided text outlines the UK public sector's plans and further development.

1 **Evolution of priorities and sector engagement**: The UK public sector has shifted its focus within the PropTech sector over the years. Starting with housing market productivity, moving on to housing market recovery, and later digital planning, this evolution signifies a responsive approach to addressing sector-specific challenges. This adaptability reflects an awareness of the dynamic nature of the property industry and the potential benefits of incorporating technological solutions. By aligning priorities with changing demands, the public sector can leverage digital technology to address pressing issues in the housing and planning sectors. This flexibility fosters an environment of innovation and responsiveness to emerging challenges.
2 **Learning from other industries**: Other industries such as finance and retail have successfully integrated digital innovation into their processes and there is a need for the property sector to catch up. The reference to open data and its transformative role in driving new services underscores the potential of data-driven approaches in enhancing service delivery. By drawing inspiration from industries and sectors that have embraced digital transformation, the property sector can also leverage technological tools to streamline operations and provide user-friendly services.
3 **Barriers to adoption and data access**: There are still barriers to the adoption of digital technology such as data access and other central and local government bottlenecks. Addressing some of these barriers can encourage both public and private sectors to

embrace innovative solutions and stimulate a more inclusive PropTech ecosystem, fostering collaboration between the public and private sectors, and potentially leading to more efficient and effective services.

4 **Government's role and PropTech advocacy**: There is still scope for the central government to raise awareness and support the drive for enhanced digital technology as this can stimulate investor interest, drive innovation, and enhance public trust in PropTech solutions, ultimately accelerating the sector's growth.

5 **Productivity and positive outcomes**: There are still opportunities for PropTech to further drive productivity, minimise risk, and improve outcomes for citizens and communities with enormous environmental, social, economic, and physical benefits. These can be achieved through improved investment in PropTech and innovation in the built environment.

In conclusion, the UK public sector's evolving approach to PropTech reflects a willingness to adapt, innovate, and embrace technology in the housing and planning sectors. Anecdotal evidence suggests that several other countries around the world are also adopting various strategies to develop the use and application of PropTech and enhance innovation in the built environment. Although these initiatives are laudable, with the potential for further benefits, mitigation measures and protocols should be in place to ensure that the public sector involvement does not evolve to the level where they become gatekeepers as this can sniffle creativity and innovation.

9.6   Chapter Summary

This chapter has provided a comprehensive analysis of the various aspects of PropTech and their impact on the built environment. By examining the enablers, stakeholders, and systems that have fostered the rise of PropTech, this chapter has shed light on the crucial factors that contribute to its continued growth and success and highlighted the potential for further advancements in innovation and digitisation within the real estate sector. Furthermore, the emphasis on calls to action from diverse stakeholder groups underscores the importance of collaborative efforts and a unified vision for the future of PropTech. The chapter further acknowledges the significance of fostering better synergy among stakeholders and systems, this chapter emphasises the need for cohesive and strategic cooperation to harness the full potential of real estate digital technologies and solutions.

The inclusion of primary data collected through semi-structured questionnaires adds depth and richness to the analysis and discussion, enhancing the understanding of the issues being discussed.

Overall, this chapter serves as an invaluable resource for a wide range of professionals and stakeholders in the built environment. Educators will find valuable insights to inform their curriculum and pedagogy, while policymakers and regulators can gain valuable guidance to shape effective strategies and frameworks. Professionals operating in the real estate sector will benefit from a heightened understanding of the transformative potential of PropTech and its implications for their businesses.

As the built environment sector continues to evolve, propelled by the ongoing advancements in digital technologies, all stakeholders are expected to remain proactive and responsive. By fostering an ecosystem of collaboration and knowledge-sharing, the industry can unlock new possibilities and achieve sustainable growth. This chapter lays the foundation for a collective action toward a digitally empowered and innovative real estate landscape. By recognising the significance of PropTech and embracing its potential, stakeholders can collectively shape a future where technology and real estate are well integrated to enhance the environmental, social, economic, and physical aspects of real estate use, operations, and services.

Notes

1 It should be noted that the content and inferences drawn by the authors do not necessarily reflect the organisational views of the respondents in the interviews.
2 Sometimes referred to as tertiary education.
3 The survey's ranking mechanism was designed to assign higher scores to the approach that was consistently preferred (e.g., placed first or second in respondents' rankings) and lower scores to an approach that was consistently ranked lower by the respondents. 3 points were given for the most preferred, 2 points for the second most preferred and 1 point for the least preferred.
4 The details of the respondents based on their backgrounds and settings are: World region: North America 48%, Europe 23%, the UK 19%, Asia Pacific 6%, the Middle East <1%, and Africa <1%; highest educational qualification: High School Diploma 9%, Bachelors degree 48%, Masters degree 41% and Doctorate or higher 2%; company size: 50 or fewer employees: 37%, 51–100 employees 14%, 101–500 employees 24%, 501–1,000 employees 8%, 1,001+ employees 17%; seniority level: Analyst 12%, Manager 28%, C-suite 17%, Director 20%, Executive 13% and VP 10%; workplace model: hybrid 59%, remote 31%, office-based 10%.
5 The report shows that by far, the widest variance in salary in the PropTech sector is among CEOs.
6 Open-ended questions are free-form questions that encourage respondents to provide in-depth, longer and well-developed answers, effectively taking them beyond yes/no responses. A semi-structured interview is a data collection method that relies on asking questions within a pre-determined thematic framework. These interview approaches were used for the other respondents in this chapter.
7 This list was compiled on 28/7/2023. It should be noted that this table does not contain all PropTech Associations in the world.
8 This role is centred on engagement with stakeholders and support for companies in the PropTech sector in order to improve transparency and efficiency in the UK housing sector. It also involves supporting specific priorities, typically involving matchmaking between PropTech companies, and policy teams, acceleration of public sector adoption and increasing PropTech awareness.
9 This role involves supporting the evolution of the real estate industry around diversity, technology and sustainability supporting a variety of next-generation focussed initiatives and start-ups. It specifically entails supporting local authorities to sift from using traditional methods for consultation and other related services to the use of digital methods in order to improve efficiency and housing delivery. It also includes the includes the delivery of the PropTech Engagement Fund which currently involves working with 45 local authorities.

References

Biesta, G. (2015). What is education for? On good education, teacher judgement, and educational professionalism. *European Journal of Education*, 50(1), 75–87. https://doi.org/10.1111/ejed.12109.
Biggs, J. (2012). What the student does: Teaching for enhanced learning. *Higher Education Research & Development*, 31(1), 39–55. https://doi.org/10.1080/07294360.2012.642839.
Cao, D., Wang, G., Li, H., Skitmore, M., Huang, T., & Zhang, W. (2015). Practices and effectiveness of building information modelling in construction projects in China. *Automation in Construction*, 49(PA), 113–122. https://doi.org/10.1016/j.autcon.2014.10.014.
Carr, D. (2003). *Making sense of education: An introduction to the philosophy and theory of education and teaching*. RoutledgeFalmer.

Fish, A. (2013). Reshaping the undergraduate business curriculum and scholarship experiences in Australia to support whole-person outcomes. *Asian Education and Development Studies, 2*(1), 53–69. https://doi.org/10.1108/20463161311297635.

Khan, M. A., & Law, L. S. (2015). An integrative approach to curriculum development in higher education in the USA: A theoretical framework. *International Education Studies, 8*(3), 66–76. https://doi.org/10.5539/ies.v8n3p66.

Krathwohl, D. R. (2002). (A REVISION OF BLOOM'S TAXONOMY) Sumber. *Theory into Practice, 41*(4), 212–219. https://doi.org/10.1207/s15430421tip4104.

LMRE and Fifth Wall. (2022). *What it's really like to work in PropTech: The industry's first salary and employee experience benchmark* (Issue September).

London PropTech Show. (2023). *Post event report: Bringing together the global PropTech industry* (Issue October).

Loo, R. (2004). Kolb's learning styles and learning preferences: Is there a linkage? *Educational Psychology, 24*(1), 99–108. https://doi.org/10.1080/0144341032000146476.

Nygaard, C., Højlt, T., & Hermansen, M. (2008). Learning-based curriculum development. *Higher Education, 55*(1), 33–50. https://doi.org/10.1007/s10734-006-9036-2.

Oladiran, O., & Nanda, A. (2021). *PropTech education integration framework (PEIF)*. https://www.ucem.ac.uk/wp-content/uploads/2021/10/Harold-Samuel-Research-Prize-HSRP-20202021.pdf.

RED Foundation. (2020). *The role of standards in enabling a data driven UK real estate market.*

RICS. (2020a). *RICS data standards*. https://www.rics.org/profession-standards/rics-standards-and-guidance/sector-standards/construction-standards/rics-data-standards.

RICS. (2020b). *The futures report 2020* (pp. 1–31). https://www.rics.org/globalassets/rics-website/media/news/news--opinion/rics-future-report-2.pdf.

RICS. (2021a). *Defining our future*. https://www.rics.org/about-rics/corporate-governance/defining-our-future.

RICS. (2021b). *RICS tech partner programme*. https://www.rics.org/networking/rics-tech-partner-programme.

RICS. (2023). *About RICS*. https://www.rics.org/about-rics.

Ronald, B. (1990). *The idea of higher education*. Society for Research into Higher Education and Open University Press.

Royal Institution of Chartered Surveyors. (2022). *Automated valuation models (AVMs): Implications for the profession and their clients* (Issue April). www.rics.org.

The UK PropTech Association. (2023). *About the UK PropTech Association*. Retrieved August 6, 2023, from https://ukproptech.com/about/.

Wolf, P. (2007). A model for facilitating curriculum development in higher education: A faculty-driven, data-informed, and educational developer–supported approach. *New Directions for Teaching and Learning, 2007*(112), 15–20. https://doi.org/10.1002/TL.294.

World Economic Forum. (2020). *What drives job satisfaction? Researchers think this is the answer*. https://www.weforum.org/agenda/2020/12/what-drives-job-satisfaction-research/.

World Economic Forum. (2023). *Future of jobs report 2023: Insight report May 2023*. https://www.weforum.org/reports/the-future-of-jobs-report-2023?gclid=Cj0KCQjw5f2lBhCkARIsAHeTvljc-d05ldRYFlfJrOuxPniw7XoYRaH0JIrmv4fx3jGSqvVTsnrW_DEaAgDtEALw_wcB.

Chapter 10

Summary and Conclusion

The Future Real Estate Digital Technology and Innovation[1]

This is the closing chapter of this book. It provides a summary of the key issues and themes discussed throughout the book, highlighting key points and themes covered. Further thoughts are provided about the evolution of the real estate sector and the role of digital technology in driving innovation. We further provide some thoughts on the prospects of PropTech and innovation in real estate over the next decade, highlighting some of the digital tools that may drive the next wave of innovation. Additionally, the chapter provides a review of "PropTech" as a buzzword and associated debates, after which concluding thoughts and a call to action are presented. This chapter also contains insights from semi-structured interviews with several PropTech and real estate stakeholders that were selected using convenience sampling techniques.

10.1 PropTech: The Past and Present

As we discussed in Chapters 1 and 3, the recent accelerated growth in real estate digitisation has created enormous opportunities for investors, entrepreneurs, professionals, and industry leaders. With close to 10,000 PropTech firms globally and a multi-billion-dollar industry, various digital tools and technologies such as the internet, mobile, cloud technology, Internet of Things (IoT), artificial intelligence (AI), blockchain technology, virtual reality (VR), and augmented reality (AR) have significantly transformed the real estate sector (Chapter 2). This transformation has been further enabled by advancements in data generation, storage, aggregation, and analytics.

A lot of the conversations around PropTech several years ago focused on these digital tools; in this book, we conceptualised that these digital tools are "mere" enablers of real estate innovation and digitisation rather than what may hitherto have been perceived as core outputs and outcomes. We therefore present PropTech as a phenomenon that is best conceptualised within the realm of real estate and the built environment, specifically through the lenses of the real estate value chain. Based on this, and following the critical analysis of existing PropTech definitions and concepts, we proposed the following definition:

> The deployment of digital and IT applications to the various sectors of the real estate ecosystem and value chain to enhance environmental, social, economic or physical efficiency of real estate use, operations and services.

DOI: 10.1201/9781003262725-13

This definition highlights the environmental, social, economic, and physical efficiency that PropTech tools have provided, capturing the problem-solving and solution-driven elements that should accompany digital technology and innovation in real estate. Based on this and previous work by Oladiran et al. (2022), the ESEP (Environmental, Social, Economic, and Physical) efficiency framework was re-introduced as illustrated below (Figure 1.4 in Chapter 1):

- The first "E" in the ESEP framework represents the environmental efficiency dimension of the framework.
- The "S" in the ESEP framework represents the social efficiency dimension with a focus on people and societies.
- The second "E" in the ESEP framework represents the economic efficiency of the framework.
- The "P" in the ESEP framework represents the physical efficiency part of the real estate, and it can also be termed "the spatial dimension".

The ESEP framework served as both an implicit and explicit basis for the analysis and discussion of the innovation and digitisation of real estate in subsequent chapters of the book and the case studies within these chapters.[2] Section 10.1.1 provides an overview of the key points and discussion in the second section of the book.

10.1.1   Digitisation and Innovation across the Real Estate Value Chain

The five chapters in Section Two covered various aspects of the real estate value chain (as illustrated in Figure 4.1 in Section Two). Each chapter of Section Two provided an overview of the use, operations, and services provided within that segment and highlighted some of the core areas of inefficiency. Furthermore, the various emergent digital tools and innovations were discussed and their roles in addressing the inefficiency in real estate use, operations, and service delivery were analysed. Each chapter concluded with a case study analysis in order to provide readers with a better understanding of the process of real estate innovation and digitisation, as well as the impact of these innovations. It is expected that students and PropTech enthusiasts would gain insight on the frontiers of real estate innovation and digitisation and use this as a springboard for creating and developing new solutions and innovations that can enhance the environmental, social, economic, and physical/spatial efficiency within the built environment. The summary of some of the key points in these chapters are:

Planning and Land Management

Planning and land management are vital for shaping sustainable and functional cities. However, traditional methods and approaches to these activities have been riddled with several inefficiencies. These include inadequate data collection, technical shortcomings, bureaucratic challenges in decision-making processes, eroding public trust, prolonged local plan adoption, and lack of emphasis on design and affordable housing negotiation.

To address the challenges associated with the traditional approaches to urban planning and land management, several digital tools have emerged. For instance, intelligent planning systems provide a blend of urban data and algorithms to automate strategic planning and decision-making. PlanTech has enabled the integration of digital technology into urban planning and design. Utilising 3D modelling, GIS, City Information Modelling (CIM), and Building Information Modelling (BIM), PlanTech enhances spatial intelligence, allowing designers to visualise, analyse, and adapt designs more dynamically. Digitised land administration systems, including blockchain, are also addressing inefficiencies in land registries by offering transparency and resilience in record-keeping. The escalating trend of urbanisation and the rise of megacities necessitate even more innovative solutions and smart cities have emerged as a response to the challenges posed by rapid urban growth. These innovative tools and systems are driving a new era of more effective, transparent, and resilient urban planning and management; they have the potential to reshape cities and address the complexities of urbanisation in the twenty-first century and generally enhance the environmental, social, economic and physical/spatial efficiency in space use and planning.

Construction, Management, and Maintenance

Real estate construction involves a range of phases from design to demolition, and this requires a high level of synergy and harmonious collaboration of architects, engineers, and other professionals in the sector to ensure structural integrity, compliance, and aesthetics. However, traditional methods and approaches to these activities are associated with several inefficiencies across diverse domains of the building's lifecycle, including material sourcing, construction, maintenance, and facilities management. Material sourcing emerges as a central challenge, as it poses environmental, social, and economic strains. Evidently, building materials production and transportation have substantial carbon footprints, and this has severe ecological implications. In the building construction phrase, there are design-construction interface challenges and inaccurate documentation, and these can further lead to complications, cost inflation, safety compromise, and reworks. Furthermore, prevailing building construction, management, and maintenance are rife with inefficiencies stemming from labour intensiveness, skill shortages, and procedural complexities.

Several innovative tools and systems have emerged in real estate construction, management, and maintenance in a bid to address some of these inefficiencies. The transformative potential of digital technologies is underscored through innovative tools and systems such as Building Information Modelling (BIM), digital twins, and smart buildings. BIM emerges as a pivotal methodology, facilitating collaboration and data-driven decision-making across stakeholders, resulting in enhanced planning, design, and construction processes. Furthermore, digital twins use sensor data to create dynamic digital replicas of physical assets, enabling real-time monitoring and optimisation, while smart buildings integrate IoT, AI, data analytics, and cloud computing to orchestrate intelligent control and resource efficiency. Additionally, digital property management and facilities management stand as pivotal paradigms, utilising digital technology to streamline tenant communication, financial tasks, maintenance, and regulatory compliance. These tools and systems are providing opportunities to achieve a greater level of environmental, social, economic, and physical/spatial efficiency across a building's lifecycle.

Funding, Finance, and Investment

Real estate investment involves ownership of real estate for income generation, and funding and finance refer to the methods used to secure capital for real estate projects and acquisition, and they are pivotal in achieving real estate investment objectives. Equity and debt financing are core approaches, with equity involving personal funds or private equity from investors, and debt finance involving the utilisation of loans for projects. Traditional real estate finance has capital, systemic, and technical constraints. Capital challenges stem from stringent lending criteria, creating obstacles in securing funds, while systemic inefficiencies are driven by macroeconomic factors, and market volatility and opaqueness; technical challenges arise from the complex nature of real estate, with diverse financial structures and strategies. Other challenges include illiquidity, high entry costs, and lack of diversification.

Digital technology is now increasingly adopted to address some of the inefficiencies in real estate finance and investment. FinTech for instance is the convergence of financial services and technology, and it is transforming real estate financial processes and transactions. It entails open banking, transactions, insurance, personal finance, online finance, and regulatory technology, enhancing efficiency and accessibility. Crowdfunding and peer-to-peer lending platforms are also reshaping real estate financing by pooling resources for projects. Tokenisation and blockchain introduce virtual assets, enhancing transparency, security, and liquidity. Virtual assets have gained prominence and they are further transforming the real estate investment landscape. Furthermore, automated investment management systems streamline processes and adapt to changing market dynamics. Additionally, virtual capital showcases the intricacies of start-up financing. These and other innovative financing and investment tools and systems are transforming the processes and approaches to real estate finance and investment, and they are enhancing the environmental, social, and economic physical efficiencies in these operations and services.

Valuation and Appraisal

The function of real estate valuation appraisal is an important aspect of determining the financial value of properties, and it is crucial for investment, finance, transactions, insurance, taxation, and litigation. Traditional approaches to valuation have challenges such as data access, subjectivity and biases in valuation judgments, contextual variation, and rigidity of processes. The heterogeneity of properties and geographical contexts often prevents direct equivalence and its scope; this is exacerbated by variation in the purpose, date, and basis of valuation.

Innovative tools and digital systems have emerged in real estate valuation, particularly in the last few years. Automated valuation models (AVMs) automate certain aspects of the valuation process, enhancing speed, consistency, and accuracy. Advanced data analytics, aided by AI and machine learning, underpin AVMs, accommodating property characteristics' variation across locations and settings. AVMs, although often misconstrued as fully automated systems, are much narrower in scope. They are currently mainly used for residential real estate valuation. We therefore propose that in order to drive a much broader scope of automation in valuation, automated valuation processes (AVPs) should be promoted as they provide a broader spectrum of the current automation in real estate valuation. AVPs go beyond modelling, automating data collection, analysis, and reporting. Although currently fragmented, they cumulatively address challenges related to speed, cost, scale, consistency, and accuracy in valuations. AVPs hold the potential to encompass various real estate asset classes, ushering in efficiency and accuracy in valuations for diverse sectors. While AVPs present solutions for various challenges in real estate valuation, their adoption for higher-risk transactions should be approached with caution due to the distribution of errors and accountability concerns. Despite these nuances, digital innovations, integrated with AI and other emerging technologies, are paving the way for a more efficient, accurate, and streamlined real estate valuation and appraisal process, across diverse asset types and markets. Scholars, professionals, and enthusiasts are encouraged to further explore the possibilities that lie in these advancements, particularly in expanding the use of AVPs for commercial properties and refining data collection and modelling methods for further enhancement in the environmental, social, economic, and spatial efficiency in real estate valuation and appraisal.

Agency, Brokerage, Marketing, and Allied Services

Property agency, brokerage, and marketing play a pivotal role in connecting property owners, developers, investors, and potential users, bridging the gap between supply and demand. However, traditional practices have long been marred by inefficiencies that hindered optimal operations. In the domain of real estate agency and brokerage, conventional models have grappled with information asymmetry, leading to fragmented property data, restricted access to pertinent information, high transaction costs, rigid market coverage, and changing consumer expectations. Inefficient processes, outdated marketing techniques, and inadequate adherence to regulations have compounded these challenges. Likewise, real estate marketing has faced hurdles due to shifting consumer behaviours, limited reach, high costs, and difficulties in measuring campaign effectiveness.

To address these and other challenges, digital technology, encompassing FinTech, blockchain, mobile technology, the internet, and AI, has emerged as a harbinger of transformation. Online property listing platforms have evolved, offering comprehensive information, advanced search functions, and streamlined transactions. Furthermore, virtual property viewing, facilitated by virtual reality and 360-degree videos, allows remote property exploration. Additionally, the shared economy leverages digital platforms for space-sharing, transforming underutilised spaces into tradeable assets. Space-time commoditisation converts physical and temporal spaces into valuable commodities, creating fluid marketplaces. Digital Customer Relationship Management (CRM) systems also have transformed real estate brokerage. These systems enable lead management, integration with marketing tools, customer profiling, transaction management, and automated communication. Such digital tools facilitate personalised experiences, efficient transactions, and targeted marketing. The emergence of these digital tools and platforms significantly improved the environmental, social, economic, and physical efficiency of real estate marketing, brokerage, and allied services.

PropTech Stakeholders, Enablers, and Systems of Growth and Development

Although the core of this book is hinged around the real estate value chain in Section Two, Chapter 9 (Section Three) provides useful insight into the core enablers, stakeholders, and systems that need to be further promoted in order for new technology and innovation in real estate to be advanced further. Learning and education are pivotal to the further growth and development of innovation in real estate. The integration of PropTech and innovation into real estate education is currently limited and some of the factors responsible for this include the fragmented nature and limited research in the domain of innovation and digital technology in real estate. The integration of PropTech in higher education is therefore advocated for its effectiveness in nurturing skills and talents necessary to enhance real estate efficiency. Furthermore, skills development and employability in PropTech are essential for fostering employability, particularly soft skills like problem-solving and self-management, along with technology use.

The roles of professional bodies, institutions, and associations in PropTech development are also enormous. These entities provide accreditation, certification, advocacy, and collaboration opportunities. The case study of the Royal Institution of Chartered Surveyors (RICS) underscores their evolving role in promoting innovation and digitisation within the sector, bridging the gap between traditional real estate and technology. Furthermore, PropTech associations are playing key roles in supporting the growth and development of PropTech and innovation globally. The UKPA, for instance, was established to drive digital transformation and offers networking, collaboration, and support for diverse stakeholders within the PropTech ecosystem. Additionally, standards, regulations, and ethics in PropTech should be emphasised and prioritised as the sector continues to evolve due to the data-intensive nature of the sector. The need for responsible and ethical data management is emphasised, with discussions on data standardisation and the challenges of "data silos". The heightened participation from the public sector is also a reflection of growing interest in digitising the built environment. Interviews with public sector entities underscore their commitment to supporting PropTech initiatives and the impact these efforts have on urban development and citizen engagement.

Although various digital tools and innovations across the real estate value chain are driving significant transformation in the built environment, there are still several prospects and opportunities for further changes to the sector. Some of these core areas of advancement are discussed in Section 10.2.

10.2   What's Next?

Several digital tools and platforms have driven digital transformation in the built environment over the last decades. Chapters 1 and 2 provided insight into the mechanics and impact of the various digital tools and systems, and the various chapters in Section Two provided insights into the application of these technologies to the real estate value chain. This sub-section highlights three core digital technologies: artificial intelligence (AI), blockchain technology, and virtual real estate, shedding light on their prospects for the future. The discussion in this sub-section is strengthened by the findings that emerged from semi-structured interviews of PropTech industry leaders. The respondents were selected using convenience sampling techniques and data were analysed using thematic and content analysis.

10.2.1   Artificial Intelligence (AI)

The real estate industry, marked by its dynamics and substantial data streams, is poised for a transformative journey with the further adoption and utilisation of artificial intelligence. The prospects for the further application of AI in real estate are vast, with the potential to reshape how professionals engage with data, customer experiences, and innovation. AI's influence is expected to continue to span across every facet of the real estate journey, sparking a transformation that is underpinned by efficiency and precision. From predictive analytics to the finesse of machine learning, AI emergence is likely to continue to simplify processes and achieve accuracy, thereby shaping an ecosystem that blends technological prowess with strategic decision-making.

AI will become increasingly important as more complex data collection techniques are adopted (such as instrumentation, logging, sensors, external data, and user-generated content), and advanced storage, transformation, and aggregation are applied. This will lead to learning and optimisation through machine learning algorithms, experimentation, and deep learning. The further development of algorithms for reading and analysing text data can lead to improved summarisation and clusterisation functions, enabling built environment professionals to gain valuable insights from several documents and extensive text through concise and summarised texts. The concept of data-centric AI will therefore continue to evolve with novel paradigms, signalling a shift towards optimising algorithms for smaller datasets. This development is particularly pertinent to real estate, where comprehensive data can be scarce, especially within the opaque realm of commercial properties. These innovative algorithms will improve the performance of models with limited data in cases where large datasets are not available. Such advancements redefine the dynamics of AI applications, emphasising data quality over sheer quantity.

In real planning, AI is expected to transform the review of citizens' views and perspectives in order to provide useful insight into citizens' expectations and thoughts about proposed projects and urban plans. For instance, AI is being used to collect and analyse social media data to understand citizens' sentiments about certain issues or programmes. Furthermore, integrating AI algorithms with generative software can provide valuable information that built environment professionals can utilise to generate information quickly and efficiently, thereby accelerating analysis and decision-making processes. This feature epitomises AI's capacity to revolutionise the way narratives are uncovered, aligning with the ultimate goal of simplifying daily activities for real estate practitioners and professionals. It is also expected that AI will remain a powerful ally in customer experience in various segments of the real estate value chain. In real estate brokerage and marketing, for instance, I-powered chatbots offer round-the-clock customer service that enhances engagement with potential buyers and renters. Furthermore, AI-powered search tools will continue to analyse a range of data including property values, school ratings, crime rates, etc. and these can be used to build specific preferences for potential renters or buyers.

AI technology is also likely to change the ways that property deals are sourced, increasing efficiency and accuracy in the process. This can manifest through lead generation, due diligence, and market analysis. AI-driven insights will empower investors, agents, and professionals with real-time data-driven recommendations, minimising guesswork, and high risk and replacing them with data-backed decisions that maximise returns and minimise risk and costs. Other benefits include personalised property recommendations and predictive trend insights. These will yield strategic benefits through improved deal evaluation and negotiation. Additionally, AI-powered tools can be used to analyse data on tenant behaviour, maintenance requests, and other factors to support property managers to identify potential issues before they become more complex.

Yet, alongside this remarkable potential, lies a set of challenges that must be acknowledged and considered. The inherent bias present in AI algorithms is a significant concern, as they can inadvertently perpetuate and exacerbate existing inequalities in the industry. If AI models are trained based on incomplete datasets or datasets with biases, they may yield results that fail to accurately reflect the needs and preferences of marginalised groups, thereby limiting the scope of application. Thus, data and curation, and algorithm development and utilisation should ensure that these issues are sufficiently mitigated. Furthermore, there are concerns that AI-enabled automation may replace certain tasks that are traditionally performed by real estate professionals. As we argued in previous chapters, AI also creates novel opportunities for those who acquire the necessary skills to harness its potential. Built environment professionals are therefore encouraged to embrace the power that it provides and the supportive tools that it presents in order for them to perform their tasks and services optimally and to reap the rewards of the next transformative wave of innovation and digitisation in the sector.

10.2.2   Blockchain Technology

The prospects of Blockchain technology to transform the future of real estate are enormous. As highlighted in Chapter 2 and subsequent chapters of this book, blockchain technology enables secure and immutable data storage, facilitates automated processes through smart contracts, and addresses several issues of land ownership in several countries. It also can streamline real estate processes by providing a shared and immutable ledger and to enhance data exchange, enabling various parties in a transaction, such as estate agents, surveyors, lawyers, and government entities, to access and share information securely and transparently. Furthermore, blockchain can enable direct transactions between landlords and tenants, improving efficiency and reducing costs in the rental market. Two core prospects are further highlighted below:

Land ownership and transparency: Many world regions grapple with the lack of secure land rights, which hampers tax collection and service provisioning while fostering disputes and corruption. Blockchain's ability to provide tamper-proof land registries is a key solution to this challenge and several public organisations and governments are likely to increase their adoption of this technology. Digital land records can enhance transparency and efficiency in land administration processes, thereby reducing disputes and ensuring more equitable access to property ownership.

Investment, funding, finance, and transactions: Blockchain technology has the potential to reduce friction in property transactions and improve overall investment efficiency. One of the major benefits of this technology is its ability to provide transparent and tamper-proof records, streamline complex real estate transactions, reduce fraud, and enhance trust among parties. Furthermore, the use of a distribution ledger, reference to a subject property, and made available to all stakeholders in a transaction can improve the due diligence, property exchange, and property search. Blockchain technology can also be used for digital asset management which registers ownership and manages and distributes property rights directly. Additionally, as smart contracts and automated funds transfers become more commonplace, the adoption of this technology may intensify. Although still in its infancy in real estate finance and investment, it can increase market transparency and improve the level of trust in the market, particularly in indirect investment and funding. It also provides enormous opportunities for cross-border capital flow and diffusion of funds.

The authors further sought insight into the prospects of blockchain technology in real estate. Ben Lerner, Managing Director, Lerner Associates PropTech M&A, John Fitzpatrick, Senior Managing Director and Chief Technology Officer at Blackstone and Nick Moore, Founder and Executive Director, Lionpoint Group shared their thoughts about this technology and the prospects it holds for real estate use, operations and services in the future, particularly for real estate investment and finance. These respondents were identified using convenience sampling techniques and semi-structured interviews were administered in the same way as Chapter 9.[3]

The question "How important is blockchain technology to the future of indirect real estate investment, funding, and finance?" prompted varied responses. Ben Lerner's response indicates a cautious optimism regarding the importance of blockchain technology in the future, particularly in indirect real estate investment, funding, and finance. He suggests that although the adoption of blockchain technology is still in its infancy, it does offer valuable opportunities in those areas as the technology matures and calls for further exploration. Ben, however, does not consider blockchain technology a top priority, nor does he see it making massive waves in real estate in the near future, although he acknowledged that this may change significantly as the technology develops further.

John Fitzpatrick's presents a more enthusiastic outlook. He considers blockchain to be a highly intriguing technology with unique use cases across the entire investment and finance value chain. John also acknowledged that blockchain's impact on real estate is in its early stages, but he anticipates it could have a considerable influence in the future. He highlighted the inefficiencies present in many real estate processes and believes that blockchain technology has the capacity to address some of these issues. However, he acknowledges that the definitive extent of blockchain's impact cannot be determined yet.

Similar to the other respondents, Nick Moore emphasises that blockchain is still in its infancy, and adds that investors in indirect real estate are relatively inexperienced when it comes to technology adoption and are primarily focused on front, middle, and back-office applications. Nick recognises that data integration and security are crucial concerns, and he anticipates a growing focus on efficiency amid changing macroeconomic conditions. While blockchain technology has been primarily utilised in back-office functions and machine learning to enhance efficiency, John points out that there are opportunities for leveraging this technology in funding and financing areas.

The responses revolve around themes of cautious optimism, potential impact, and varying degrees of development in the integration of blockchain technology in real estate, particularly investment, funding, and finance. Overall, the responses collectively underscore the evolving nature of blockchain's role in the real estate sector. While opinions differ on the immediate significance of blockchain, there is a shared recognition of its potential to reshape various aspects of real estate use, operations, and services as the technology continues to advance.

In conclusion, the prospects of blockchain technology to transform the real estate industry are dependent on the development and wider adoption of this technology across other non-real estate sectors. This technology has several merits such as enhanced transparency, increased collaboration, improved property rights and land ownership, and advancement in other technologies such as internet, computing, data, and analytics will further enhance its potential in the transformation of real estate markets.

10.2.3   Virtual Real Estate and the Metaverse

In recent years, the concepts of virtual real estate have emerged as a transformative force that is reshaping how individuals interact, work, socialise, and even engage with the real estate industry. Virtual real estate refers to digital properties, spaces, or parcels that exist within virtual environments. Virtual environments may include various platforms, such as online games, virtual reality simulations, and blockchain-based virtual worlds. Virtual real estate can be thought of as digital land that users can own, develop, and interact with. It is typically represented by Non-Fungible Tokens (NFTs) on blockchain networks, which provide proof of ownership and authenticity. Virtual real estate can serve multiple purposes, including social interactions, entertainment, gaming, commerce, and even business activities. For example, users might purchase virtual land within a game or virtual world, and then develop structures, buildings, or businesses on that land. Virtual real estate can therefore have value both within the digital environment and in terms of potential monetisation.

The metaverse can be defined as a virtual platform that uses extended reality and other emerging technologies to allow real-time interactions and experiences in ways that are not possible in the physical world (Koohang et al., 2023). It is not a single technology or device, and it is not a service provided by a single company (Deloitte, 2023); rather, it is a broader concept that encompasses interconnected virtual worlds and environments, forming a collective digital universe. It represents a shared space where users can engage, interact, create, and transact, often through the use of avatars or digital identities. "Metaverse" is the combination of "meta" which refers to experiences that transcend a physical setting and "universe" which can also represent the "world". This term was developed by Neal Stepheson's 1992 novel "Snow Crash" where he envisioned a convergence of the physical and virtual realms of human engagement and interaction. In this narrative, the Metaverse represents a shared 3D virtual space where users, represented as avatars, engage socially, conduct business, and explore environments beyond the limitations of reality. Stephenson's concept laid the foundation for discussions on immersive digital experiences which echoes aspects of today's emerging metaverse. The term continues to shape contemporary discourse on interconnected virtual worlds and their potential to reshape human engagement across physical and digital boundaries.

The Metaverse represents a digital realm where immersive experiences simulate the real world or create entirely new environments beyond physical reality. The metaverse is not limited to a single platform or virtual environment; instead, it encompasses a network of interconnected digital spaces. Users can move seamlessly between different virtual worlds, each with its own rules, designs, and purposes. The metaverse aims to create a comprehensive digital existence that mirrors aspects of the real world while also introducing new possibilities for interaction and commerce. It is a dynamic and evolving concept that involves a range of technologies including augmented reality (AR), virtual reality (VR), blockchain, artificial intelligence (AI), 3D reconstruction, and the Internet of Things (IoT) underpin the trading of virtual real estate assets using Non-Fungible Tokens (NFTs).

Extended Reality (XR), encompassing AR and VR technologies, plays a pivotal role in creating immersive experiences within the Metaverse. Through XR, users can engage with virtual environments that mimic real-world settings or introduce fantastical worlds. This technology offers potential applications in real estate, such as virtual property viewings and immersive property development simulations. XR bridges the gap between the physical and virtual worlds, enhancing the user experience and expanding the possibilities of real estate transactions.

The metaverse is increasingly garnering significant attention from technology giants, with Facebook, now Meta, being a key player in shaping its development. It offers a platform for digital interaction, commerce, and connection through avatars and digital identities. This phenomenon has the potential to revolutionise the real estate industry by providing unique opportunities for interaction, business, and property ownership.

Metaverse environments allow individuals to create and interact with bespoke spaces that overcome limitations of physical space and costs. This could transform business meetings, workshops, and even social interactions into immersive experiences with spatial sound and unique designs. Furthermore, the Metaverse introduces the concept of digital ecosystems, complete with their own currency, property rights, and possessions, granting users a novel form of digital identity.

The metaverse is creating several investment opportunities, and this is likely to become more significant over the coming years. Investment in virtual real estate within the Metaverse is gaining significant traction, with early adopters focusing on the big four platforms: Sandbox, Decentraland, Cryptovoxels, and Somnium. It has been reported that metaverse mortgages are being issued to buy virtual land, and one of the pioneering deals was recently completed for a property in Decentraland (Rosen, 2021). The evolving concept of the Metaverse introduces both opportunities and challenges to the real estate industry.

While investment in virtual real estate assets is yielding positive returns, concerns persist (Koohang et al., 2023). One of these relates to the impact on physical real estate markets and the sense of real community. Furthermore, investment in this landscape requires a unique set of skills and competencies, often referred to as "digital intelligence". Investors must navigate complexities such as identifying virtual lands with high traffic, understanding geospatial digital footprints, and comprehending the nuances of digital asset trading. These skills are essential for successful business transactions in the metaverse, ensuring informed investment decisions. Furthermore, it is not clear how this will affect physical socialisation and the impact on physical real estate. There are concerns that the decline experienced in the physical retail sector due to the increase in e-commerce can have similar effects on physical real estate use.

The success of the Metaverse relies on overcoming feasibility, privacy, user safety, and social issues. As technology giants like Meta and others invest substantial capital, the future of the Metaverse remains uncertain but full of potential. Marketing, education, tourism, and healthcare sectors are some of the potential areas that will significantly impact over the next few years (Dwivedi et al., 2022). As the digital landscape continues to evolve, research into the scope, opportunities, challenges, risks, and long-term impact of this phenomenon would be valuable. Extensive research in this area will support investors and other stakeholders to understand the opportunities and risks presented by the metaverse and virtual real estate; thus, enabling them to maximise the benefits presented.

10.2.4 Other Opportunities and Challenges

Although Section 10.2 has focused primarily on the transformative potential of AI, blockchain, and the metaverse, other digital tools are likely to further transform the real estate sector. This sub-section discusses some of the issues that relate to the future of real estate innovation, the prospects for further enhancement, and potential challenges. We sought insight into this from the three respondents mentioned in Section 10.2, using the same sampling and analytical techniques. In addition to interviews with Ben Lerner, John Fitzpatrick, and Nick Moore, we also interviewed other industry stakeholders: Brad Greiwe, Co-Founder and Managing Partner at Fifth Wall, Aaron Block, Co-Founder and Managing Partner, MetaProp, Raj Singh, Managing Partner at JLL Spark, Michael Beckerman, CEO CREtech and CREtech Climate and Zach Aarons, Co-Founder and General Partner, MetaProp.

10.2.4.1 Opportunities for Further Development and Diversification in Real Estate Innovation and Digitisation

In Chapter 3, we revealed that PropTech funds and companies are currently concentrated in the United States. We sought the views of some of the respondents on the necessity and strategies for a more diversified PropTech landscape – beyond the United States.

Ben Lerner acknowledged the US's dominance in the PropTech investment landscape due to its large market size and willingness for early-stage investments. He suggests that the US's investment criteria should not be solely focused on US businesses, rather, it should also consider solutions that can attract benefits to the industry at a global scale. Ben proposes a broader approach to innovation through partnerships with local investment companies, which can support the allocation of funds for the exploration of digital solutions in other world regions. John Fitzpatrick also acknowledged the US' dominance on the PropTech landscape but pointed out that there are several promising PropTech solutions emerging from other regions, like Europe and Singapore. John sees potential for solutions addressing region-specific use cases and suggests that it is unlikely that the US's dominance will decrease drastically in the short term.

In Nick Moore's view, the US is a magnet for PropTech investment capital and entrepreneurship partly because it is also a core real estate investment destination, and because of its ability to provide scaling opportunities. Nick points out that while technologies may originate in other parts of the world, their entry into the US often increases their acceptability and wide usage. In addition to this, the US market provides a diverse investment base with substantial capital for companies with proven business models, and macroeconomic conditions may also contribute to the attraction of funds to a country.

The responses centre around the need for a broader spread of PropTech investment to further advance innovation and digitisation in real estate. It is also important that country and regional-based innovative solutions digital tools are developed for specific use cases to enhance localised growth and development in the sector. It should be noted that while the diffusion of capital may have global benefits, the allocation of capital is typically a function of several macroeconomic, geopolitical, demographic, and socio-cultural fundamentals, and the mere desire to diffuse these funds will need to be driven by the right investment, innovation, and entrepreneurial climate.

Zach Aarons anticipates significant growth in international emerging markets for PropTech. He cites Nigeria as an example, with its rapidly growing population and expanding financial technology adoption. Zach notes that PropTech trends often mirror FinTech trends in mature markets and given the surge in FinTech adoption in West Africa, he expects a similar trajectory for PropTech. He highlights the need for digitisation in markets like Nigeria, where traditional challenges like obtaining mortgages in person are prevalent. Aarons predicts similar dynamics in other emerging markets like Mexico, Colombia, and Brazil, emphasising the inevitability of digitisation in these regions.

The respondents also discuss the recent trend where trade and industry buyers are engaging in acquisitions to address product gaps, acquire talents and resources, and expand into new markets to achieve scale. Ben, however, pointed out that the market has seen a proliferation of overlapping point solutions, which makes consolidation inevitable. He further predicts steady growth in acquisitions by larger incumbents, challenging start-ups developing technological solutions, and private equity (PE) firms focusing on smaller businesses for organic and inorganic growth. Ben identifies office/workplace technology, ESG (Environmental, Social, Governance), construction technology, and property management tools for efficiency and cost savings as areas with significant change potential. John also emphasises the anticipation of continued innovation in the ESG and climate tech space. He opines that construction technology, historically underinvested and with low adoption rates, is likely to receive heightened attention.

John also foresees increased investment in areas like lending, leasing, and financing, noting the potential for investment in both legacy PropTech providers and public companies that may have faced market challenges. He suggests that PropTech providers offering multiple use cases, as opposed to point solutions, will be particularly attractive to investors as the market evolves. Furthermore, companies that focus on holistic solutions, spanning multiple aspects of the property lifecycle, are more likely to thrive. Aaron Block further pointed out the exponential growth of real estate finance and construction. He also pointed out that some of the transformations in residential real estate such as single-family and multifamily living have been underpinned by digital technology; he further noted that areas of real estate that could benefit from further innovation are healthcare and logistics. Brad Greiwe echoed this sentiment. He added that single-family rentals and other residential segments have been particularly active due to consumer expectations and demand. This segment, enabled by digital technology, has shown effective outcomes.

Other themes that emerged in relation to the prospects for further development of innovation in the interview centre around education and upskilling. The pathway to a successful PropTech career involves a combination of education, hands-on experience, networking, and continuous learning. Real estate students should pursue a balanced understanding of real estate and digital technology through formal and informal educational opportunities, internships, participation in events, building a peer support network, and other opportunities to enable them to understand the complexities of the property sector and effectively navigate and create meaningful opportunities. Aaron also echoes this sentiment. Real estate education should encompass courses that bridge the gap between real estate and digital technology, catering to a new generation of professionals with diverse skill sets. Brad further pointed out that courses in built environment education are typically wrapped around the environmental sciences, leading to an insufficient understanding of how businesses operate within the industry. He advocates for a nuanced understanding of real estate operations and finances, coupled with the ability to connect industry knowledge to emerging technologies. The emphasis is on being able to operationalise solutions and effect real-world change. Raj Singh also highlighted the importance of applying existing skill sets to the real estate industry and making it more attractive to younger generations. He emphasised that while the necessary skill sets exist, they are often being directed towards other industries. Raj suggests that creating a "coolness factor", and emphasising the environmental factors can attract younger talent to the real estate industry.

Additionally, Raj discusses the concept of the metaverse and virtual real estate and their potential impact on real estate. While the metaverse is seen as aspirational and capable of attracting investment and attention, he suggests that its practical application and adoption may require further technological advancement and time. He predicts that the metaverse will likely experience cycles of hype and progress, aligning with other technological trends.

As the industry continues to evolve, staying updated on current trends and maintaining a serious and informed outlook will be crucial for those aspiring to make a meaningful impact in the PropTech ecosystem. Brad particularly emphasised the importance of context in understanding industry intricacies and bridging the gap between technology and real estate. The role of intrapreneurship, entrepreneurship, understanding industry operations, and fostering collaborations between real estate and technology experts were underscored as crucial for driving innovation.

10.2.4.2 Climate Change Technology

There is an increasing imperative for the real estate industry to adopt technology to achieve carbon neutrality. Brad specifically highlights the influence of the various governments through legislation and policy, whilst also acknowledging the role of industry partners in this drive, which has significantly enhanced real estate innovation and further development. Climate tech and environmental, social, and governance (ESG) solutions stand out as crucial areas for continued innovation, with the conversation now shifting to climate tech, positioning it as a crucial area of future growth. Brad emphasises that addressing carbon impact is imperative, especially as regulations shift towards a government-mandated approach. Michael Beckerman added that the acceleration of technology adoption in response to the pandemic has also facilitated a greater focus on environmental, social, and governance (ESG) considerations. Brad reveals that the intersection of real estate, technology, and climate presents a significant opportunity for retrofitting and sustainability efforts. The industry's responsibility to achieve carbon neutrality is underscored by the need for bottom-line impacts and compliance.

Michael Beckerman provides further thoughts. He emphasised the importance of addressing the impact of the real estate industry on the environment, highlighting that the built environment is both the largest asset class globally and the single largest contributor to greenhouse gas (GHG) emissions. This theme underscores the urgency of decarbonisation efforts to mitigate the industry's substantial environmental footprint. Michael also points out that this has the potential to attract commercial opportunities. He posits that reframing the climate crisis as a commercial opportunity could drive the real estate and technology sectors to prioritise climate technology adoption. He links the potential for financial return on investment (ROI) to galvanising the industry, aligning with the notion that climate technology ventures hold significant growth potential, akin to the rise of unicorns in the technology sector.

Michael also considers the intrinsic link between real estate and various industries, impacting global urbanisation trends, energy consumption, and overall societal well-being. He highlights the role of real estate in fostering change through innovation, particularly in response to challenges such as the climate crisis and the COVID-19 pandemic. Furthermore, he discusses the role of technology in addressing climate challenges, drawing parallels to Tesla's success in the automotive industry. This theme underscores the transformative potential of technological innovations in advancing the decarbonisation agenda within the real estate sector.

The views of the respondents highlight the dual nature of the built environment being both an asset and a major emitter of GHGs. Although current climate change actions and policies are primarily aimed at reducing real estate-related emissions, they are also driving opportunities for commercialisation as seen in climate technology companies. It argued that the industry's potential for financial returns should drive engagement with climate technology, as seen in the success of Climate Tech companies.

10.3 Buzzwords

The term "PropTech" has emerged as a prominent buzzword, signifying the fusion of property and technology within the real estate sector for close to one decade (Chapter 1 provides an overview of the evolution of PropTech as a buzzword and a phenomenon). While the term has garnered widespread usage, concerns about its use, overuse, and limitations have arisen. Buzzwords, like "PropTech", are embraced due to their succinctness and accessibility. They act as encapsulations of complex concepts, facilitating communication and understanding. While concerns about overuse and dilution of meaning exist, it is important to recognise that buzzwords play a crucial role in initiating conversations, raising awareness, and fostering a common vocabulary within an industry.

Indeed, some of the critiques of the buzzword "PropTech" are valid. They often revolve around its perceived inability to capture the full spectrum of technological innovations within the built environment, the intersection with other tech-driven phenomena, and the perception that the buzzword may become redundant in the near future. In separate interviews, John Fitzpatrick and Nick Moore point out that when the PropTech buzzword emerged, it was mainly associated with physical real estate, and that the phenomenon has evolved beyond the physical asset, now encompassing various aspects such as acquisition, management, and finance. In another interview, Andrew Knight, the Global Data and Tech Lead at the RICS, argues that the scope of the current buzzword "PropTech" is inadequate, and its significance might wane as the real estate sector burgeons. This sentiment is also reflected in Nicoara (2022) and Oladiran and Nanda (2021). Moreover, the term's intersection with other technological domains like "LawTech", "FinTech", "InsureTech", and "ClimateTech" undermines its specificity. Andrew also opines that the emergence of "PropTech" reflects the lack of an internal drive for innovation and digitisation in the built environment sector, highlighting that the forces of change have emerged from external forces.

Despite the criticisms and limitations, all four respondents agree that the term "PropTech" is currently relevant and a reasonable encapsulation of innovation and digitisation in the built environment. They acknowledge that the term has become engrained in the real estate vocabulary and is currently the most recognisable and widely used label. This aligns with our approach in this book to conceptualise PropTech and its application beyond the application of digital technology to the physical asset to every area of the built environment value chain spanning planning to brokerage. The persistence of "PropTech" as a buzzword is indicative of the real estate industry's acknowledgement of its transformative journey. The term has acted as a conduit for acknowledging the integration of technology into real estate operations, and it highlights the industry's movement towards technological advancements, prompting stakeholders to explore and adopt innovative solutions.

One of the merits of a widely recognised buzzword like "PropTech" lies in its consistency and coherence across discussions, literature, and conversations. Although developing new buzzwords such as BETech (built environment tech) or RETech (real estate technology) may be perceived as more all-encompassing, this effort is unnecessary, and it may risk fragmenting the industry's collective vocabulary and impeding effective communication. "PropTech" has become a unifying term that bridges the gap between technical experts and traditional real estate professionals and is currently fit for purpose. Another merit of the "PropTech" buzzword is that it encourages an ongoing dialogue about the evolving relationship between technology and real estate. By maintaining a consistent term, the industry fosters a space for sustained discussions on innovations, challenges, and opportunities. The continuity of the term ensures that the conversation about technology remains at the forefront of real estate discussions. Andrew however points out that the target should be for the buzzword to diminish, as this would be a good sign that the built environment sector is increasingly integrating innovation, data, and digitisation.

The complexity of digital integration in real estate can be overwhelming for stakeholders unfamiliar with technical jargon, particularly those in traditional real estate. "PropTech" acts as a simplification tool, allowing a diverse range of professionals to engage in conversations without feeling alienated. It serves as a bridge that connects different facets of the industry. "PropTech" also serves as a catalyst for collaboration between technology experts and real estate professionals. As we have tried to advocate all through this book, PropTech needs to be conceptualised from problem-solving, efficiency enhancement, flexibility, agility, and sustainability points of view. It should also be seen to span the entire built environment value chain, and the goal should be for it to be perceived as the mechanical component of a vehicle rather than the driver. By presenting a shared point of reference, it fosters a collaborative environment where both sides can leverage their expertise to drive meaningful advancements in the industry. A new buzzword might disrupt this collaborative dynamic.

The gravitas of a label, exemplified by "PropTech", transcends mere linguistic semantics. It encapsulates a dynamic interplay between industry maturation, technological integration, and cultural acceptance. The quest for a novel appellation signifies the anticipation of a sector's evolution from its embryonic phase of technology adoption to its seamless integration. This paradigm shift signifies the transformation of technology from a novel notion to a foundational tenet in real estate operations. However, this quest may be an effort in futility, as the current buzzword remains fit for purpose and relevant in the current landscape.

10.4 Conclusion

In drawing the curtains on this comprehensive review of PropTech and the broader innovation within the real estate sector, this book concludes with some reflections and expectations. Various areas of the built environment industry have been critically analysed, with emphasis on their unique and shared inefficiencies, as well as the digital technological tools that have been used to address some of these inefficiencies. These perspectives have covered real estate planning, land management, construction, property management, building maintenance, facilities management, investment, finance, valuation, brokerage, agency, and marketing.

One fact remains apparent: PropTech has transcended the realm of buzzword and found a firm foothold in the very fabric of the present and future of real estate use, operations, and service. Based on prospective analysis, it is evident that there is still much more to come. Digital technology, often portrayed as an enabler, is assuming the role of a visionary, shaping the trajectory of real estate. The evolution from physical spaces to intelligent environments, from static assets to dynamic ecosystems, is underway. As we contemplate this evolution, it is important to remember that the heart of innovation is not solely in the adoption of individual technologies but in the transformation mindset that propels action. As Souza et al. (2021) point out, true technological transformation does not stem from embracing isolated technologies but from embracing radical thought translated into action. As we peer ahead to the next decade, we look forward to new tools poised to redefine PropTech: blockchain, artificial intelligence, augmented reality, and the seamless integration of data promise to be the torchbearers of this evolving era.

Throughout this book, we've grappled with the label "PropTech" itself – a term that encapsulates and defines the digital real estate phenomenon. As scholars, students, and practitioners, we are encouraged to acknowledging its fundamental role as a catalyst for further innovation in the sector. Beyond the buzzword, our focus should be on the continuous dialogue it engenders, the amity it nurtures, and the shared purpose it instils. The interviews with stakeholders reflect optimism for the future and a consensus that "PropTech" is not a fading star but a North Star, guiding the real estate industry toward uncharted realms of possibility.

Finally, we should embrace the prospect of unseen disruptions in the built environment, driven by innovation and digital technology. The emerging tools and platforms in the case studies exemplify the unexpected directions that the integration of technology and property can take. We are reminded that "PropTech" is not a terminus but an evolution. Our actions should therefore resonate with this philosophy as we continue to evolve towards a sector where innovation isn't an outcome but the norm.

Notes

1 It should be noted that the content and inferences drawn by the authors do not necessarily reflect the organisational views of the respondents in the interviews.
2 Although we illustrate scales of efficiency (Figure 1.5 in Chapter 1), we do not apply this to the analysis; scholars are encouraged to explore the potential of utilising these scales or to propose other scaling approaches.
3 The data from this interview is also used in the discussion of other sub-sections in this chapter.

References

Deloitte. (2023). *A whole new world? The metaverse and what it could mean for you Metaverse technology and its implications for business leaders.* Perspectives. https://www2.deloitte.com/us/en/pages/technology/articles/what-does-the-metaverse-mean.html (accessed 24 October 2023).

Dwivedi, Y. K., Hughes, L., Baabdullah, A. M., Ribeiro-Navarrete, S., Giannakis, M., Al-Debei, M. M., Dennehy, D., Metri, B., Buhalis, D., Cheung, C. M. K., Conboy, K., Doyle, R., Dubey, R., Dutot, V., Felix, R., Goyal, D. P., Gustafsson, A., Hinsch, C., Jebabli, I., ... Wamba, S. F. (2022). Metaverse beyond the hype: Multidisciplinary perspectives on emerging challenges, opportunities, and agenda for research, practice and policy. *International Journal of Information Management*, 66(July), 102542. https://doi.org/10.1016/j.ijinfomgt.2022.102542.

Koohang, A., Nord, J. H., Ooi, K. B., Tan, G. W. H., Al-Emran, M., Aw, E. C. X., Baabdullah, A. M., Buhalis, D., Cham, T. H., Dennis, C., Dutot, V., Dwivedi, Y. K., Hughes, L., Mogaji, E., Pandey, N., Phau, I., Raman, R., Sharma, A., Sigala, M., ... Wong, L. W. (2023). Shaping the metaverse into reality: A holistic multidisciplinary understanding of opportunities, challenges, and avenues for future investigation. *Journal of Computer Information Systems*, 63(3), 735–765. https://doi.org/10.1080/08874417.2023.2165197.

Nicoara, B. (2022). *Why the term "PropTech" will disappear.* https://www.forbes.com/sites/forbestechcouncil/2022/03/16/why-the-term-proptech-will-disappear/.

Oladiran, O., & Nanda, A. (2021). *PropTech education integration framework (PEIF).* https://www.ucem.ac.uk/wp-content/uploads/2021/10/Harold-Samuel-Research-Prize-HSRP-20202021.pdf.

Oladiran, O., Sunmoni, A., Ajayi, S., Guo, J., & Abbas, M. A. (2022). What property attributes are important to UK university students in their online accommodation search? *Journal of European Real Estate Research.* https://doi.org/10.1108/JERER-03-2021-0019.

Rosen, P. (2021). *Metaverse mortgages are being issued to buy virtual land — And one of the first ever was just signed for a property in Decentraland.* Business Insider. https://markets.businessinsider.com/news/currencies/metaverse-mortgage-terrazero-decentraland-virtual-land-real-estate-crypto-finance-2022-2.

Souza, L. A., Koroleva, O., Worzala, E., Martin, C., Becker, A., & Derrick, N. (2021). The technological impact on real estate investing: Robots vs humans: New applications for organisational and portfolio strategies. *Journal of Property Investment and Finance*, 39(2), 170–177. https://doi.org/10.1108/JPIF-12-2020-0137.

Index

Note: **Bold** page numbers refer to tables; *italic* page numbers refer to figures and page numbers followed by "n" denote endnotes.

For Product Safety Concerns and Information please contact our EU representative GPSR@taylorandfrancis.com Taylor & Francis Verlag GmbH, Kaufingerstraße 24, 80331 München, Germany